Agricultural Production and Forestry

Agricultural Production and Forestry

Editor: Pat Jordan

R CALLISTO REFERENCE

www.callistoreference.com

Callisto Reference,
118-35 Queens Blvd., Suite 400,
Forest Hills, NY 11375, USA

Visit us on the World Wide Web at:
www.callistoreference.com

ISBN: 978-1-63239-950-2 (Hardback)

Cataloging-in-Publication Data

Agricultural production and forestry / edited by Pat Jordan.
   p. cm.
Includes bibliographical references and index.
ISBN 978-1-63239-950-2
1. Agriculture. 2. Forests and forestry. I. Jordan, Pat.
SB98 .A375 2018
630--dc23

Table of Contents

Preface

Agricultural production refers to the animal and plant production that is made available for human consumption. Whereas, the management and preservation of the forests is known as forestry. Forestry also aims at utilizing the forest resources for commercial production of timber, fuel wood, etc. Some of the sources used in increasing agricultural productivity are mechanization, fertilizers, irrigation, pesticides, animal feed and genetic engineering. This book provides comprehensive insights into the field of agricultural production and forestry. Those with an interest in the agricultural production and forestry would find this book helpful.

After months of intensive research and writing, this book is the end result of all who devoted their time and efforts in the initiation and progress of this book. It will surely be a source of reference in enhancing the required knowledge of the new developments in the area. During the course of developing this book, certain measures such as accuracy, authenticity and research focused analytical studies were given preference in order to produce a comprehensive book in the area of study.

This book would not have been possible without the efforts of the authors and the publisher. I extend my sincere thanks to them. Secondly, I express my gratitude to my family and well-wishers. And most importantly, I thank my students for constantly expressing their willingness and curiosity in enhancing their knowledge in the field, which encourages me to take up further research projects for the advancement of the area.

Editor

Effect of Provenance and Storage Agroecology on Duration of Yam (*Dioscorea rotundata* Poir.) Tuber Dormancy

Elsie Ihuakwu Hamadina[1], Robert Asiedu[2]

[1]Crop and Soil Science Department, Faculty of Agriculture, University of Port Harcourt, Port Harcourt, Nigeria
[2]International Institute of Tropical Agriculture, IITA, Ibadan, Nigeria

Email address:
elsieile@yahoo.com (E. I. Hamadina)

Abstract: Crop improvement in yam is slow due to poor understanding of tuber dormancy. Tuber provenance and storage agroecology are thought to affect the duration to sprouting in yam, but systematic studies on the role of these factors are rare. The objective of this study was to determine the effects of tuber provenance and storage agroecology on the duration to sprouting in *D. rotundata*. Twenty landraces [comprising 7 originating from the Guinea savanna (GS), 5 from the Forest/transition (TS), and 8 from the Humid forest (HF)] were collected and multiplied at a location in their respective agroecologies: Abuja (GS), Ibadan (TS) and Onne (HF). Thereafter, 100 tubers of each of the 20 landraces were stored at each of the three sites, and dates of the appearance of shoot bud (ASB)/ sprouting were recorded. The results showed that provenance did not significantly affect the duration from planting to sprouting or the duration from date in storage to sprouting in *D. rotundata*. The duration to ASB varied by up to 21 days (d) for landraces originating from HF and TS, and 37 d for landraces originating from GS. Variations among landraces within a provenance group were greater than between provenance groups. This suggests that the provenance of a landrace is not a major factor controlling the duration to ASB. All landraces responded to storage agroecology/ environment in a similar manner. There was no interaction between provenance and storage environment. Tubers stored at Onne and Ibadan sprouted about 10 d earlier than those at Abuja, and this was associated with slightly higher temperature and RH at Onne and Ibadan. Storage agroecology is an important factor controlling the duration to sprouting but provenance is not.

Keywords: Provenance, *Dioscorea Rotundata*, Humid Forest, Guinea Savanna, Forest Transition

1. Introduction

Dormancy is an adaptive trait associated with distinct seasons that vary in time and space ([1], [2]). In yam, the timing of the end of the growing season (marked by the commencement of vine senescence) coincides with the start of the dry season. Also, these seasons are associated with the start of yam tuber dormancy while the onset of the rainy season is associated with the onset of sprouting. The association of these two events with distinct seasons of the year highlights their ecological importance.

Duration of dormancy is regulated by both endogenous and environmental mechanism(s) ([3], [4]). In yams, tuber dormancy is long (90 to 150 d) and variable. Variability in the duration of dormancy has been attributed to differences in species, variety, crop management, date of tuber harvest, growth, and storage environment ([5], [6]). Many studies

have been conducted to understand the endogenous and environmental mechanism(s) that regulate the duration of yam dormancy. These studies have been successful at identifying the roles of endogenous plant growth regulators (PGRs) and storage temperature on dormancy. While it is clear that PGRs and temperature can affect dormancy, their effects have been minimal [7].

One aspect of environmental control of dormancy that has been poorly studied is the effect of provenance/adaptation on the duration of dormancy. A very early indication of a possible role of agroecology of origin on differences in duration of tuber dormancy was reported by Coursey, 1976 [8] and Passam, 1982 [6]. In Coursey, 1967 [10] tubers of *D. elephantipes* grown in semi-desert area were observed to exhibit dormancy that is as long as the dry period. Also, the work by Ile, 2004 [11] suggests that *D. cayenensis* that is found more in the high rainfall ecological zone (where the growing season is long and dry season is short) has a long growing period and tends to sprout slightly earlier than *D. rotundata* is found more in the moist savanna. However, this assertion has not been systematically studied. Within species differences in duration of dormancy is also common but the reason for this is unclear. We ask therefore, whether the duration of dormancy in *D. rotundata* is under some ecological control determined by adaptation to growing and storage condition. Therefore, the objectives of the study were to determine the effects of tuber provenance and storage agroecology on the duration to sprouting in *D. rotundata*.

2. Materials and Methods

2.1. Experimental Locations

The experiments were conducted in three IITA stations located at Abuja, Ibadan, and Onne in Nigeria. These locations cut across three yam producing agroecological zones from a mid-altitude northern part to a low altitude southern part of Nigeria. Abuja, Ibadan and Onne represent the Guinea savanna (GS), Forest/ savanna (TS) and the Humid rainforest (HF), agroecological zones respectively (Fig. 1).

Selected ecological characteristics of the above locations are presented in Table 1. The locations are different in altitude, annual total rainfall (total, duration and distribution pattern) and hence in the duration of the growing season. They are also different in annual mean relative humidity and less clearly in annual mean air temperature. However, the annual mean values for average minimum and maximum air temperatures indicate that night and day temperatures are about 1°C warmer and cooler respectively from the north (Abuja) to south (Onne). Abuja is distinctly less humid compared to Ibadan or Onne. Rainfall pattern at Abuja and Onne are similar with a peak in September. However, it is distributed over much of the year at Onne compared to Abuja, resulting in a longer growing season at Onne (approx. 300 d) than at Abuja (approx. 210 d). At Ibadan, rainfall peaks in June and again in September with a break in August and the duration of the rainy/ growing season is about 270 d

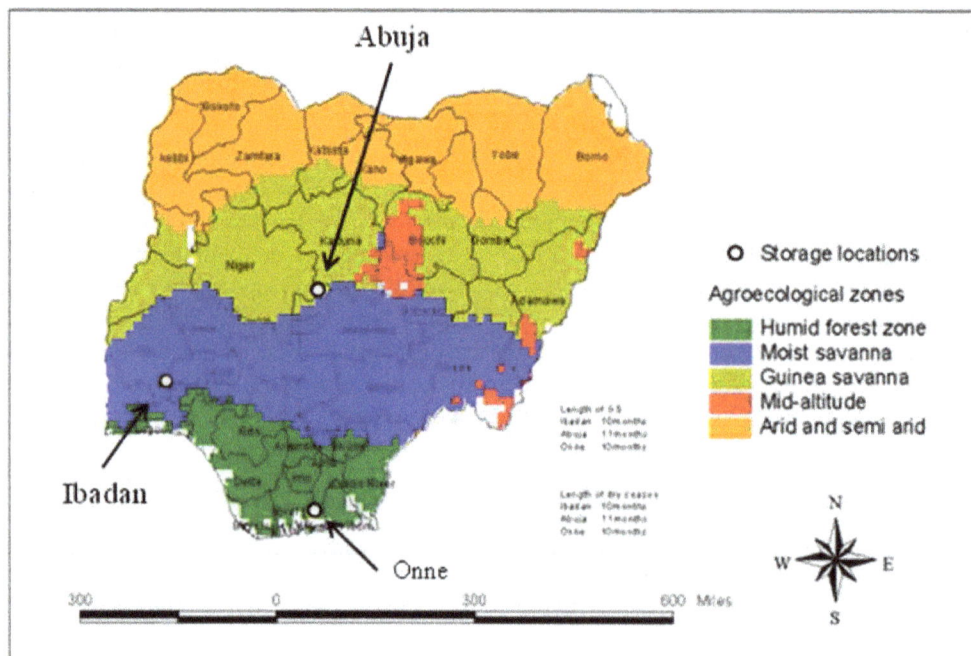

Figure 1. *Agro-ecological zones in Nigeria showing study locations. Source: IITA, Ibadan, Agroecological Unit- Modified.*

Table 1. *Ecological characteristics of three selected locations representing the three agroecological zones in Nigeria.*

Location	Co-ordinates (Long., Lat.)	Altitude (m)	Annual rainfall (mm)	Rainfall pattern	Annual average temperature min/max (°C)	Annual average relative humidity min/max (%)	Duration of rainy season
Abuja	9°16'N; 7°20'E	300	1302	Unimodal	21.3 / 33.3	14.4 / 37.4	Apr.-Oct.
Ibadan	7°26'N; 3°54'E	210	1253	Bimodal	22.0 / 32.0	55.6 / 87.0	Mar.-Nov.
Onne	4°46'N; 7°10'E	10	2400	Unimodal	23.0 / 31.0	62.8 / 94.5	Feb.-Nov.

Source: Jagtap (1993)

2.2. Collection of Landraces and Tuber Multiplication

Twenty-two yam varieties (landraces) indigenous to the GS, TS and HF agroecologies were collected from farmers in each agroecology. Eight of these landraces were collected from the GS, five from the TS and seven from the HF (Table 2). Tubers were then multiplied (by planting mini-setts); landraces from GS were multiplied at Abuja while those from TS and HF were multiplied at Ibadan and Onne respectively.

They were multiplied within their agroecology of origin to avoid any possible effects of growing location.

At Abuja, planting was done in May while at Ibadan and Onne planting was done in April. Tubers were harvested at vine senescence on December at Abuja, Onne and Ibadan respectively.

Table 2. Dioscorea rotundata landraces collected across three agro-ecological zones of the yam belt of Nigeria.

Agro-ecological zone	Acc. No.	Local name	Co-ordinates
Guinea savanna	Jan-99	Giwa	9o 22'N;6o 18'E
Guinea savanna	Feb-99	Suba	9o 05'N;6o 38'E
Guinea savanna	Mar-99	Akwuki	9o 05'N;6c 38'E
Guinea savanna	Apr-99	Maisaki	10o 85'N;7o 33'E
Guinea savanna	May-99	Kpakogi	9o 18'N;6o 15'E
Guinea savanna	Jun-99	Yar-ganye	10o 52'N;7o 34'E
Guinea savanna	Jul-99	Chikakwudu	11o 08'N;7o 34'E
Guinea savanna	Aug-99	Amara	11 o 05'N;7 o 31'E
Forest/ savanna	Sep-99	Lasirin	7o 39'N;3o 39'E
Forest/ savanna	Oct-99	Ajelanwa	7o 39'N;3o 39'E
Forest/ savanna	Nov-99	Ayin	7o 40'N;3o 45'E
Forest/ savanna	Dec-99	Ehuru	7o 38'N;3o 40'E
Forest/ savanna	99-13	Omi-efu	7o 38'N;3o 40'E
Humid rainforest	99-14	Ekpe	6o 20'N;6o 50'E
Humid rainforest	99-15	Adaka	6o 20'N;6o 50'E
Humid rainforest	99-16	Nwopoko	6o 40'N;7o 22'E
Humid rainforest	99-17	Abi	6o 40'N;6o 48'E
Humid rainforest	99-18	Obiaturugo	5o 49'N;7o 26'E
Humid rainforest	99-19	Azia	4c 43'N;7o 18'E
Humid rainforest	99-20	Okom	5o 19'N;7o 21'E
Humid rainforest	99-21	Bilazia	4o 40'N;7o 23'E
Humid rainforest	99-22	Agbaka	5o19'N;7o 21'E

2.3. Reciprocal Transfer of Tubers from Agroecology of Origin to Storage Locations

Two 99-8 (from the GS) and 99-22 (from the HF) out of the 22 landraces were not included in the sprouting/storage study, thus twenty landraces were used in this study. Tubers of each landrace were divided into three groups. Each third of the groups was stored at one of the three storage locations. These transfers were completed within five days of harvest and tubers are considered as being in their various growth agroecologies until they arrive at the storage location. Thus, at each storage location, there were 20 landraces replicated 100 times; there were 100 tubers/ landrace and each tuber was considered a replicate. Individual tubers were weighed, numbered and placed in open wooden or plastic boxes, and placed on shelves in the yam barn.

2.4. Data Collection and Analysis

Temperature and relative humidity (RH) in the yam barns were recorded every 2 h using Tinytalk® (2 nos.) temperature and relative humidity sensors. These readings were taken in duplicate. Individual tubers were observed every 7 d for the appearance of shoot bud on the tuber surface (ASB or sprouting). The data analyzed was duration from harvest to sprouting/ASB. Data was analyzed as a complete 32 factorial experiment, consisting of three provenances (20 landraces nested in provenances) and three storage locations with individual tubers as replicates. SAS V8, GLM procedure was used to run this analysis.

3. Results

3.1. Temperature and Relative Humidity in the Yam Barns

The major difference between storage locations was minimum temperature (^{o}C). Abuja had the coolest night temperature (14.6 ^{o}C) while the nights were warmer by 2 and 5 ^{o}C at Ibadan and Onne respectively. Maximum and average temperatures were similar across storage locations, average temperature values were 26.2 and 26.8 ^{o}C. On the other hand, average RH at Abuja was 42 % while it was 55 % at Ibadan.

3.2. Effects of Provenance and Storage Location on Duration from Planting to Sprouting and Storage to Sprouting

Since planting (multiplication) date was different at Abuja compared to Onne, it was necessary to determine whether the effects of provenance and storage location on date of sprouting were confounded with the duration from planting to sprouting. Further because tubers arrived at the storage locations on slightly different dates, it was also necessary to determine whether the effects of provenance and storage location on the date of sprouting can be accounted for by differences in the date of the start of storage or the duration from storage to sprouting.

The duration from planting to sprouting was significantly affected by provenance (p=0.001) and storage location (p=0.001) but not their interaction. The duration from planting to sprouting was 288 d for tubers originating from GS. The duration from planting to sprouting was 22 d longer for tubers originating from HF than those from GS (Table 2) while it was 23 d longer for tubers from TS compared to those from GS. There was no significant difference in the duration from planting to sprouting in HF tubers compared to TS tubers. The fact that planting was done 24 d later at GS compared to HF or TS indicated that the 22 d shorter duration to sprouting observed at GS was associated with the duration from planting to sprouting. Thus the duration from planting to spouting in this case may not clearly reflect the effects of provenance.

The effect of provenance on the duration from storage to sprouting showed that tubers from the GS had the longest (74 d) mean duration from storage to sprouting (Table 3). Compared to tubers from the GS, the duration from storage to sprouting was significantly shorter by 4 and 8 d for tubers originating from TS and HF respectively. The fact that storage began 6 d later for tubers originating from HF compared to those from GS suggests that storage date was confounded in provenance.

In contrast, the effect of storage location on the duration from planting to sprouting showed that the duration was significantly (p=0.001) shorter by 10 d if tubers were stored at Onne (HF) compared to storage at Abuja (GS) (Table 4). Although planting was done on the same dates at Onne and Ibadan, the duration from planting to sprouting was 5 d shorter at Onne (p=0.05) compared to Ibadan. Similarly, the effect of storage location on the duration from storage date to sprouting showed that storage at Abuja resulted in the longest duration to sprouting, with sprouting occurring after a mean of 74 d (Table 4). Compared with storage at Abuja, storage at Ibadan and Onne resulted in 4 and 9 d shorter durations to sprouting respectively. Date of start of storage was not confounded in storage location since the average start date of storage varied by only 1 to 2 d.

Thus, storage location affected the duration from planting to sprouting by 4 and 8 d while storage location affected the duration from storage to sprouting by the 4 to 9 d. in contrast, provenance did not significantly affect the duration from planting to sprouting or the duration from storage to spouting.

Table 3. Effect of provenance on the duration from planting to sprouting and the duration from storage to sprouting.

Provenance	Duration from planting to sprouting			Duration from storage to sprouting		
	Planting date (DOY)	Mean (d)	SE	Average days in storage (DOY)	Mean(d)	SE
GS	136	288	0.62	351	74	0.62
TS	112	311	0.66	354	70	0.66
HF	112	310	0.51	357	66	0.51

Table 4. Effect of storage location on the duration from planting to sprouting and the duration from storage to sprouting.

Provenance	Duration from planting to sprouting			Duration from storage to sprouting		
	Planting date (DOY)	Mean (d)	SE	Average days in storage (DOY)	Mean(d)	SE
Abuja	136	308	0.62	355	74	0.62
Ibadan	112	303	0.61	354	70	0.61
Onne	112	298	0.57	353	66	0.57

4. Discussion

The yam belt of Nigeria, cuts across, from the North to the South, three distinct agro-ecological zones; Guinea savannah, Forest savannah transition and the Rainforest [12], which vary in latitude, rainfall pattern/start and duration of the rainy and dry seasons, solar radiation and photoperiod. This study followed a carefully planned method to determine whether the differences in duration to sprouting within the species of *D. rotundata* was caused by differences in tuber provenance and storage agroecology

This study is one study that evaluated the sprouting date of a very large number (over five thousand tubers) of *D. rotundata* tubers, and it was clear that the duration to sprouting which is an important indication of the duration of visible dormancy is highly plastic. This duration could be as long as < 311 d (as in this study) when considered as a duration from planting, between the months of April and early May, to sprouting or as short as < 74 d when considered as a duration from storage in mid-December to sprouting. Furthermore, the fact that a small (less than 6 d) difference in start dates can lead to statistical significant result emphasizes the need to have uniform start dates across treatments. Thus, yam dormancy studies that evaluate the duration of visible dormancy must provide adequate information on the start date, preferably in days of the year (DOY), for the evaluation of duration of dormancy.

This study also found that differences in the duration from a defined planting date to sprouting or from a defined start of storage to sprouting in *D. rotundata* is not significantly influenced by tuber provenance. Indeed, the variation in date of sprouting was greater among landraces (up to 37 d) within a provenance than among provenances (1 to 2 d). The mean date of sprouting among provenance groups ranged from 55 to 57 DOY and it followed the order GS > TS > HF. The uniformity in sprouting date among provenance and presence of landraces with early and late sprouting times indicated that, though the 20 landraces used in this study were collected from different agroecologies, which vary in the timing of the start and length of growing seasons, there was no clear evidence to suggest that they were developed due to adaptation to the lengths of the wet-growing and dry-storage period at the various agroecologies. Thus while variations in sprouting date among yam species may be due to differences in agroecology of origin ([6], [8], [13]), the difference among landraces within the species of *D. rotundata* may not be related to differences in agroecology of origin within Nigeria. The next question would be are the landraces used in this study truly specific to the respective agroecologies? This question can best be ascertained through molecular analysis. However, this study has assumed that they are specific to the respective agroecologies because their local names depict their origin. Further, their popularity among farmers in the various agroecological zones suggested that they are well known. In support of this, is the report of Okoli, 1980 [14], which said that the landrace 'Ekpe' is adapted to Anambra State in South-east Nigeria, and in this study 99-14 is described as 'Ekpe' and was collected from Anambra State in the HF. The presence of both early and late sprouting landraces among provenances supports the suggestion of Alexander and Coursey, 1969 [15] that farmers have selected for varieties with varied sprouting dates to meet their short and long term need of the tuber. The evolution of *D. rotundata* suggests that it is the result of hybridization between a rain forest species, *D, cayenensis* with short dormant period due to short dry season, and a wild species from the Guinea savanna with slightly longer dormant period due to longer dry season [8]. Therefore, it is proposed that the presence of landraces with varying dates of sprouting may have developed due to differences in the genes for length of dormancy or difference in the dominance of the genes selected for by farmers.

Results from thousands of tubers evaluated in this study showed that the earliest time of sprouting occurred in late December, but mostly within the first three months of the year ([16], [17], [18]). The occurrence of sprouting at a defined period or season (at end of the Harmattan season to the beginning of the rains) in the year in Nigeria such as between late December of previous year and April of the next year, suggests that landraces have developed life cycles/stages that fit the annual seasonal cycle. More specifically, the fact that sprouting is not restricted to beginning in April or May which marks the onset of the rains in the GS for example, negates the suggestion that the time of sprouting coincides with the onset of the rainy season and the time of planting ([19], [20]) per se. It is therefore suggested that it is not the actual start of the growing season that is the cue to start sprouting in Ibadan or Abuja particularly but rather that landraces have adapted or developed life cycles that fit the annual seasonal cycle and that sprouting always occurs before the onset of the rains because they have internal mechanism(s) that determine the onset of the next growing season. Okoli, 1980 [14] and Degras, 1993 [13] have long ago suggested also that the timing of the end of dormancy/start of sprouting is regulated by an endogenous biological 'clock'.

The effect of storage environment (in the barn or controlled environment facility) on yam dormancy is known but the effect of storing tubers in agroecologies other than their adapted one have not been investigated. In contrast to provenance, storage agroecology affects the duration of sprouting. Sprouting was earliest at Onne and latest at Abuja with Ibadan being intermediate. Generally, sprouting was 10 d earlier if tubers were stored at Onne compared to storage at Abuja Therefore, date of sprouting was clearly dependent on inherent genetic makeup of the landraces and the agroecology where they are stored, with Onne being more favorable for early sprouting. Earlier sprouting was probably encouraged at Onne due to the 1 to 2 ^{o}C warmer average storage temperatures [21] and higher relative humidity than that at Abuja [22]. The rate of progress of plant developmental events has been shown severally to increase with increasing temperature up to the optimum temperature for the process [1]. In the past, temperature has also been shown to affect the duration to sprouting in yam. It has been shortened by 10 to 16 d by storing tubers at 30 ^{o}C compared to 20 ^{o}C [17]. The effect of temperature is further enhanced when high temperatures of 30 to 35 ^{o}C are combined with high relative humidity in the range 70 to 80-85 % ([22], [17]). Therefore, the small (10 d) degree of effect of storage agroecology on date of sprouting indicated that the effects of storage location on date of sprouting may be related to their ability to affect the progress of Phase 2 of dormancy [23].

5. Conclusion

In summary, this study has shown that the differences in the date of sprouting between landraces within the species of

D. rotundata were not be due to differences in their agroecology of origin, because landraces with early and late sprouting time are present at all agroecologies, but rather due to their storage agroecologies.

References

[1] Roberts, E. H. and Summerfield, R. J. (1987). Measurement and Prediction of Flowering in Annual Crops. In: *Manipulation of Flowering* (eds J. G. Atherton), pp.17-50. Butterworths, London.

[2] Roberts, E. H. and Summerfield, R. J. (1987). Measurement and Prediction of Flowering in Annual Crops. In: *Manipulation of Flowering* (eds J. G. Atherton), pp.17-50. Butterworths, London.

[3] Leopold, C. A. (1996). Natural History of Seed Dormancy. In: *Plant Dormancy: Physiology, Biochemistry and Molecular Biology* (ed G. A. Lang), pp.3-16. CAB International, Wallingford.

[4] Lang, A. (1952). Physiology of flowering. *Annual Review of Plant Physiology,* 3, 265-306.

[5] Lang, G. A., Early, J. D., Martin, G. C. and Darnell, R. (1987). Endo-, para-, and ecodormancy: physiological terminology and classifications for dormancy research. *HortScience,* 22, 371-377.

[6] Passam, H. C. (1982). Dormancy of Yams in Relation to Storage. In: *Yams- Ignames* (eds J. Miege and N. Lyonga), pp.285-293. Oxford: OUP.

[7] IITA, (1997). Project 13- Improvement of Yam-base Systems. Annual Report 1997. IITA Ibadan, Nigeria.

[8] Barker, D. J., Keatinge, J. D. H. and Asiedu, R. (1999). Yam dormancy: potential mechanisms for its manipulation. *Tropical Science,* 39, 168-177.

[9] Coursey, D. G. (1976). The Origins and Domestication of Yams in Africa. In: *Origins of African Plant plant Domestication* (ed. J. R. Harlan), pp.383-408. Mouton, La Hague.

[10] Coursey, D. G. (1967). *Yams*. Longmans, Green and Co. Ltd, London

[11] Ile, E. I. (2004). Control of Tuber Dormancy and Flowering in Yam (Dioscorea rotundata Poir.) tuber. PhD thesis, The University of Reading, Reading UK.

[12] Orkwor, G. C. (1998). The importance of yams. In: *Food Yams: Advances in Research* (eds G.C. Orkwor, R. Asiedu and I. J. Ekanayake), pp. 1-12. NRCRI and IITA, Ibadan, Nigeria.

[13] Degras, L. (1993). *The Yam: A Tropical Root Crop.* (ed R. Costo). MacMillan, London.

[14] Okoli, O. O. (1980). Dry matter accumulation and tuber sprouting in yams Dioscorea spp.). *Experimental Agriculture,* 16, 161-167.

[15] Alexander, J. and Coursey, D. G. (1969). The Origins of Yam Cultivation. In: *The Domestication and Exploitation of Plants* (eds P. J. Ucko and G. W. Dimbleby), pp. 405- 425. London: Duckworth.

[16] Arnolin, R. (1982). Vegetative Cycle of the Yam *D. alata* cv. Tahiti and Belep: Influence of Spaced Planting. In: *Proceedings 18th Annual Meeting, Caribbean Food Crops Society, Mayaguez, Puerto Rico University, Puerto Rico.* pp.146-169. Caribbean Food Crops Society

[17] Swanell, M. C., Wheeler, T. R., Asiedu, R. and Craufurd, P. Q. (2003). Effect of harvest date on the dormancy period of yam (*Dioscorea rotundata*). *Tropical Science* 2003, 43, 103-107.

[18] Shiwachi, H., Ayankanmi, T., Asiedu, R. and Onjo, M. (2003). Induction of germination in dormant yam (*Dioscorea* spp.) tubers with inhibitors of gibberellins. *Experimental Agriculture,* 39, 209-217.

[19] Njoku, E. (1963). The propagation of yams *Dioscorea spp.* by vine cuttings. *Journal of West African Science Association,* 18, 29-32.

[20] U.S.D.A, (United States Departments of Agriculture) (1972). Yam Production Methods. In: *Production Research Report* No. 147. U.S.D.A. USA.

[21] Passam, H. C. (1977). Sprouting and apical dominance of yam tuber. *Tropical Science,* 19, 29- 39.

[22] Mozie, O. (1984). Influence of ventilation and humidity during storage on weight and quality changes of white yam tubers *Dioscorea rotundata* Poir. *Journal of the University of Puerto Rico,* 68, 341-348.

[23] Ile, E. I. P. Q. Craufurd, N. H. Battey and R. Asiedu (2006). Phases of tuber dormancy in yam (*Dioscorea rotundata* Poir.), *Annals of Botany 97;* 497-504.

2

Farmers' Awareness of the Effects of Climate on Growth and Yield of Potato (Solanum Tuberosum) in Jos-South Local Government Area of Plateau State, Nigeria

Wuyep Solomon Zitta[1, *], Samuel Akintayo Akinseye[2], Yakubu Pwajok Mwanja[3]

[1]Department of Geography, Plateau State University Bokkos, Nigeria and Department of Geography, Environmental Management and Energy Studies, University of Johannesburg, South Africa

[2]Department of Geography, Environmental Management and Energy Studies, University of Johannesburg, South Africa

[3]Department of Microbiology, Plateau State University Bokkos, Nigeria and Department of Botany and Plant Biotechnology, University of Johannesburg, South Africa

Email address:

wuyepsol@yahoo.com (S. Z. Wuyep), sammiesam31@live.co.za (S. A. Akinseye), ubukay30@yahoo.com (Y. P. Mwanja)

Abstract: This study assessed farmers' knowledge on the effect of climate on growth and yield of potato in Jos -South Plateau State. The instrument of data collection used for this study include structured questionnaire. Purposive sampling design was followed in the selection of 200 farmers. The study was undertaken in four districts of Jos- South (Du, Vwang, Kuru and Gyel). One village was selected from each district using the simple random technique to avoid bias. Primary data collected from the farmers include socio-economic characteristics such as gender, age, marital status, years of farming experience as well as relevant questions in order to assess farmers' knowledge on the effect of climate on growth and yield of potato. Descriptive techniques of data were employed such as simple percentages to describe the knowledge of the respondents. Findings indicate that farmers have good knowledge of the effect of climate on growth and yield of potato. The effect identified includes reduced yield due to excessive rainfall during tuber bulking stage. The disease by late-blight is the most important disease that reduces the yield in the study area. It reduces between 40-80% of the total yield. However, there is need for adequate knowledge of the effect of climate on potato and adaptative strategies.

Keywords: Climate, Farmers, Potato, Knowledge, Awareness, Yield

1. Introduction

In recent years, there has been a growing awareness that scientific knowledge alone is inadequate for solving climate crisis. Thus, the indigenous local farmers have been recognised as powerful knowledge holders on climate change and key factors for developing policy to mitigate and cope with its effects. The knowledge of the local and indigenous farmers is increasingly recognised as important source of climate knowledge and adaptation strategies (Natural Sciences, 2012). Traditional knowledge, innovations and adaptation practices embody local adaptative management to the changing environment and compliment scientific research, observations and monitoring (International indigenous People's Forum on Climate Change, 2009).

Most local communities possess traditional and local knowledge that may help them better adapt to the impact of climate. Some communities are using traditional knowledge to record their observations of climate and its impact on the environment as a result of their close relationship with land and their dependence on natural resources for their livelihoods and have long been observing and noting the impact of climate conditions (UNESCO, 2010).

Unless appropriate mitigative and adaptive measures are taken, climate change will frustrate farmers' efforts to achieve sustainable agriculture production and food security. However, developing such strategies will require information from the farmers since the ability to adapt and cope with climate change depends on the knowledge, skills, experiences and other socio-economic factors (Maharjan *et al.*, 2011). It is against this background that this study seeks to assess farmers' awareness on the effect of climate on growth and yield of Potato (Solanum tuberosum) in Jos-South Local Government Area of Plateau State, Nigeria.

2. Study Area and Methods

This study assesses farmers' knowledge on the effect of climate on growth and yield of potato. Jos- South-Local Government Area of Plateau State Nigeria is one of the seventeen local governments in Plateau State. It is made up of four districts: Vwang, Du, Gyel, and Kuru. The local government area has its headquarters in Bukuru. It lies on latitude 8° 43° N and longitude 8° 46° N with an altitude of 1293.2m above sea level. The local government area is bounded by Barkin- Ladi local government to the South, Riyom local government to the South West, Jos-East local government to the East and Bassa Local Government to the West (Figure. 1). The local Government has a population of 650,835 (National population Commission, 2006) with an average land area of 1,037km^2.

Figure 1. Location map of study areas

This paper makes use of two types of data. The first data is secondary sourced materials from past studies e.g. books and journals. The second data is from administering of questionnaire. Purposive sampling design was followed in the selection of 200 farmers. The study was conducted in four districts of Jos-South (Du, Vwang, Kuru and Gyel).One village was selected from each district using the simple random technique. Descriptive techniques of data were employed such as simple percentages to describe the knowledge of the respondents.

3. Results and Discussion

The number of respondents per location sampled reveals that the Kuru location has the highest with Vwang location

recording the lowest number of respondents (Figure 2). The sex distribution ratio per each location was depicted in table 1. The results revealed that male and female constitute 47% and 54% respectively. This implies that more females are into farming potato than males.

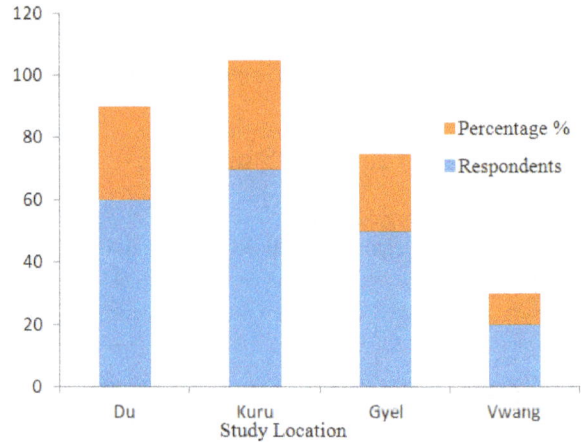

Figure 2. Number of respondents.

Table 1. Sex distribution of respondents.

Location	Du		Kuru		Gyel		Vwang			
Sex	F	%	F	%	F	%	F	%	Total	%
No of Males	29	48.30	30	42.9	22	44.0	12	60	93	46.5
No of Females	31	51.7	40	57.1	28	56.0	8	40	107	53.5
Total	60	100	70	100	50	100	20	100	200	100

Key: F: Frequency Source: Field Survey, 2014

Table 2 shows both young and old people are involved in farming. The distribution shows that about 46% of the respondents were between 31-40 years of age. Respondents that were over 41years of age constituted about 14%.This imply that most of the respondent (about 86%) was relatively young and physically active. This has a direct bearing on the availability of able-bodied manpower for agricultural production and also, age influence the ability to seek and obtain off-farm jobs and income which could increase.

Table 2. Age range of respondents.

Location	Du		Kuru		Gyel		Vwang			
Age group	F	%	F	%	F	%	F	%	Total	%
20-30	30	50	38	54.3					92	46
31-40	23	38.3	25	35.7	22	44	10	50	80	40
41-50	7	11.7	7	10	10	20	4	20	28	14
Total	60	100	70	100	50	100	20	100	200	100

Key: F: Frequency Source: Field Survey, 2014

The literacy level of farmers which is an important factor that determines the ability of a farmer to understand policies and programmes relating to climate was employed in this research. The distribution of the respondents per each location is presented in Table 3. The table revealed that 8% of the respondents had no formal education, 33% attained primary education, 43% had secondary education while 17% attained tertiary education. Thus, 92% of the respondents have some formal education. This study has revealed that literacy level is high amongst the respondents and this could have implications for agriculture production.

Table 3. Educational level of famers.

Location	Du		Kuru		Gyel		Vwang			
Educational level	F	%	F	%	F	%	F	%	Total	%
Primary School	18	30	17	24.3	20	40	10	50	65	32.5
Secondary School	22	36.7	40	57.1	18	36	5	25	85	42.5
Tertiary Education	18	30	10	14.3	2	4	4	20	34	17.0
No Formal Education	2	3.3	3	4.3	10	20	1	5	16	8
Total	60	100	70	100	50	100	20	100	200	100

Key: F: Frequency Source: Field Survey, 2014

Information in table 4 indicates that 49% of the respondents have been farming potato for 11-20 years. 30% of the respondents have been farming potato for 1-10 years while 22% have been farming Potato for 21-30 years. This implies that knowledge of the respondents on the effects of climate that affect the growth and yield of potato develop as they put more years in potato farming. They become more matured and conscious thereby gaining more experience on the understanding of the environment. This statement corroborates with the study of Boulding (1956) who suggested that over time, individuals developed mental impressions of the world through their everyday contacts with the environments, with these impressions knowledge acting as the basis for their behaviour.

Table 4. Years of farming experience by respondents.

Location	Du		Kuru		Gyel		Vwang			
Year	F	%	F	%	F	%	F	%	Total	%
1-10	18	30	22	31.4	11	22	8	40	59	29.5
11-20	30	50	28	40	30	60	9	45	97	48.5
21-30	12	20	20	28.6	9	18	3	15	44	22.0
Total	60	100	70	100	50	100	20	100	200	100

Key: F: Frequency Source: Field Survey, 2014

Table 5 reveals that 29% of the respondents said soil temperature affects the germination and growth of potato, 18% said air temperature affects the germination and growth of potato while 54% said rainfall affects the germination and growth of potato. This farmers' observation on climatic

elements corroborated the findings of Wuyep *et al.*, 2013 that precipitation effectiveness is important for good germination or sustained growth of potato which have effect on the final yield. It also confirmed the assertion made by Ifenkwe and Okonkwo (1983) that during the rainy season, time of planting depends on the onset of rain when rain becomes stable usually between the last week of April and first week of May.

Table 5. Respondents knowledge on climatic elements affecting germination of Irish potato.

Location	Du		Kuru		Gyel		Vwang			
Climatic elements	F	%	F	%	F	%	F	%	Total	%
Rainfall	30	50	39	55.7	29	58	9	45	107	53.5
Air temperature	20	33.3	11	15.7	2	4	2	10	35	17.5
Soil temperature	10	16.7	20	28.6	19	38	9	45	58	29
Total	60	100	70	100	50	100	20	100	200	100

Key: F: Frequency Source: Field Survey, 2014

Respondents' knowledge with regards to time of planting and harvesting of potato was analysed. It was revealed that 76% planted potato in April while 76% of the farmers also harvested potato in July Table 6. This farmers' observation corroborated the findings of Ifenkwe and Okonkwo, 1983 that in Jos–South, potato is planted when rain becomes stable usually between the last week of April. Also, Ifenkwe, 1989 reported that yield declined with delay in date of planting probably as a result of premature killing of plants by late-blight disease.

Table 6. Respondents knowledge on the time of planting and harvesting of irish potato.

Location	Du		Kuru		Gyel		Vwang			
	F	%	F	%	F	%	F	%	Total	%
April planting period	39	65	50	71.4	46	92	16	80	151	75.5
May planting period	21	35	20	28.6	4	8	4	20	49	24.5
Total	60	100	70	100	50	100	20	100	200	100
July harvesting period	39	65	50	71.4	46	92	16	80	151	75.5
August harvesting period	21	35	20	28.6	4	8	4	20	49	24.5
Total	60	100	70	100	50	100	20	100	200	100

Key: F: Frequency Source: Field Survey, 2014.

Table 7. Respondents knowledge on increase or decrease in temperature.

Location	Du		Kuru		Gyel		Vwang			
	F	%	F	%	F	%	F	%	Total	%
Temperature increased	40	66.7	62	88.6	30	60	19	95	151	75.5
Temperature decreased	20	33.8	8	11.4	20	40	1	5	49	24.5
Total	60	100	70	100	50	100	20	100	200	100

Key: F: Frequency Source: Field Survey, 2014

Table 7 shows that 76% noticed an increase in temperature while 24.5% of the respondents stated that temperature trend in the study area has decreased. The farmers' assessment agreed with the expert report Zemba et al., (2013) that temperature has increased in the Jos-Plateau and the yield of tuber is on the decline due to knobbiness and secondary growth at emergence/vegetative stage.

Table 8. *Respondents knowledge on decreased and increased rainfall trend in the last five years.*

Location	Du		Kuru		Gyel		Vwang			
	F	%	F	%	F	%	F	%	Total	%
Decrease in rainfall	25	41.7	20	28.6	8	16	7	35	60	30
Increase in rainfall	35	58.3	50	71.4	42	84	13	65	140	70
Total	60	100	70	100	50	100	20	100	200	100

Key: F: Frequency Source: Field Survey, 2014

The distribution of the respondents according to their assessment of rainfall trend in the area is represented in Table 8. Majority of the respondents 70% noticed an increase in rainfall while 30% opined that rainfall trend has been on a decrease. This implies that the more the rain at sprouting to emergence/vegetative and tuber set/initiation stage, the better the growth and yield of potato. This finding agrees with Zemba et al., (2013) that the total rainfall correlate significantly (r=0.470) with potato at 5% probability level.

Table 9 shows that 74% of the respondents indicated that rainfall trend affect the final yield of potato, 16% said minimum temperature affect the final yield of potato while 10% indicated that maximum temperature affect the final yield of potato. This implies that the higher the rainfall at tuber bulking stage, the lower the yield. This finding agrees with Wuyep et al., (2013) that high rainfall during tuber initiation of potato is not healthy to the crop as it causes poor aeration and subsequently poor development of tubers. Also, this finding corroborate with Zemba et al., (2013) that rainfall amount has negative correlation coefficient of -0.665 at 1% level of significance. This implies that the higher the amount of rainfall in July, the lower will be the yield. This is not surprising because the month of July coincides with tuber bulking/ripening stage.

Table 9. *Respondent knowledge on variable affecting the final yield of irish potato.*

Location	Du		Kuru		Gyel		Vwang			
Climatic elements	F	%	F	%	F	%	F	%	Total	%
Minimum temperature	4	6.7	25	35.7	1	2	2	10	32	16
Maximum temperature	6	10	3	4.3	10	20	1	5	20	10
Rainfall trend	50	83.3	42	60	39	78	17	85	148	74
Total		60	100	70	100	50	100	20	100	200	100

Key: F: Frequency Source: Field Survey, 2014

Table 10 reveals that 9% are of the opinion that maximum temperature causes lates-blight disease of potato, 14% opined that minimum temperature causes late-blight disease of potato while 76% said rainfall causes late-blight disease of potato. This implies that the more the rain during tuberization, the lower the yield of potato. Thus, the farmers assessment corroborate with the findings of expert Nwakocha (1987) that blight causes between 40-80% reductions in yield. The peak incidence is between July and August when the haulm of most susceptible varieties are destroyed by inciting pathogen phythopthora. This disease is accompanied by high relative humidity, dew and frequent rainfall (Hienfling, 1987).

Table 10. *Respondent knowledge on the causes of late-blight disease of irish potato.*

Location	Du		Kuru		Gyel		Vwang			
Climatic elements	F	%	F	%	F	%	F	%	Total	%
Rainfall trend	50	83.3	50	71.4	37	74	18	90	155	77.5
Maximum temperature	4	6.7	10	14.3	3	6	1	5	18	9
Minimum temperature	6	10	10	14.3	10	20	1	5	27	13.5
Total			60	100	70	100	50	100	20	100	200	100

Key: F: Frequency Source: Field Survey, 2014

Table 11 results shows that 35% of the respondents don't know how to eradicate this disease while 66% are of the opinion that early planting will halt the late-blight disease. This confirms the work of Ifenkwe (1989) that harvesting is carried out in early July to avoid destruction of tubers by inciting pathogen phythopthora.

Table 11. *Respondents knowledge on overcoming late-blight disease of irish potato.*

Location	Du		Kuru		Gyel		Vwang			
	F	%	F	%	F	%	F	%	Total	%
Early planting	43	71.7	40	57.1	37	74	11	55	131	65.5
No knowledge	17	28.3	30	42.9	13	26	9	45	69	34.5
Total	60	100	70	100	50	100	20	100	200	100

Key: F: Frequency Source: Field Survey, 2014.

4. Conclusion and Recommendation

The conclusion drawn from the findings is that there is a good knowledge on the effect of climate on the growth and yield of potato among farmers in the study area. Also, the study depicts that farmers are experiencing the negative effect of climate in form of reduced crop yield. Based on these findings, the following recommendations are proffered

1. The present planting period for potato is found suitable and should be maintained. This will help to maximize the positive effects of climate.
2. Information on climatic data should be collected all over the study area to provide information for long term planning.
3. Late-blight resistance varieties seed of potato should be developed in order to eradicate late-blight disease which reduces 40-80% yield of potato.
4. Seminars and workshop should be organized by the relevant authorities to enlighten the farmers more on effects of climate on growth and yield of potato

References

[1] Ahmed, S.A. (1980). Potato production in Bangladesh in proceeding of third international symposium on Potato production for South-Asia and pacific region held at Bandung, Indonesia.

[2] Boulding, J. (1956) in Adebayo A.A (1989). Man environment relations: The Geographers new points, pp15 unpublished lecture notes.

[3] Burton, W.G. (1989). The potato. Veenman and Zonen, Wageningen. Netherlands, 382.

[4] Hienfling, J.W. (1987). Late-blight of potato. Technical Bill. International potato center (IPC), 12.

[5] Ifenkwe, O.P & Okonkwo,J.C. (1983). Determination of the most suitable time to plant Potato, taking into account the onset of rain. Annual report.of National Root Crop Research Institute, Umudike, Nigeria.

[6] Ifenkwe, O.P. (1989a). Comparison of flat and ridges for dry season planting of potato. Annual report of National Root Crop Research Institute, Umudike, Nigeria.

[7] International Indigenous People's Forum on Climate Change (IIPFCC, 2009). Policy paper on climate change. A policy paper finalized at the IIPFCC meeting in Bangkok, Thailand. September 26-27, 2009.

[8] Kowal, J.M & Andrew, D.J. (1973). "Patterns of water availability and water requirement for grain sorghum production at Samaru, Nigeria. Tropical Agriculture (Trinidad), 50:89-100.

[9] Lopez, D.F, Boe A.A, Johnsen R.H & Jansky,S.H. (1987). Genotype X environment interactions, correlations and coping ability of six traits in potato. American potato, 64:44.

[10] Maharjan, S.K, Sidjel E.R, Sthapit B.R & Regmi, B.R. (2011). Tharu community's perception on climate change and their adaptative initiations to withstand its impacts in Western Terai of Nepal. International N.G.O, 6(2): 35-42.

[11] Natural Sciences (2012). Turning tables on change: Indigenous assessments of impacts and adaptation.http://www.unesco.org/new/en/natural sciences.12/07/2014

[12] Nwokocha, H.N. (1987). Weed interference studies in potato. Annual report, National Root crops Research Institute, Imudeke, 88-93.

[13] UNESCO (2010). The role of traditional and local knowledge on climate adaptation. A session at the 5th Global conference on oceans, coasts and islands. Ensuring survival preserving life and improving governance. May 3-7, 2010 at UNESCO in Paris.

[14] Wuyep, S.Z, Zemba A.A & Jahknwa C.J. (2013). Effects of precipitation effectiveness on the yield of potato (Solanum Tuberosum) in Jos-Plateau, Nigeria. International Journal of Research in Applied Natural and Social Science, 1, 5: 27-32.

[15] Zemba, A.A, Wuyep S.Z. Adebayo A.A & Jahknwa, C.J (2013). Growth and yield response of potato (Solanum Tuberosum) to climate in Jos-South, Plateau State, Nigeria. Global Journal of Human Social Science, xiii, 1: 13-18.

3

Study on Coppice Management of *Acacia nilotica* Tree for Better Woody Biomass Production

Abrham Tezera Gessesse*, **Tesfaye Teklehaymanot Gezahegn, Hailie Shiferaw Wolle**

Amhara Region Agricultural Research Institute, Debre Birhan Agricultural Research Center, Debre Birhan, Ethiopia

Email address:

abrhamtezera@gmail.com (A. T. Gessesse)

Abstract: Over 90% of the energy consumed in the country is depending on woody biomass. The rising demand for tree products and expanding population pressure resulted in decline of forest cover and consequently the demand for tree products exceeded the supply. This study was conducted at Armania Kebele with the aim of to evaluate different cutting time and management practices yielding more wood biomass volume. One indigenous tree species, preferable by the farmers, namely *Acacia nilotica* was selected and the experiment design in 3x4 factorial experiment with randomized complete block design arrangement and replicated three wise. Eight trees per plot were planted in two rows with 2 meter spacing. The result shows that, *Acacia nilotica* could be coppiced well with 13 to 29 numbers of coppices per stump at all stages of cutting time. Root collar diameter and plant height were highly correlated with growth period. Hence, Leaving of two and three number of coppices could give higher woody biomass volumes as compare to leaving one and all number of coppices and control (uncut). Therefore it is necessary that farmers should allow two and three number coppice per stump to get high woody biomass volume for fuel-wood consumption.

Keywords: Acacia, Coppicing, Biomass, *nilotica*

1. Introduction

The rising demand for tree products and expanding population pressure resulted in the decline of forest cover and consequently the demand for tree products exceeded the supply. The supply of wood and woody biomass products in the north Shewa zone comes from different forest and vegetation types and production systems, including natural forests, woodlands, bush lands, community woodlots, and farm forestry. The share of the total domestic energy is: fuel wood and tree residues 70%, dung 8%, agricultural residues 7% and the rest comes from other sources (EFAP, 1994). Hence, 90% of the total energy is coming from biomass. The trend in north Shewa is not different from the national, especially for the lowlands. According to EFAP, 1994 report a large deficit wood has occurred since 1992 (33 million m^3), and fuel-wood deficit amounted to 32.5 million m^3. This deficit has been the main cause for the mining of forest resource base. Based on assumed per capita consumption requirements, in 1992 total requirements for wood products have been estimated to be 47.4 million m^3, of which fuel wood demand was 45 million m^3 (EFAP, 1994). The demand for construction poles and fence posts for construction of new houses and renovation was estimated to be 2.1 million m^3. The total deficit is expected to be 84.2 million m^3 by the year 2014. Thus, being aware of these deficit alternative sources of energy, afforestation (massive plantation) program and efficient utilization and management of forest land should continue to receive attention based on research results for a greater biomass and quality production.

The existing experience on plantation species in northern Shewa zone indicates that most species used are exotics, dominantly Eucalyptus. There is a danger in overlying on a few species. Pest or disease epidemics could wipe the entire species with devastating effects. This will become serious when there is no alternative species to rely on. Focus has given to exotic species due to the information available and their fast growth rate. However, the indigenous species have been blamed for their slow growth rate. As a general truth, for most of the indigenous species there is no information at all and some of the indigenous species are known for their slow growth rate but there are species with fast growth rate, even faster than the exotics.

Study at Sirinka agricultural research centers revealed that *Acacia polyacantha* produces four times more biomass than *Eucalyptus camaldulensis* at the end of three years growth (Yigardu 2002). Farmers also indicated their preference to *Acacia nilotica* for it produces a good charcoal and fuel wood. Moreover, survey report of the center at various weredas indicated that increasing demand for fuel-wood and the decreasing supply trend. The indigenous species which have fast growth rate also have additional desirable attributes over the exotics (e.g. termite resistance/adaptation to the environment). Wood and charcoal sale has been becoming the main stay of some farmers in northern Shewa zone specially when there is food shortage/drought. The accessibility of some weredas to the main road made current and future wood market promising. Though farmers have their own management practices of the above mentioned indigenous species, research efforts made to improve the growth rate/performance through better management is not significant.

Acacia Nilotica occurs in woodlands and scrub in dry and moist kola agro climatic zones, 600-1700 m.a.s.l (Azene Bekele, 2007). It is Medium to fast growing large shrub or small tree, usually 2-6m but can reach 14m, branching from the base to make a rounded crown. Literatures indicate that *Acacia nilotica* do not coppice well (Tree data base CD, Azene Bekel, 2007,) but farmers in the lowlands of north Shewa say it can coppices well. Therefore, the trial was conducted with the aim of evaluating *Acacia nilotica* tree species for a greater woody biomass and quality production under different management regimes.

2. Materials and Methods

One indigenous tree species, known for its relative fast growth rate and preferable by the farmers, namely *Acacia nilotica* was selected for evaluating its coppice potential. The field experiment was conducted at Armania Kebele, Tarma-Ber district, North Shewa zone and the experiment design in 3x4 factorial experiments with randomized complete block design arrangement and replicated three wise. The dimension of a plot was 6.5mx10m (65 m^2) and eight trees per plot were planted in two rows with 2 meter spacing. At the age of three years old, four years old and five years old trees were cut close to the ground. And then after rising of sprouts from the stump Leaving one number of coppice , Leaving two number of coppice , Leaving three number of coppice and leaving all coppice per stump per cutting time were applied with one satellite control uncut trees. Growth parameters (root collar diameter, height and DBH), survival rate and fresh and dry weight data were collected.

Data Analysis

Data were analyzed using SAS 9.0 followed by least significant difference (LSD) test, was applied for detecting significant differences among means.

3. Result and Discussion

3.1. Root Collar Diameter and Height

The growth parameter result shows that root collar diameter (RCD) and height were increased by R^2 98 and 97 percent each month, respectively. The result indicated that root collar diameter and height were highly correlated with growth period (Fig 1 and 2).

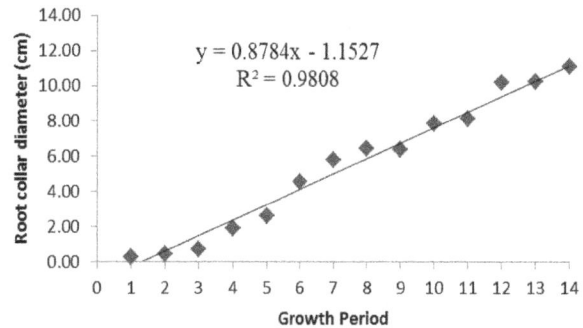

NB:- 0=0MAP, 1=3MAP, 2=6MAP, 3=12MAP, 4=15MAP, 5=24MAP, 6=30MAP, 7=36MAP, 8=42MAP, 9=48MAP, 10=54MAP, 11=60MAP, 12=66MAP, 13=72MAP and 14=78MAP

Fig. 1. *Root collar diameter of Acacia nilotica at 36 month after planting*

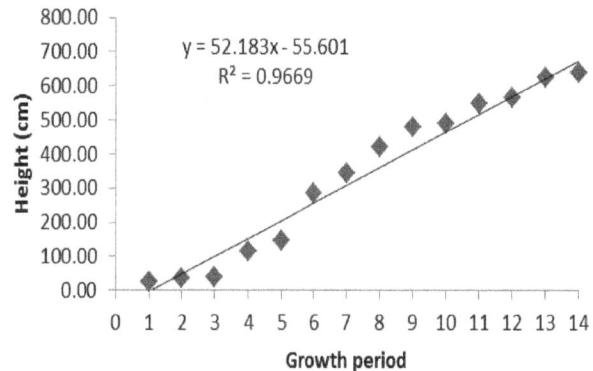

NB:- 0=0MAP, 1=3MAP, 2=6MAP, 3=12MAP, 4=15MAP, 5=24MAP, 6=30MAP, 7=36MAP, 8=42MAP, 9=48MAP, 10=54MAP, 11=60MAP, 12=66MAP, 13=72MAP and 14=78MAP

Fig. 2. *Height of Acacia nilotica at 36 month after planting*

3.2. Fresh and Dry Woody Biomass

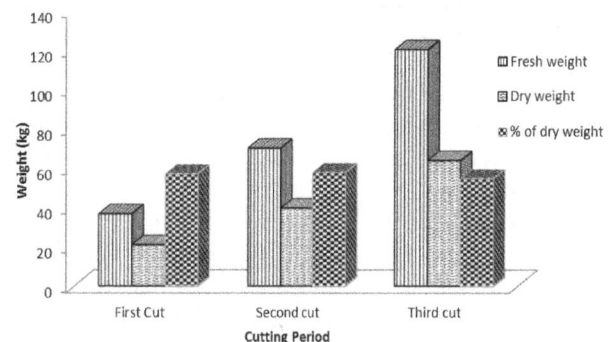

Fig. 3. *Fresh and Dry weight of Acacia nilotica*

The result of dry wood biomass of *Acacia nilotica* ranges from 55 to 58 percent.

3.3. Coppiced Potential and Woody Biomass Volume of Acacia nilotica

The results of this study disproof the previous findings that says *Acacia nilotica* did not coppice (Tree data base CD, Azene Bekele, 1993,). However, the present study confirmed that *Acacia nilotica* could be coppiced well at the age of three, four and five years after planting like other coppiced

tree species (Chirwa *et al*, 2003). From the first cutting time at the age of three years after planting, number of coppice per stump ranges from m 13 to 29 and gave higher volume of woody biomass. The result of cutting period and leaving of number of coppice indicates that leaving of two coppices number of *Acacia nilotica* could gave higher woody biomass volumes at 36 MAP cutting period as compare to leaving one coppice at 48MAC and leaving all coppices, three coppices and one coppice at 84 MAP (Table 1).

Table 1. *Woody biomass volume comparison of different cutting period and leaving of different number of coppice of Acacia nilotica*

| Treatment | Cutting periods | | | | | |
| | 36MAP | | 48MAP | | 60MAP | |
	48MAC m³ha⁻¹	84MAP m³ha⁻¹	36MAC m³ha⁻¹	84MAP m³ha⁻¹	24MAC m³ha⁻¹	84MAP m³ha⁻¹
All coppice	4830a	6190ab	3268ab	6504ab	1688ab	6412ab
One coppice	1412b	3013b	1387b	4518b	380b	5262b
Two coppice	5245a	6316a	5546ab	9982ab	1770a	8657a
Three coppice	3794ab	5327ab	8434a	12172a	1070ab	6993ab
CV	49.46	33.55	51.72	38.65	69.05	21.39

Columns with same letter are not significantly different, MAP (months after plant), MAC (months after cutting) and CV (coefficient of variance).

On the other hand, at 48MAP the result shows that, leave three coppices can harvest 12,172 m³ha⁻¹ volume of woody bio-mass and followed by leaving two coppices (9982m³ha¹) and all coppices (6504 m³ha⁻¹) than leaving one coppice (4518m³ha⁻¹) at 84MAP. Leaving three, two and all coppices increases the biomass yield by 169%, 120% and 44%

respectively as compared to leaving one coppice at 84MAP. The total woody biomass volume obtained at 84 MAP by leaving two and three number of coppices had higher than leaving one coppice.

3.4. Woody Biomass Volume at Coppicing Levels and Cutting Period of Acacia nilotica

Table 2. *Volume comparison of coppicing levels and cutting periods of Acacia nilotica with control*

| Treatment | Total woody bio-mass volume (m³ha⁻¹) | | |
	84MAP (36MAP+48 MAC)	84MAP (48MAP+36MAC)	84MAP (60MAP+24 MAC)
All coppice	6190a	6504ab	6412ab
One coppice	3013a	4518b	5262ab
Two coppice	6316a	9982ab	8657a
Three coppice	5327a	12172a	6993ab
Control (Un-cut)	4142a	4142b	4142b
CV	38.24	46.59	33.43

Columns with same letter subscription are not significantly different, MAP (months after plant), MAC (months after cutting), CV (coefficient of variance).

Woody biomass volume comparison of different coppice levels with control (uncut tree) at different cutting period result indicated that there was no significant different among coppice level treatments on woody biomass volume at 84MAP (volume at 36MAP+48MAC) compared to control (Table 2). But higher volume was obtained by leaving two numbers of coppices per stump. The result indicated that, 52% higher woody biomass volume was obtained by leaving two coppices as compare to uncut (control) followed by leaving all and three coppices by 49.4% and 28.6%, respectively.

Leaving three coppices of *Acacia nilotica* at 48MAP cutting

period was obtained significantly higher woody biomass volume than as compare to uncut tree (control) at 84MAP. Leaving three numbers of coppices at 48MAP was gives 12,172 m³ha⁻¹ volume of woody biomass which is 193% higher than that of control (un-cut tree). Leaving two and all coppices was also increased woody biomass volume by 140% and 57% per hector as compare to uncut tree, respectively.

3.5. Woody Biomass Volume at Different Cutting Periods

The result of woody biomass volume at different cutting periods indicated that significant difference was observed among treatments (Table 3). Cutting of *Acacia nilotica* at 48 MAP and 60 MAP periods was essential and preferable to obtain higher woody biomass volume of tree as compare to uncut tree (Harmer and Howe, 2003). The above two cutting periods could be increase woody biomass volume by 104% and 65%, respectively as compared to uncut.

Table 3. *Woody biomass volume at different cutting periods*

Treatment	Total Woody Biomass Volume at 84MAP (m³ha⁻¹)
36 MAP (Three Years old)	5211bc
48 MAP (Four Years old)	8457a
60 MAP (Five Years Old)	6831ab
84 MAP (Control)	4142c
CV	32.86

Columns with the same letter subscriptions was not significantly different, MAP (months after plant), MAC (months after cutting), CV (coefficient of variance).

3.6. Woody Biomass Volume of Different Coppice Levels

Woody biomass of different coppice levels result indicated that significant difference was observed among treatment (Table 4). Leaving of two and three number of coppices were increased the woody biomass volume of *Acacia nilotica* by 200.82%, and 197% respectively as compare to uncut tree.

Table 4. *Woody biomass volume of different coppice levels*

Treatment	Total Woody Biomass Volume at 84MAP (m³ha⁻¹)
All coppice	6352b
One coppice	4264c
Two coppice	8318a
Three coppice	8164ab
Un-cut (Control)	4142c
CV	33

Columns with the same letter subscription was not significant different, MAP (months after plant), MAC (months after cutting), CV (coefficient of variance).

4. Conclusion and Recommendation

Acacia nilotica could be coppiced well and it is a medium to fast growing tree species, at every three months the root collar diameter and height of this tree increased by 1.5cm and 100cm respectively. Therefore, Cutting of *Acacia nilotica* tree at 36 MAP and 60 MAP with leaving two and three number of coppices gives higher woody biomass volume and increases the woody biomass volume by 270% and by 365% respectively over leaving one number of coppice and control. Despite of the present findings on *Acacia nilotica* coppice management there is needs to do more research on other ages in order to overcome the problems of fuel wood and other wood demand shortage in the district.

References

[1] Azene Bekele, 2007. Useful trees and shrubs for Ethiopia. Identification, propagation and management for 17 agroclimatic zones. world agroforestry centre, East Africa region, Nairobi, Kenya.

[2] Chirwa, T. S., Mafongoya, P. L. and Chitu, K. 2003. Mixed planted fallow using coppicing and non-coppicing trees for degraded acrisols in Eastern Zambia. Journal of Agroforestry Systems 59(3): 243 – 251

[3] EFAP (Ethiopian forestry action programme), 1994. Ethiopian forestry action program. Volume II. The challenges for development. EFAP secretariat, Addis Ababa, Ethiopia.

[4] Harmer, R, and Howe, J. 2003. The silviculture and management of coppiced woodlands. Forestry Commission, Edinburgh.

[5] Yigardu, M. 2002. Aboveground Biomass of the Dominant Tree Species on Farmlands in Sirinka Catchment, North Wollo, Ethiopia. MSc thesis, Wondo Genet Collage of Forestry and Swedish University of Agricultural Sciences, Sweden

4

Effects of land use practices on soil organic carbon, nitrogen and phosphorus in river Nzoia drainage basin, Kenya

Wabusya Moses[1], Humphrey Nyongesa[2], Martha Konje[3], Humphrey Agevi[3], Mugatsia Tsingalia[1]

[1]Department of Biological Sciences, Moi University, Eldoret, Kenya
[2]Department of Sugar Technology, Masinde Muliro University of Science and Technology, Kakamega, Kenya
[3]Department of Biological Sciences, Masinde Muliro University of Science and Technology, Kakamega, Kenya

Email address:
mugatsi2005@ygmail.com (M. Tsingalia), wabusyam@yahoo.com (W. Moses)

Abstract: Land use activities along River Nzoia Drainage Basin, Kenya, include cultivation along the river banks, over grazing, deforestation, draining of wetlands for horticulture, harvesting of sand and brick-making. These activities have brought about changes in soil properties in the drainage basin adversely affecting farming output and the ecosystem in general. Consequently, it is important to understand how the different land use activities influence the soil properties in order to design and implement effective soil management strategies. This study examined the effects of land use practices on selected soil nutrients in Nzoia River Drainage Basin in Bungoma County. Cultivation and grazing were identified as important land use practices, while undisturbed sites were treated as controls. Land use practices along the river were identified by actual surveying of the study area. Secondary data on land use practices were obtained from technical reports, from local authorities and government offices. Soil samples were collected from different land use areas using randomly placed 5mx5m quadrats. Solis were collected at depths of 15cm in zigzag grid layout in each sample quadrat using soil auger. A total of 72 soil samples were collected in the study sites and analyzed for total nitrogen (N), available phosphorus (P) and organic carbon (C). Analysis of variance and correlation were performed to determine the significant land use practices affecting soil N, C and P. Cultivation had a significant effect on soil organic C mean value of 1.91 but negatively correlated with total Nitrogen and soil C while undisturbed sites exhibited positive correlation with C ($P\leq 0.05$). On the basis of our findings, it is recommended that conservation agriculture be practiced in the River Nzoia and its drainage system.

Keywords: Cultivation, Grazing, Organic Carbon, Nitrogen, Phosphorus, Conservation Agriculture

1. Introduction

Land-use changes are widely recognized as key drivers of global carbon dynamics [13, 29] and grasslands have received much attention for their substantial potential to act as carbon sinks in the recent past (4, 7, 9, 21, 23). With improved management practices, such as soil fertilization, promotion of native vegetation, sowing of legumes and replanting perennial grasses, most grasslands worldwide are considered to be important carbon sinks [12]. However, overgrazing and poor pasture management have led to significant losses of carbon from soils [6, 10, and 11]. Deforestation causes increased losses of carbon, nitrogen, phosphorus, and sulphur from terrestrial ecosystems. Where

deforestation is followed by conversion to other land uses, the effects of deforestation are magnified. The major causes of organic carbon losses are harvesting and burning of forest residues, accelerated decomposition, decreased production of wood and roots, and erosion. Nitrogen and sulphur are lost through the same pathways, and additionally by leaching to stream- and ground-water, and by the anaerobic production of nitrogen and sulphur containing gases. Phosphorus is lost primarily through harvest and erosion. More than half of the carbon and nitrogen and somewhat less phosphorus and sulphur can be lost in sites where forests are converted to other uses [16]. Losses of these elements following deforestation are most rapid in sites with high decomposition rates, especially in the tropics and on fertile soils. The interactions of the carbon, nitrogen, phosphorus, and sulphur cycles affect losses of any element through nutrient limitations to biological transformations, ratios of element availability, which cause either biological mobilization or immobilization, and anion/anion interactions in the soil solution [16].

Intensified cultivation practices often lead to widespread increases in the levels of nitrogen and phosphorous in lowland watercourses resulting in subsequent nuisance growths of algae and other aquatic plants [3]. In New Zealand for instance, the introduction of nitrogen fixing clovers, use of nitrogen fertilisers, including the practice of spreading animal wastes on pastures, and direct addition of stock urine and faeces in pasture have increased the amounts of nitrate leaching from pastoral catchments [3].

Reduction in vegetation cover reduces the amount of soil organic carbon in the soil. Soil organic carbon has been shown to be adequate in forest and bush lands in the upper zones but deficient in the lower zones which was attributed to the reduced plant cover and high rate of decomposition and mineralization of organic matter in the lower zones [17]. The results also showed that soil organic carbon was higher in annual crops, pasture and fallow as a result of the addition of farmyard manure or use of inorganic fertilizers.

Livestock grazing has been proposed to be the greatest source of riparian habitat degradation [25]. Cattle consume streamside vegetation, disturb soils, destabilise stream banks, and churn up channel sediments, and deposit manure and urine. Livestock damage to riparian vegetation and soils destabilizes the banks and lead to mobilization of fine sediments that in turn cause sedimentation in the channel and reduced stream clarity [26]. In addition, more runoff of sediments occurs from soil disturbed and compacted by livestock trampling [18]. The resulting increased sediment load is accompanied by particular nutrients that may contribute to stream enrichment as well as eutrophication of lakes and estuaries downstream (e.g., 15, 27] estimated that bank erosion contributed 32% of sediment discharge and 10% to the export of phosphorus from agriculture catchments in Canada.

A study on the effects of increasing grazing pressure on soil carbon and nitrogen storage in temperate grasslands of northern China revealed that carbon and nitrogen storage in both 0-10cm and 10-30cm soil layers decreased linearly with increasing stocking rate [30]. Carbon storage in the 0-10cm soil layer was significantly higher in lightly grazed than in heavily grazed grasslands after a five year grazing treatment. Their findings suggest that there is an underlying transformation from soil carbon sequestration under light grazing to carbon loss under heavy grazing. It was demonstrated that grazing increased the bulk density and moisture content through compaction and exposure of the soil to the sun, but reduced most soil nutrients through feeding and subsequent erosion due to the reduced ground cover [30].

This study sought to determine the effects of land use practices on soil organic Carbon and total Nitrogen in the Nzoia River Drainage Basin. The current agricultural practices that involve deliberately maintaining ecosystems in a highly simplified, disturbed, and nutrient rich state by supplying limiting factors, especially water, mineral nitrogen, and mineral phosphate in excess and active control of pests in order to maximise production turns is a major problem of nutrient management in Nzoia River Drainage Basin. There is therefore the need to reclaim the soils and improve on its properties in order to increase agricultural productivity in the Nzoia River Drainage Basin in Bungoma County.

2. Materials and Methods

2.1. Study Site

Nzoia River Basin lies between latitudes 10 30'N and 00 05′ S longitudes 34 0' W and 35 45' E. The Nzoia River originates from the Cherengani Hills at a mean elevation of 2300 m above sea level and drains into Lake Victoria at an altitude of 1000 m [19]. It runs approximately south–west and measures about 334 km with a catchment area of about 12,900km^2, with a mean annual discharge of 17777×106 m^3 year. The study site in Bungoma East sub county extended between longitudes 34°34'00" to 34°51'30"E and latitudes 0°23'00" to 0°37'30"N. The climate of the basin is mainly tropical humid characterised by the day temperatures varying between 16^{0}C in the highland areas of Cherengani and Mt Elgon to 28^{0}C in the lower semi–arid areas on annual basis. The mean annual night temperatures vary between 4^{0}C in the highland areas to 16^0 C in the semi–arid areas. Mean annual rainfall varies from a maximum of 1100 to 2700mm and a minimum of 600 to 1100mm. The area experiences four seasons in a year as a result of inter-tropical convergence zone. There are two rainy seasons, namely, short rains (October to December) and the long rains (March and May) and two dry seasons (If I understand well, is this rearrangement correct?). The dry seasons occur in the months of January to February and June to September ([19].

Among the main features in the area are the Nzoia Sugar Company plantations, Malava Forests, the Nandi Escarpment, and the Pan Paper Milling Factory at Webuye [19]. The riparian communities engage in sugarcane agriculture, fishing and livestock keeping along with subsistence farming of maize, beans, sweet potatoes and cassava. Three main sites, Kakimanyi, Nambalayi and Kuywa were sampled in the study area. Grazing and horticulture were the main land use practices while certain areas that were undisturbed (without any land use practice), were treated as control sites during the study.

2.2. Land Use Practices Data

Primary data was obtained through actual survey and mapping of the study area to identify the land use practices. Secondary data on land use practices was obtained from research technical reports by the Nzoia River Basin Management Initiative and local and government authorities such as Webuye District Agricultural offices.

2.3. Soil Collection and Analysis

Soil samples were randomly collected at depths of 0-15 cm in a zigzag grid layout using an auger across the sample plots of 5m x 5m in study sites under different land use practices during rain and dry seasons. The individual samples of soil were bulked and mixed thoroughly in buckets. A total of 72 soil sub samples were obtained from the bulks and packed in clear labelled plastic bags. The samples were placed in cooler boxes with ice for transportation to the laboratory. The samples were air-dried, ground to pass through a 2-mm mesh sieve for nitrogen, phosphorus and carbon analyses. Soil total nitrogen was determined by semi micro-Kjedal digestion method where 4.4ml of the digestion mixture (selenium, lithium sulphate, hydrogen peroxide and concentrated sulphuric acid) was added to 0.3g soil and digested for two hours at $360^{\circ}C$. Later, 25ml of distilled water was added, the solution cooled and aliquot of 5ml taken for distillation for total nitrogen. Another aliquot of 5ml was taken and available phosphorus determined by spectrophotometer method at wavelength 880nm as described by [20]. Total organic carbon was determined by oxidizing 0.3g of soil with a solution of 5ml of 0.0667M potassium dichromate and 7.5ml of concentrated sulphuric acid in the digestion block. The concentration of potassium dichromate remaining after digestion was titrated with 0.033M ferrous ammonium sulphate as described by [20]. Analysis of variance (ANOVA), with the aid of General Linear Model, and correlation analysis were performed between land use practices and soil nutrients using SAS software to determine the significant land use practices on soil nitrogen, phosphorus and carbon soils were randomly collected in study sites based on the land use practices using a randomized block design. .

3. Results

Figs 1, 2, and 3 show the levels of soil nutrients in the different sites of the study area under different land use practices. In all the study sites, total soil organic carbon was highest in areas that were undisturbed followed by areas that were under intense grazing (Fig. 1). Total nitrogen on the other hand differed both by study site and land use practice being highest in land under grazing, especially in Kuywa (Fig. 2). Phosphorus differed at different study sites under different land use practices without showing any specific pattern (Fig 3).

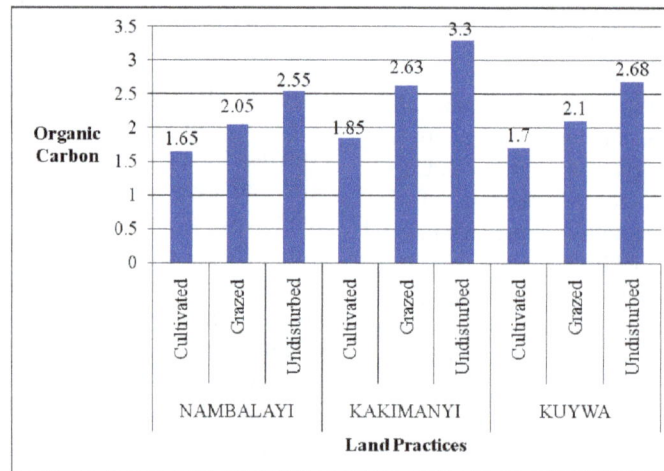

Figure 1. *Levels of total soil organic carbon in various study sites with different land use practice.*

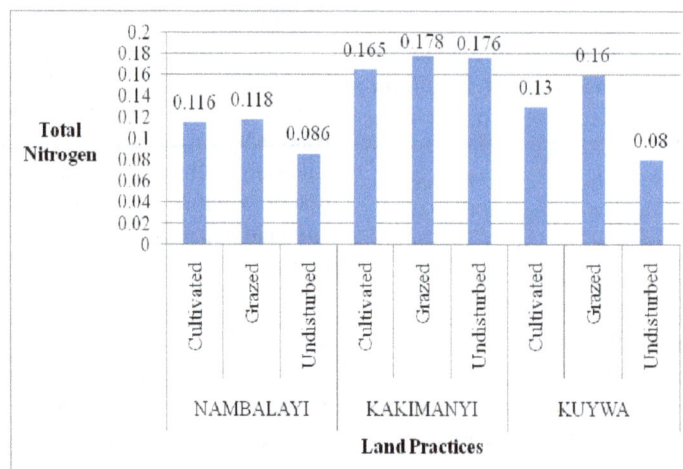

Figure 2. *Levels of total soil nitrogen in various study sites with different land use practice.*

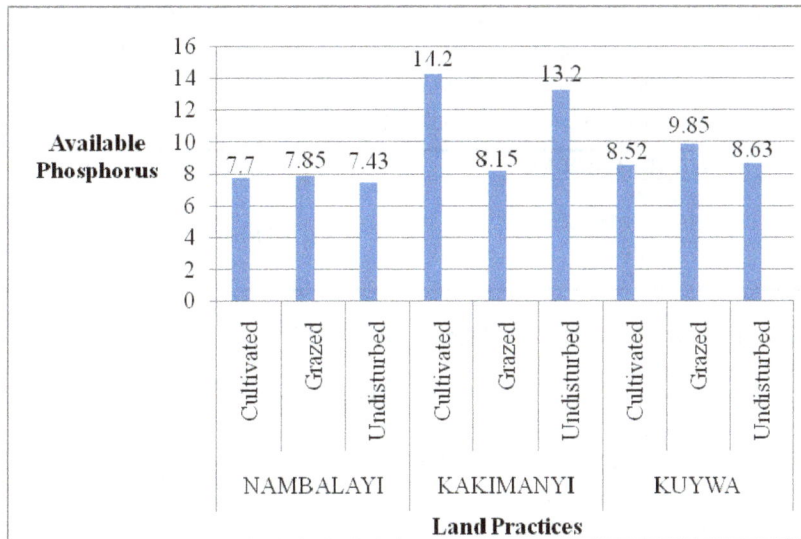

Figure 3. Levels of available soil phosphorus in various study sites with different land use practice.

Selected results of ANOVA are shown in table 1. These results clearly show that cultivation significantly decreased soil organic C.

Table 1. *Effects of land use practices on the mean soil total nitrogen (N), available phosphorus (P) and carbon (C) in the River Nzoia Drainage Basin. Means followed by different letters within a row are significantly different (ANOVA; P ≤ 0.05).*

Land use practice	Total N %	Available P ppm	Organic C %
Grazed	1.64a	2.60a	1.97a
Cultivated	1.64a	2.66a	1.91b
Undisturbed	1.64a	2.69a	1.99a

There were no significant differences between grazed and undisturbed sites for total nitrogen and phosphorus ($P> 0.05$). Similarly, cultivation did not significantly affect N and ($P > 0.05$).

At significant level of $P ≤0.05$) undisturbed sites positively correlated with soil organic carbon (r=0.50785; p=0.0001). In addition, cultivation negatively correlated with total N and soil organic carbon (r= -0.29010 and P=0.0222; and r= -0.41052 and P=0.0009 respectively).

Table 2. *Pearson Correlation Coefficients between land use practices and soil nutrients (number in parenthesis is the probability (P)-value).*

Practices	Grazed	Cultivated	Undisturbed	Nitrogen	Phosphorus	Carbon
Grazed	1.00000	-0.40394 (0.0011)	-0.34941 (0.0054)	0.04866 (0.7072)	-0.13928 (0.2803)	0.19000 (0.1391)
Cultivated		1.00000	-0.40891 (0.0010)	-.29010 (0.0222)	0.11236 (0.3846)	-0.41052 (0.0009)
Undisturbed			1.00000	-.08292 (0.5217)	0.07712 (0.5513)	0.50785 (0..0001)
Nitrogen				1.00000	-0.13940 (0.2799)	0.18120 (0.1587)
Phosphorus					1.00000	-0.16806 (0.1917)
Carbon						1.00000

4. Discussion

4.1. Land Use Practices

The observed land use practices in the study area are attributed to residents in these areas who rely on farming for both commercial and subsistence reasons. The other reason is the existence of the two sugar industries (Nzoia Sugar Company and Mumias Sugar Company) and the Webuye maize milling plant. These industries have encouraged farmers to cultivate more sugar canes and maize so as to earn a better living. Undisturbed areas have been preserved for purpose of cultural practices especially circumcision.

4.2. Cultivation and Soil Organic Carbon

Correlation between land use practices and soil organic Carbon was negative (r= -0.44, P= 0.0004). This is probably because under heavy cultivation practices, the soil is broken and highly churned thus improving the aeration. It also exposes soil organic matter to micro-organisms that later on enhances decomposition of soil organic matter hence a decline in soil organic carbon. More so, the crops in cultivated areas are usually harvested by removing both the fruit and the foliage which finally leads low addition of organic matter levels and hence a decline in soil organic carbon [29].

Studies by [14] demonstrated a decline in soil organic

carbon over time while trying to compare the conversion of land from grass land to arable cultivation land. Studies by [1] have shown that the highest carbon stocks are found in natural forest, artificial forest and artificial grassland ecosystems, and that continuous cropping reduced carbon stock by about 70%.

Table 3. *Standard levels of Soil Nutrients (N, P and C) [20] .*

ORGANIC CARBON	TOTAL NITROGEN	AVAILABLE PHOSPHORUS
> 3.0 High	> 0.25 very high	
1.5 – 3.0 moderate	0.125 – 0.25 moderate	< 10ppm very low
0.5 – 1.5 low	0.05 – 0.125 low	
< 0.5 very low	< 0.05 very low	

Grazing did not have any significant effect on soil N, P and C. This contradicts the findings by other researchers that have shown that grazing depresses soil organic carbon content and storage [24 and 28]. It has been demonstrated that there is a range of potential indirect mechanisms through which soil carbon may be affected by increased grazing [2, 5, and 22]. It appears that grazing depresses soil organic carbon storage through reducing plant biomass especially by removal of palatable grasses and sedges that built soil carbon pools. Palatable plant species generally produce litter of high quality for decomposers than do unpalatable species [9]. The results of our findings can be attributed to the system of grazing that is practiced in the basin. Grazing is small scale with very few animals as most farmers prefer the cultivation of maize and sugar cane to rearing of livestock. Secondly most of their grazing grounds are usually used for growing some crops that are harvested after a short season.

The total Nitrogen and available Phosphorus were selectively affected by the land use practices. Cultivated and grazed sites had higher total N unlike undisturbed sites. This fact takes place probably because of the chemical and organic fertilizers used, waste products such as dung and urine from animals [3].

5. Conclusions and Recommendations

In conclusion, cultivation and fallow demonstrate an inverse significant influence of the soil carbon. Whereas cultivation reduced soil carbon, leaving the land under fallow increased the soil carbon. Soil nitrogen and phosphorus were unaffected by land use practices point to their resilience to land use changes. In all the study sites, available phosphorus was below the standard levels [20] except in one study site (Kakimanyi) where the levels were higher under cultivation and undisturbed sites. Organic carbon and nitrogen were moderately high in all the study sites under different land use practices.

On the basis of our findings, we recommend that conservation agriculture be practiced in the River Nzoia Basin and its drainage system. Interspersing different land use practices can help in recovery of nitrogen, carbon and phosphorus. On the basis of the effects on the three nutrients studied, grazing and fallow are recommended as the preferred land use practices because of their minimal effects It is also recommended that more studies need to be carried out given the extensive nature of the Nzoia River Basin.

References

[1] Anikwe M. (2010) Carbon storage in soils of South-eastern Nigeria under different management practices. *Carbon Balance and Management 5:5*

[2] Bardgett R.D., and Wardle D.A. (2003) Herbivore-mediated linkages between aboveground and below ground communities. *Ecology*, 84: 2258–2268.

[3] Baudry J. and Thenail, C. (2003) Instit National de la Recherché Cedex, France. Last accessed 9/24/014.

[4] Conant R T, Paustian K, Elliott E T, 2001. Grassland management and conversion into grassland: Effects on soil carbon. *Ecological Applications*, 11: 343–355.

[5] De Deyn, G.B., Cornelissen, J.H.C. and Bardgett, R.D. 2008. Plant functional traits and soil carbon sequestration in contrasting biomes. *Ecology Letters*, 11, 516–531.oration of (?)

[6] Elmore A J, Asner G P, 2006. Effects of grazing intensity on soil carbon stocks following deforestation of a Hawaiian dry tropical forest. *Global Change Biology*, 12: 1761–1772.

[7] Fang J Y, Guo Z D, Piao S L *et al.*, 2007. Terrestrial vegetation carbon sinks in China. *Science in China* (Series D), 50: 1341–1350

[8] Grime J.P., Cornelissen J.H.C., Thompson K. and Hodgson J.G.(1996): Evidence of a causal connection between anti-herbivore defence and the decomposition rate of leaves. *Oikos*, 77: 489–494.

[9] Guo L B, Gifford R M, 2002. Soil carbon stocks and land use change: A meta-analysis. *Global Change Biology*, 8: 345–360.

[10] He N P, Yu Q, Wu L *et al.*, 2008. Carbon and nitrogen store and storage potential as affected by land-use in a *Leymus chinenis* grassland of northern China. *Soil Biology & Biochemistry*, 40: 2952–2959.

[11] He N P, Zhang Y H, Yu Q *et al.*, 2011. Grazing intensity impacts soil carbon and nitrogen storage of continental steppe. *Ecosphere*, 2, art8, doi:10.1890/ES1810-00017.00011.

[12] Jones M B, Donnelly A, 2004. Carbon sequestration in temperate grassland ecosystems and the influence of management climate and elevated CO2. *New Phytologist*, 164: 423–439.

[13] IPCC (Intergovernmental Panel on Climate Change), 2007. Fourth Assessment Report, Climate Change 2007: Synthesis Report. Cambridge, UK: Cambridge University Press.

[14] Johnson A. E, Poulton P. R, and Coleman K. (2009) Soil organic matter: Its importance in sustainable agriculture and carbon dioxide fluxes. *Advances in Agronomy* 101, 1-57

[15] Kondolf G. M, Piegay and N. landon (2007). Changes since 1830 in riparian zone of the lower eygues river. France landscape Ecology 22:367-384

[16] Lobe J. (2004). "*Hamburger Consumption Spurs Amazon Deforestation*" *Common Dreams* (http://www.commondreams.org/headlines04/0409- 05.htm) Last accessed 8/29/014.

[17] Mehlich A, Bellis E, and Gitau J.K (1964) *Fertilizing and liming in relation to soil chemical properties.* Scott Laboratories, Department of Agriculture, Nairobi.

[18] Nguyen M. L; Sheath G W; Cooper A b (1998) Impact of cattle treading on hill land 2 Soil physical properties and contaminant runoff. Newzealand journal of agricultural Research 41:279-290

[19] Nzoia River Basin Management Initiative Technical Report (2006-2011). A public private partnership between Water Resources Management Authorities and Civil Societies, Learning Institutions and Communities. pp.24.

[20] Okalebo, J.R., K.W. Gathua, and P.L. Woomer. (1993). *Laboratory Methods of Soil and Plant Analyses: A Working Manual.* Technical Publication, no. 1. Nairobi,

[21] Post W M, Kwon K C, 2000. Soil carbon sequestration and land-use change: Processes and potential. *Global Change Biology*, 6: 317–327.

[22] Semmartin M., Bella C.D., and de Salamone I.G. (2010): Grazing-induced changes in plant species composition affect plant and soil properties of grassland mesocosms. *Plant and Soil*, 328: 471–481.

[23] Soussana J F, Loiseau P, Vuichard N *et al.*, 2004. Carbon cycling and sequestration opportunities in temperate grasslands. *Soil Use and Management*, 20: 219–230.

[24] Snyman H.A., and Du Preez C.C. (2005) Rangeland degradation in a semi-arid South Africa-II: influence on soil quality. *Journal of Arid Environments,* 60: 483–507.

[25] Trimble S. W. (1994) Stream channel erosion and change Resulting from riparian forests. *Geology* 25:467–469.

[26] Waters T. F. (1995) Sediment in streams. *American Fisheries Society Monograph 7. American Fisheries Society, Bethesda, Maryland,* 251 pp.

[27] Williamson, M. (1996) *Biological Invasions.* Chapman and Hall, London

[28] Wu G.L., Liu Z.H., Zhang L., Chen J.M. and Hu T.M. (2010). Long-term fencing improved soil properties and soil organic carbon storage in an alpine swamp meadow of western China. *Plant and Soil,* vol.332: 331–337.

[29] Yang Y S, Xie J S, Sheng H *et al.*, 2009. The impact of land use/cover change on storage and quality of soil organic carbon in mid-subtropical mountainous area of southern China. *Journal of Geographical Sciences,* 19:49–57.

[30] Zhang Y. H, Yu Q, Chen Q. S, Pan Q. M, Zhang G. M and Han X. G (2011).Grazing intensity impacts soil carbon and nitrogen storage of continental steppe. *Ecosphere www.esajournals.org 2(1): art 8.* Pp. 1- 8

5

Analyzing risks related to the use of pesticides in vegetable gardens in Burkina Faso

Rayim Wendé Alice Naré[1, *], Paul Windinpsidi Savadogo[1], Zacharia Gnankambary[1], Hassan Bismarck Nacro[2], Michel Papaoba Sedogo[1]

[1]Laboratoire Sol-Eau-Plante. Institut de l'Environnement et de Recherches Agricoles (INERA), ouagadougou, Burkina Faso
[2]Laboratoire d'Etude et de Recherche sur la Fertilité des Sols (LERF), Université Polytechnique de Bobo Dioulasso (UPB), Bobo Dioulasso, Burkina Faso

Email address:

alice.nare@gmail.com (R-W. A. Naré.), paul.savadogo@gmail.com (P. W. Savadogo), gnank_zach@hotmail.com (Z. Gnankambary), nacrohb@yahoo.fr (H. B. Nacro), michel_sedogo@yahoo.fr (M. P. Sedogo)

Abstract: In West Africa, the uncontrolled use of pesticides by vegetable farmers leads to contamination of soils as well as surface and ground water. Farmers also use various sources of organic amendments which could impact the fate of the pesticides in soils. This study was conducted to identify the type of pesticides and organic amendments used in the main vegetable gardens in Ouagadougou, Ouahigouya and Bobo-Dioulasso three cities of Burkina Faso. Farmers were interviewed individually on their practices regarding organic amendments and pesticides. Sixty one percent (61%) of farmers do not know the instruction regarding the application of pesticides. Fifty three percent (53%) of farmers did never receive training on pesticide application. We found that pyrethroid-based insecticides like lambda-cyhalothrin and delthametrin were the most used by farmers. About 69%, 59% and 100% of the farmers apply the pesticides periodically in their fields respectively in Ouagadougou, Ouahigouya and Bobo-Dioulasso. All the farmers interviewed attested that they do not respect the recommended doses of pesticides. Manure was the organic amendment mostly used in the three cities (41-75%), followed by household garbage (15-41%). In Ouagadougou most of farmers (69%) apply organic amendment at the recommended rate or more while in Bobo-Dioulasso (69%) and in Ouahigouya (57%), the majority apply low rates. There is a correlation between the social status (sex and the education level) and the pesticide and organic amendment management.

Keywords: Soil, Pesticides, Organic Amendment, Environment Pollution, Burkina Faso

1. Introduction

In Africa, the use of pesticides accounts for 2–4% of the world pesticide use [1].Various active ingredients like organophosphates, carbamates, pyrethroids and organochlorins are commonly used in urban vegetable production [2, 3]. In Burkina Faso, approximately one hundred active ingredients are used [4] and around 75% of these active ingredients are insecticides, acaricides or nematicides [5].

Numerous studies in Africa showed that pesticides are used incorrectly. Those studies revealed that famers use unauthorized or banned products and also have inappropriate practices like the non-use of protection material, [6-9]. Inappropriate use of pesticides negatively impact on farmers and consumers health [10, 11], and leads to a contamination of soil [12], water [13-15], air [16-18], and crops [19].

The persistence of endosulfan, profenofos and others organo-chlorin pesticides in the cotton fields in Burkina Faso was reported [20] as well as their effect on microbial activities [21-23].

Soil amendment with animal manure or crop residues is a common practice in in the sub Saharan African agricultural systems. This practice has beneficial effects on soil nutrient status and stimulate soil biology, particularly in degraded and arid environments [24, 25].Organic amendments also play an important role on the fate of pesticides in soil [26]. [27] studied the effects of composts on the adsorption-desorption of three carbamate pesticides in different soils. The authors found that the adsorption capacity was positively and significantly correlated with soil organic carbon and the cation exchange capacity and negatively correlated with soil pH. They also found that desorption was more in unamended soil than amended soil.

In Burkina Faso [28] showed that manure applied at the rate of 3.33 mg/kg of soil increased organo-chlorin pesticide degradation 1.5 time more than unamended soil. .

Whereas numerous studies on pesticides use were conducted in the cotton systems in Burkina Faso, such investigations on vegetable gardens are not well documented. This paper reports on farmers practices regarding the use of pesticides and organic amendments in vegetable gardens in three main cities of Burkina Faso, Ouagadougou, Bobo-Dioulasso and Ouahigouya, and analyses social environment and the risks of environment and human contamination by pesticides. We hypothesized that (i) the types of pesticides used in vegetables farming are not recommended, and they are not managed properly (ii) the organic amendments are not managed properly (iii) farmers social status affects pesticides and organic amendment management.

2. Materials and Methods

2.1. Study Sites

The survey was conducted in September 2012 in the three cities of Burkina Faso, Ouagadougou in the centre, Ouahigouya in the north and Bobo-Dioulasso in the west. In these cities the biggest vegetable gardens were selected. In those gardens, tomato, onion, cabbage, lettuce and potato are the vegetables mostly grown by farmers [29]. The vegetables are all produced with irrigation with waste water from canals and/or water from well and dams. For soil fertility management farmers apply organic amendments and inorganic fertilizers. They use various types of pesticides for pest and diseases control.

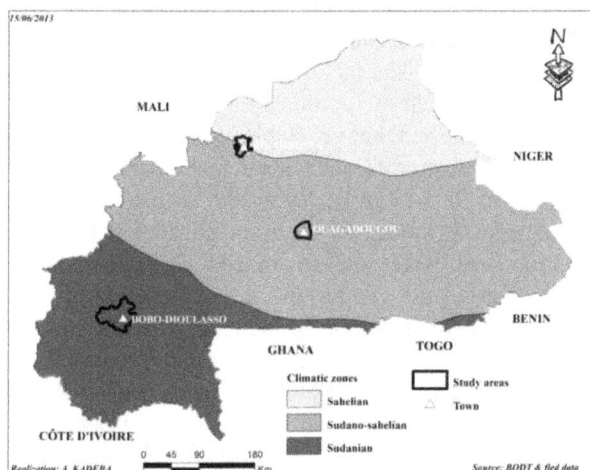

Figure 1. Map of Burkina Faso showing the location of the study areas.

The number of farmers interviewed was 109, 101 and 100 respectively for Ouahigouya, Ouagadougou and Bobo-Dioulasso. The farmers were interviewed on their level of instruction, their knowledge on pesticides and organic amendments. The three studied areas are located in different climate zones of the country (fig1).

2.2. Survey

In each zone, a team of surveyors visited farmers. The studied farmers were selected randomly in the three climate zones. The investigations were conducted in four vegetables sites in Ouagadougou (Tanghin, Boulmiougou, Kossodo, and Wayalgin), in BoboDioulasso eight sites (Dogona, Ko-Deni, Sector 25, Sagaby, Kunima, Sabarydougou, Legeuma and Kwa) and in Ouahigouya, 5 sites (Barrage Oumarou Kanazoé, Thiou, Pèla, Gouinré, secteur 05, secteur 14). For the vegetables farms were farmers are organized in associations, we contacted the head of the association to mobilise the others farmers for the interview. The selection of farmers was voluntary basis. A total of 310 farmers were interviewed. We interviewed farmers individually regarding the types and rate of organic amendments and pesticides used. We also collected information on the frequency of application and the equipment used. The information regarding the social status of the farmers was also collected namely sex, age, education level and the affiliation to organizations.

2.3. Calcul and Statistics

The percentages and frequencies were calculated using the software Sphinx lexica. The PCA (principal components analysis) was performed using XLSTAT version 6. The names of the variables were shortened as follow: M: Member of association; M0: not member of association; 0 illeterate; I: primary school; II: secondary school; S0: no sprayer Spest: sprayer; Pa: Periodical pesticides application; A0: application in case of attack; OM1: one time organic amendment; OM2: two time organic amendment; OM3: amendment depending of the availability; OMl: application of organic amendment at less than recommended rate; B: Broadcasting of organic amendment; OMh: application of organic amendment at more than recommended rate; L: localise application of organic amendment; OMu: quantity of organic amendment unknown; LP: localise application of organic amendment followed by ploughing

3. Results-Discussions

3.1. Social Status of Vegetable Farmers

The majority of farmers were men in Ouahigouya and Bobo Dioulasso while in Ouagadougou, only 46% of farmers were men (table 1). The average age of farmers was almost the same in the three cities (43 in Ouagadougou ; 37 in Ouahigouya and 39 in Bobo Dioulasso). The highest proportions of illiterate farmers were encountered in Ouagadougou and Bobo-Dioulasso, the two main cities, while in Ouahigouya the highest proportion of farmers having primary and secondary education level was recorded. Our results corroborate with those of [30] who reported that in Cameroun, urban agriculture is practiced mostly by youth, illiterate and that urban farmers more often are not organized in association. The high proportion of illiterate farmers in Ouagadougou could be explained by the fact that in big cities educated people are employed in other sectors where a certain level of education is required. Besides, most of the vegetable farmers in big cities are migrants from rural areas who did not have access to

education. [31] reported that the majority of vegetable farmers in Burkina Faso are illiterate. The young people involved in vegetable garden secure the labor. The education level (illiterate) could have some consequences on pesticides use. Indeed it might be that illiterate people cannot read the instruction regarding the pesticide use. Besides, they will not be able to write the trainings note for their own use or share with other famers.

Table 1. Social status of vegetable farmers in Burkina Faso (survey conducted in three cities (Ouagadougou, Ouahigouya and Bobo-Dioulasso).

	Social Status of farmers	Ouagadougou	Ouahigouya	Bobo Dioulasso
Age	Average	43.03	36.69	39.37
	Minimum	18	18	19
	Maximum	80	55	70
	SD	12.39	9.66	12.46
Sex	Male (%)	46	96	89
	Female (%)	54	14	11
Level of education	Illiterate	87	42	57
	Primary	6	54	28
	Secondary	7	13	13
Member of farmers associations	Yes	56	47	48
	No	46	53	51

3.2. Pesticides

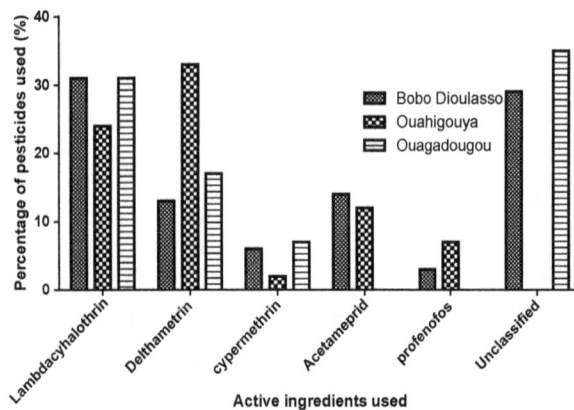

Figure 2. Types of pesticides used by vegetable farmers in three cities of Burkina Faso (Ouagadougou, Ouahigouya and Bobo-Dioulasso)

In the three cities, pyrethroid-based insecticides such as lambda-cyhalothrin and delthametrin were used by a high proportion of the farmers (respectively 24-31% for lambda-cyhalothrin and 17-33% for delthametrin) to control insects in vegetable crops (fig 1). Other insecticides like profenofos, cypermethrin, acetamiprid, were used in Ouahigouya and Bobo Dioulasso. Acetamiprid and profenofos were not used in Ouagadougou. The pyrethroids are probably the most available pesticides, explaining the importance of their use as farmers use the products they can easily find. The pyrethroids are highly hydrophobic as shown by their low water solubility and low KOC [32, 33]. Some authors [19,

34-38] showed that those type of insecticides are highly toxic to fresh water fish and to bees even at low concentrations (<0.5 lg/l).The pesticides with low Koc like the pyrethroids are more able to bind to soil particles than pesticides with high Koc value [39, 40]. All the famers interviewed apply pesticides themselves in their farms.

In Ouahigouya and BoboDioulasso almost all the farmers use a spray for pesticides application, whereas in Ouagadougou, only 41% use this material. The 59% of farmers in Ouagadougou use other materials like leaves or brooms to apply pesticides. This might be due to the low education level of most of farmers in this city. In the three provinces, farmers spray pesticides periodically (every 3 days or weeks) to prevent insect attack and plants diseases with 59% in Ouahigouya, 69% in Ouagadougou, and 100 % in Bobo-Dioulasso. The pesticides quantities used were unknown (table 2). The fact that most of farmers did not get training on pesticide use could explain that situation. In this line [41, 42] reported that vegetable farmers have the highest pesticides application frequencies in many African countries. According to [11, 43, 44], inappropriate use of pesticides leads to negative impact on users and consumers health, and vegetation, soil, water and air. Indeed, [20] showed that in Burkina Faso, the soils in the cotton fields were contaminated by endosulfan in the level of 1 to 22 µg/kg in farmer fields. [12] reported that three months after application, endosulfan and profenofos concentration varied in the range of 10-30µg/kg in cotton field in Burkina Faso. The impact of pesticides on human health was also reported. Approximately three million people are poisoned and 200,000 die each year around the world from pesticide poisoning and a majority of them belongs to the developing countries [45, 46].

Table 2. *Practices of pesticides application by vegetable farmers in three cities of Burkina Faso (Ouagadougou, Ouahigouya and Bobo-Dioulasso).*

		Percentage of responses (%)		
Pesticides management		**Ouagadougou**	**Ouahigouya**	**Bobo-Dioulasso**
Mode of pesticides application	Spraying	41	99	100
	Other	59	0	0
Frequency of application	Periodically	69	59	100
	In case of attack	27	24	0
Pesticides dosage	Precise	0	0	0
	Not Precise	100	100	100
Training on the use of pesticides	Yes	36	34	54
	No	64	66	46

3.3. Organic Amendments

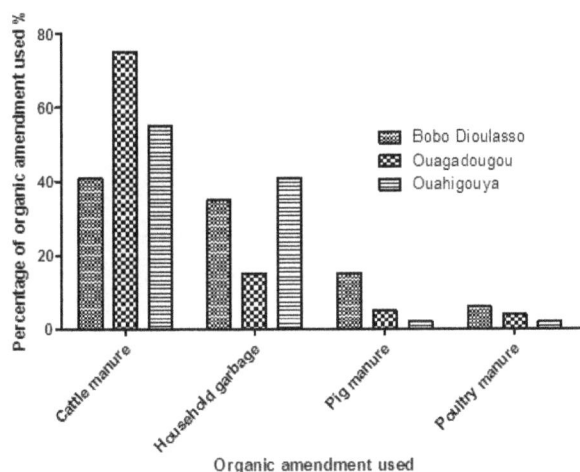

Figure 3. *Types of organic amendment applied in vegetable gardens in three cities of Burkina Faso (Ouagadougou, Ouahigouya and Bobo-Dioulasso)*

Manure was the most used organic amendment in all the three cities (41-75%), followed by the household garbage (15-41%) (fig 3). In Ouagadougou, the cattle manure was used by 75% of famers, while in Ouahigouya and Bobo Dioulasso it was 55% and 41% respectively. Household garbage was used by 41% of farmers in Ouahigouya, followed by Bobo-Dioulasso (35%) and Ouagadougou (15%).The other organic amendments (pig manure, poultry manure and urban compost) were used but not significantly (less than 20% of farmers). In many West Africa countries, organic amendments are used in vegetable gardens. In Lomé (Togo), organic amendments are used in all the city gardens [2]. In Dakar (Senegal), it has been estimated that 25% of the nutrients for horticultural crops come from compost, 25% from animal manure [47].

Table 3. *Quantities of organic amendment applied by vegetable farmers in three cities of Burkina Faso (Ouagadougou, Ouahigouya and Bobo-Dioulasso)*

	Percentage of responses (%)		
Quantities (T/ha)	**Ouagadougou**	**Ouahigouya**	**Bobo Dioulasso**
<10	8	47	20
10-19	25	10	49
20-40	55	39	7
>40	4	1	6
Unknown	8	3	18

The quantities of organic amendments applied by farmers varied between the three cities (table 3). In Ouagadougou most of the farmers (69%) used organic amendments at 20t/ha which is the recommended rate or even more. However in Ouahigouya and Bobo-Dioulasso, the majority (respectively 57% and 69%) use less than the recommended rate. In Ouagadougou and Bobo-Dioulasso, most of farmers apply organic amendment one time during the plant cycle (respectively 76% and 56%). In Ouahigouya and Ouagadougou the organic amendments are spread on the fields (77% and 98% respectively) while in Bobo-Dioulasso 58% of farmers practice localise application and 14% of those farmers plough in the amendments after application.

Table 4. *Methods of application of organic amendments by vegetable farmers in three cities of Burkina Faso (Ouagadougou, Ouahigouya and Bobo-Dioulasso).*

	Ouagadougou	Ouahigouya	Bobo-Dioulasso
Frequency of application			
One-time	76	13	56
Two-times	10	5	20
Depending of avaibility	14	80	21
Application methods			
Broadcast	98	77	15
Localise	0	10	58
Localise-ploughing in	0	0	14

The use of organic amendments can lead to pesticide adsorption or to their degradation. The sorption of uncharged organic compounds like pyrethroids in soils has been shown to be highly correlated with soil organic matter content [48, 49]. In the three cities, the use of organic amendments could be an opportunity to reduce surface and groundwater pollution as soil sorption is one of the most important processes affecting the fate of pesticides in the environment [50]. Many studies highlighted the effects of soil organic amendments on pesticides retention and degradation. They showed that pesticides are adsorbed on both organic and inorganic soil constituents [51, 52]. In general, addition of organic matter, including soluble and insoluble fractions, increases the pesticides adsorption and decreases their subsequent mobility in the soil profile [53, 54]. According to [50] for all those pesticides which have low polarity and water solubility like the pyrethroids soil organic matter will be the important sorbent, simply because the solvent is water and hydrophobic interactions are the driving force. Lambda-cyhalothrin because of its high hydrophobicity is sorbed more strongly by soil particles than deltamethrin and cypermethrin [55]. Regarding the sorption affinity the three pyrethroids are ranged as follow lambda-cyhalothrin>deltamethrin>cypermethrin.

In Bobo Dioulasso and Ouagadougou where lambda-cyhalothrin is the most used active ingredient, and cattle manure is used by most of farmers, water pollution probably will be reduced. In Ouahigouya where deltamehrin is the most used pesticide, water is likely to be polluted by pesticide compared to the other cities. Hydrophobic sorption to mineral surfaces may decrease the rate of chemical and biological degradation [55]. However, pyrethroids when used as soil insecticides are not selective and may also kill beneficial soil microorganisms [56].

3.4. Impact of the Social Status of Famers on Pesticides and Organic Amendment Management

The principal component analysis (PCA) show that the first axis F1(65% of the variability) was positively correlated with female farmers, age, illiteracy, affiliation to farmers' association, pesticides application in case of pest attack, the broadcast of organic amendment, the application of high rate of organic amendment (fig 4).

This axis was negatively correlated with the male farmers, primary and secondary education level, not member of farmers' cooperative, use of organic amendments according to the availability, and application of organic amendments at less than recommended rate, localise application of organic amendments.

The second axis F2 was positively correlated with age, illiteracy and secondary education level, periodically application of pesticides, one and two time of organic amendment application, application of organic amendment at low rate, localise application of organic amendment. This axis was negatively correlated with farmers sex (male and female), other methods for pesticides application, application in case of pest attack, application of organic amendment at high rate.

The individual PCA showed positive correlation with two provinces (Ouahigouya and Bobo Dioulasso) and negative correlation with Ouagadougou. The PCA revealed a strong correlation between the sex, the affiliation of farmers to association, the education level and the management of pesticides and organic amendments. In fact women are illiterate, member of farmers' association, and they use pesticide in case of attack and use material other than sprays for pesticides application. They broadcast organic amendment at a rate higher than recommended. Men have primary education level, not member of farmers association, and they use organic amendments according to the availability. In most of the sub Saharan countries including Burkina Faso the education rate is low [57] mostly for women. It was reported that women education level in Burkina Faso was 37.2% for primary school in 2002 and

8.51% for secondary school in 1999 [58]. This low rate corroborate with our results. Although women are member of association meaning that they could benefit of trainings on pesticides use, they do not adopt a good practice regarding the application of pesticides. This could be explained by the fact that the materials are inaccessible for women because men are the manager most of the time. This confirms a social influence on women activities. [59] showed the relation between technology adoption and farmers' socio-economic status in Burkina Faso. They reported that the gender and the age are the factors influencing the technologies adoption by farmers. The same finding was reported by [60] for compost adoption by farmers in Cameroon.

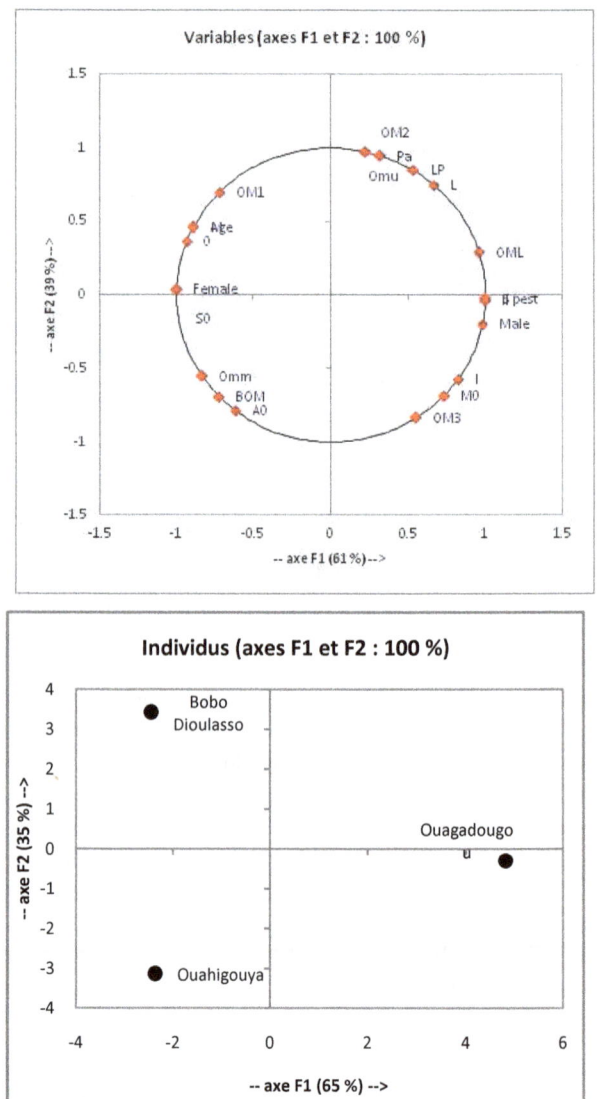

Figure 4. *Principal Component Analysis showing the impact of the social status of farmers on the use of pesticides and vegetables in three cities (Ouagadougou, Ouahigouya and Bobo Dioulasso) in Burkina Faso.*

4. Conclusion

The survey showed that nearly all farmers in the three cities use pyrethroids based insecticides. Most of the farmers in these cities are not instructed and did not receive training on pesticide use. The majority use leaves or broom (brushes or twigs) for pesticides application in Ouagadougou. All farmers use organic amendments, but it is only in Ouagadougou where organic amendments were used at the recommended rate or more. The use of organic amendment use could contribute to pesticides retention. There is urgent need for more training on pesticides and organic amendment use so that farmers will be able to use pesticide correctly and improve their efficiency. This will contribute to environment, users and consumers health protection and to a sustainable agriculture. From our results we conclude that for a safe use of pesticides in vegetable gardens, women education level need to be increased and more training on pesticides and organic amendment use need to be provided. The researches on the use of bio-pesticides are also needed.

Acknowledgements

This work was financed by International Foundation for Science (IFS) through grant number C/5002-1 awarded to the first author. We are grateful to IPICS and RAPFE for the publication fee, and Dr Kiba Delwendé Innocent Agronomist at INERA (Burkina Faso) for comments on the draft of the manuscript.

References

[1] Agrow. World agchem market steady. AGROW 4979 June 2006,. 2006:17.

[2] Tallaki K. The pest-control system in the market gardens of Lomé, Togo. in: Mougeot LJA (Ed), Agropolis: the social, political and environment dimensions of urban agriculture, IRDC, Ottawa, Canada, . 2005:pp. 51-88.

[3] Cissé I, Fall ST, Akinbamijo OO, Diop YM. L'utilisation des pesticides et leurs incidences sur la contamination des nappes phréatiques dans la zone des Niayes au Sénégal. in: Akinbamijo OO, Fall ST, Smith OB (Eds),Advances in crop-livetsock integration inWest African cities, ITC, ISRA, CRDI, Ottawa, Canada, . 2002: pp. 85-99.

[4] Toé MA. Limites maximales de résidus de pesticides dans les produits agricoles d'exportation dans trois pays du CILSS. Etude du Burkina Faso. . Projet Gestion des pesticides au Sahel http://wwwinsahorg/pdf/cataloguePublicationpdf. 2003:pp. 1-3.

[5] Orou Guidou G. La campagne cotonnière 97/98 au Bénin. In: Symposium sur l'utilisation des intrants en culture cotonnière et maraîchère. Dakar, 25-28 janvier. 1998:pp. 49-52.

[6] Gomgnimbou APK, Savadogo PW, Nianogo AJ, Millogo-Rasolodimby J. Usage des intrants chimiques dans un agrosystème tropical : diagnostic du risque de pollution environnementale dans la région cotonnière de l'est du Burkina Faso. Biotechnol Agron Soc Environ. 2009;13(4):499-507.

[7] Addo S, Birkinshaw LA, Hodges RJ. Ten years after the arrival of Larger Grain Borer: farmers' responses and adoption of IPM strategies. Int J Pest Manage. 2002;48(4): 315-25.

[8] Dinham B. Growing vegetables in developing countries for local urban populations and export markets: problems confronting small-scale producers. Pest Manage Sci. 2003;59:575-82.

[9] Matthews G, Wiles T, Baleguel P. A survey of pesticide application in Cameroon. Crop Prot 2003;22:707-14.

[10] Kishi M. The health impacts of pesticides: what do we now know? In:Pretty, J (Ed), The Pesticide Detox Towards a More Sustainable Agriculture Earthscan, London. 2005:pp. 23-38.

[11] Ahouangninou C, Fayomi BE, Martin T. "Evaluation des Risques Sanitaires et Environnementaux des Pratiques Phytosanitaires des Producteurs Maraîchers dans la Commune Rurale de Tori-Bossito (Sud-Bénin)" Cahiers Agricultures. 2011;Vol. 20(No. 3): pp. 216-22.

[12] Ondo Zue Abaga N, Alibert P, Dousset S, Savadogo PW, M. S, M. S. Insecticide residues in cotton soils of Burkina Faso and effects of insecticides on fluctuating asymmetry in honey bees (Apis mellifera Linnaeus). Chemosphere 2011;83:585-92.

[13] Leu C, Singer H, Stamm C, Muller S, Schwarzenbach R. Variability of herbicide losses from 13 fields to surface water within a small catchment after a controlled herbicide application. . Environ Sci Technol. 2004;38(14):3835-41.

[14] Panuwet P, Siriwong W, Prapamontol T, Barry Ryan P, Fiedler N, Robson MG, Boyd Barr D. Agricultural pesticide management in Thailand: status and population health risk. environmental science & policy 2012;17 72 - 81.

[15] Tapsoba HK, Bonzi-Coulibaly YL. Production cotonnière et pollution des eaux par les pesticides au Burkina Faso. J Soc Ouest-Afr Chim. 2006;21:87-93.

[16] Aulagnier F, Poissant L. Some pesticides occurrence in air and precipitation in Québec, Canada. . Environ Sci Technol. 2005;39:2960-7.

[17] Manirakiza P, Akinbamijo O, Covaci A, Pitonzo R, Schepens P. Assessment of Organochlorine Pesticide Residues in West African City Farms: Banjul and Dakar Case Study. Arch Environ Contam Toxicol 2003;44:171-9.

[18] Adu-Kumi S, Kareš R, Literák J, Boru˚ vková J, Yeboah PO, Carboo D, Akoto O, Darko G, Osae S, Klánová J. Levels and seasonal variations of organochlorine pesticides in urban and rural background air of southern Ghana. Environmental Science and Pollution Research 2012;19 (6):1963-70.

[19] Ahouangninou C, Martin T, Edorh P, Bio-Bangana S, Samuel O, St-Laurent L, Dion S, Fayomi B. Journal of Environmental Protection. 2012;3:241-8.

[20] Savadogo WP, Traoré O, Topan M, Tapsoba KH, Sédogo PM, Bonzi-Coulibaly LY. Variation de la teneur en résidus de pesticides dans les sols de la zone cotonnière du Burkina Faso. Journal Africain des Sciences de l'Environnement. 2006;1:29-39.

[21] Naré R-WA, Savadogo WP, Gnankambary Z, Sedogo PM. Effect of Endosulfan, Deltamethrin and Profenophos on Soil Microbial Respiration Characteristics in Two Lands Uses Systems in Burkina Faso. Research Journal of Environmental Sciences. 2010;4(3):261-70.

[22] Ouattara B, Savadogo PW, Traoré O, Coulibaly B, Sedogo PM. Effet des pesticides sur l'activité microbienne d'un sol ferrugineux tropical au Burkina Faso. Cameroun Journal of Experimental Biology. 2010;Vol 06 N°01:11-20.

[23] Naré R-WA, Savadogo WP, Gnankambary Z, Nacro HB, Sedogo PM. Effect of Three Pesticides on Soil Dehydrogenase and Fluorescein Diacetate Activities in Vegetable Garden in Burkina Faso. Curr Res J Biol Sci. 2014;6(2):102-6.

[24] Ouédraogo E, Mando A, Zombré NP. Use of compost to improve soil properties and crop productivity under low input agricultural system in West Africa. . Agr Ecosyst Environ. 2001 84:259-66.

[25] Ros M, Hernandez MT, Garcia C. Soil microbial activity after restoration of a semiarid soil by organic amendments. Soil Biol Biochem. 2003;35:463-9.

[26] Kumar A, Singh S. Adsorption and Desorption Behavior of Chlorotriazine Herbicides in the Agricultural Soils. J Pet Environ Biotechnol. 2013;4(5):154.

[27] Bansal OP. The Effects of composts on adsorption-desorption of three carbamate pesticides in different soils of Aligarh district. J Appl Sci Environ Manage. 2010;14 (4):155 - 8.

[28] Savadogo PW, Lompo F, Bonzi-Coulibaly YL, Traoré AS, Sedogo PM. Influence de la Température et des Apports de Matière Organique sur la Dégradation de l'Endosulfan dans trois types de Sols de la Zone Cotonnière du Burkina Faso. J Soc Ouest-Afr Chim 2008;026:79 - 87.

[29] Bassolé D, Ouédraogo L. Problématique de l'utilisation des produits phytosanitaires en conservation des denrées alimentaires et en maraîchage urbain et péri urbain au Burkina Faso : cas de Bobo Dioulasso, Ouahigouya et Ouagadougou. IFDC 2007.

[30] Sotamenou J, Parrot L. Sustainable urban agriculture and the adoption of composts in Cameroon. International Journal of Agricultural Sustainability. 2013;11(3):282-95.

[31] Smith OB. Overview of urban agriculture and food security in West Africa. International Development Research Centre, Ottawa, Canada. 2001.

[32] Wauchope RD, Buttler TM, Hornsby AG, Augustun Beckers PWM, Burt JP. The SCS/ARS/CES pesticides properties for environmental decision-making. . Rev Environ Contam Toxicol. 1992;123:1-155.

[33] Tomlin CDS. The Pesticide Manual 11th ed. British Crop Protection Council, UK. 1997.

[34] Edwards R, Millburn P. The metabolism and toxicity of insecticides in fish. . In: Hutson, DH, Roberts, TR (Eds), Insecticides, vol 5 John Wiley and Sons, New York. 1985:pp. 249-69.

[35] Murty AS. In: Toxicity of Pesticides to Fish. CRC Press, Inc, Florida. 1986a;vol. 1.

[36] Murty AS. In: Toxicity of Pesticides to Fish. CRC Press, Inc, Florida. 1986b; vol. 2.

[37] Wise, Ltd. LIS. The e-Pesti-cide Manual 2000-2001 (Twelve Edition). Version 20 Editor: CDS Tomlin; British Crop Protection Council.

[38] Hertfordshire Uo. Base de Données FOOTPRINT PPDB sur les Propriétés des Pesticides. FOOTPRINT: Des Outils Innovant Pour l'Evaluation et la Réduction du Risque Pesticides. 2008:http://www.eu-footprint.org/fr/ppdb.html.

[39] Elliott JA, Cessna AJ, Nicholaichuk W, Tollefson LC. Leaching rates and preferential flow of selected herbicides through tilled and untilled soil. Journal of Environmental Quality 2000;29:1650-6.

[40] Vereecken H. Mobility and leaching of glyphosate: a review. Pest Management Science 2005;61:1139-51.

[41] Williamson S, Ball A, Pretty J. Trends in pesticide use and drivers for safer pest management in four African countries. Crop Protection 2008;27:1327- 34.

[42] Ouédraogo M, Tankoano A, Ouédraogo T, Guissou I. Environ Risque Sante. Risk factors for pesticide poisoning among users in the cotton-production region of Fada N'Gourma in Burkina Faso / Etude des facteurs de risques d'intoxications chez les utilisateurs de pesticides dans la région cotonnière de Fada N'Gourma au Burkina Faso. 2009;8:343-7.

[43] Pretty J, Hine R. Pesticide use and the environment. In: Pretty, J (Ed), The Pesticide Detox Towards a More Sustainable Agriculture Earthscan, London,. 2005:pp. 1-22.

[44] Tariq M, Hussain. I, S. A. Policy measures for the management of water pollution in Pakistan. Pak J Environ Sci 2003;3:11-5.

[45] WHO. Public health impact of pesticides used in agriculture. World Health Organization, Geneva. 1990.

[46] FAO. Project concept paper. HEAL:health in ecological agricultural learning. Prepared by the FAO programme for community IPM in Asia. [http://wwwfaoorg/nars/partners/2nrm/proposal/ 9-2-6doc]. 2000.

[47] Fall ST, Cissé I, Akinbamijo OO, Adediran SA. Impact de l'intégration entre l'horticulture et l'élevage sur la productivité des systèmes périurbains dans l'espace sénégambien in: Akinbamijo OO, Fall ST, Smith OB (Eds),Advances in crop-livetsock integration inWest African cities, ITC, ISRA, CRDI, Ottawa, Canada,. 2002:pp.69-84.

[48] Clausen L, Fabricius I, Madsen L. Adsorption of pesticides onto quartz, calcite, kaolinite, and α-alumina. J Environ Qual. 2001;30:846-57.

[49] Wauchope RD, Yeh S, Linders JB, Kloskowski R, Tanaka K, Rubin B. Pesticide soil sorption parameters: theory, measurement, uses, limitations and reliability. . Pest Manage Sci. 2002;58(5):419-45.

[50] Tang X, Zhu B, Katou H. A review of rapid transport of pesticides from sloping farmland to surface waters: Processes and mitigation strategies.Journal of Environmental Sciences. 2012;24(3):351-61.

[51] Wang H, Huang B, Shi X, Darilek JL, Yu D, Sun W, Zhao Y, Chang Q, Öborn I. Major nutrient balances in small-scale vegetable farming systems in peri-urban areas in China. Nutr Cycl Agroecosyst 2008;81:203-18.

[52] Li L-L, Huang L-D, Chung R-S, K-H. F, Zhang.Y-S. Sorption and Dissipation of Tetracyclines in Soils and Compost. . Pedosphere. 2010;20(6):807-16.

[53] Briceño G, Palma G, Durán N. Influence of organic amendment on the biodegradation and movement of pesticides. Critical Reviews in Environmental Science and Technology. 2007;37: 233-71.

[54] Si YB, Zhang J, Wang SQ, Zhang LG, Zhou DM. Influence of organic amendment on the adsorption and leaching of ethametsulfuron-methyl in acidic soils in China. Geoderma. 2006;130:66-76.

[55] Oudou HC, Hansen HCB. Sorption of lambda-cyhalothrin, cypermethrin, deltamethrin and fenvalerate to quartz, corundum, kaolinite and montmorillonite. Chemosphere. 2002;49:1285-94.

[56] Matsumura F. In: Toxicology of Insecticides. Plenum Press, New York. 1985; vol. 2.

[57] INSD. (RGPH, Enquête prioritaire, Enquête burkinabé sur les conditions de vie des ménages 2003 et Enquête QUIBB), Direction des Etudes et de la Planification du Ministère de l'Enseignement de Base et de l'Alphabétisation -DEP/MEBA. Direction des Etudes et de la Planification du Ministère des Enseignements Secondaire, Supérieur et de la Recherche Scientifique -DEP/MESSRS 2003;annuaires.

[58] Paré-Kaboré A. La problématique de l'éducation des filles au Burkina Faso GUFNU-Dijon 1/10. 2003.

[59] Somda J, Nianogob J, Nassa S, Sanou S. Soil fertility management and socio-economic factors in crop-livestock systems in Burkina Faso: a case study of composting technology. . Ecological economics. 2002;43:175-83.

[60] Sonnemann I, Baumhaker H, Wurst S. Species specific responses of common grassland plants to a generalist root herbivore (Agriotes spp. larvae). Basic and Applied Ecology. 2013;13(7):579-86.

6

Compatibility of *Jatropha Curcas* with Maize (*Zea Mays* L.) Cv. Obatampa in a Hedgerow Intercropping System Grown on Ferric Acrisols

Abugre S.[1,*], Twum-Ampofo K.[2], Oti-Boateng C.[3]

[1]Department of Forest Science, School of Natural Resources, University of Energy and Natural Resources, Sunyani, Ghana
[2]Department of Environmental Management, School of Natural Resources, University of Energy and Natural Resources, Sunyani, Ghana
[3]Department of Agroforestry, Faculty of Renewable Natural Resources, KNUST, Kumasi, Ghana

Email address:
simon.abugre@uenr.edu.gh (Abugre S.)

Abstract: Skeptics are talking about the impact of the biofuel crop on food production. It is important that the compatibility of *Jatropha curcas* in agroforestry systems is investigated to provide answers to some of these problems being advanced. The Randomized Complete Block Design (RCBD) with three hedgerow spacings of 2 m x 1 m, 3 m x 1m, 4 m x 1 m of *Jatropha curcas* and a control (No hedgerow) was used to lay out the experiment. This was replicated 3 times. The study showed that in the second year, plant height and plant diameter at first node differed significantly between the treatments. Maximum stover weight was 11.9 tons/ha and 7.5 tons/ha in the first and second year respectively for 4 m x 1 m spacing. Generally yields were lower in the second year in all the treatments compared to the first year. Maximum grain yield of maize was 4.47 tons/ha and 2.99 tons/ha in the first and second year respectively at the control treatment. Chemical properties of the soil did not record any significant decline after two years of cultivation. pH, organic Carbon, total nitrogen, organic matter, exchangeable cations, total exchangeable bases, exchangeable acid and base saturation did not show significant difference between the treatments. The highest Land Equivalent Ratio (LER) of 1.6 and 1.2 was recorded at 4 m x 1 m for both years, making it the most suitable plant spacing for *Jatropha curcas* with maize.

Keywords: *Jatropha Curcas*, Growth, Yield, Land Equivalent Ratio, Nutrient Status

1. Introduction

The exploitation of bio-energy sources of fuel has recently been given much prominence by the scientific community and commercial entrepreneurs as a way to solve the energy crisis. Bio-diesel is the most valuable form of renewable energy that can be used directly in any existing unmodified diesel engine [1]. It is an alternative fuel that can be used in diesel engines and provides power similar to conventional diesel fuel. Biofuel can help reduce the country's dependence on foreign oil imports. Recent environmental and economic concerns (Kyoto protocol) have prompted a resurgence in the use of biodiesel throughout the world. In 1991, the European Community, (EC) proposed a 90% tax reduction for the use of biofuels, including biodiesel [1]. Biofuels create new markets for agricultural products and stimulate rural development because they are generated from crops. They hold enormous potential for farmers. The long term challenge is the ability to supply feedstock to keep up with growing demand. The supply of feed stock from maize, soya beans will be limited by competition from other uses and land constraints. The key to the future of biofuels therefore is finding inexpensive feed stocks that can be grown by farmers. *Jatropha curcas* proves to be one of the many plants that hold great promise as a biofuel crop. *Jatropha curcas* is more recently being cultivated as a bio-diesel plant. Soybean and rapeseed have a relatively low oil yield compared with *Jatropha curcas*. A yield of 375 kg/ha (280 gallons/acre) is reported for soybean in the United States. In Europe, yield for rape seeds is said to be 1000 kg/ha (740 gallons per acre) whilst in India, *Jatropha curcas* is reported to have yield of 3000 kg/ha (2226 gallons/acre).

The world's population has grown from almost 5 billion 1980 to over 6 billion in 2000 [2], and finite availability of fertile land makes meeting energy needs for this growing population difficult. Ghana had a population of about 12.4 million in 1984. This figure increased to 18.8 million in 2000 with a growth rate of 2.6%.

The increase in population has caused a corresponding rise in food and fuel consumption, straining the earth's natural resources [3]. In developing countries, the pressure on natural resources is more acute because nearly 70% are subsistence-based and live in rural communities [4]. Heavily reliant on natural resources for food and energy, people by their basic instinct to survive derive their diet from their surroundings [5, 6].

Maize (Zea mays L.) belongs to the family Poaceae (Gramineae) and the tribe maydeae. Based on area and production, maize is one of the most important cereal crops in Ghana. It is rich in calories and forms part of the staple diet of every Ghanaian. It is Ghana's number one staple crop followed by rice. The yield in Ghana is low compared to other maize producing countries and there is an average short fall in domestic maize of 12%. There is concern that the use of land for the cultivation of bio-fuels could further jeopardize Ghana's self- sufficiency in maize production. Increasing grain yield per unit area and increasing the corn cultivable area are recognized as better solutions to solving the gap between consumption and production. The total land area of Ghana is about 23.8 million hectares of which 35% is cultivated. Ghana's agriculture is predominantly small holder, traditional and rainfed.

Agricultural production is undertaken by about 2 million, predominantly small holder subsistence farmers who account for about 80% of food in the country. The mean farm size is less than 1.2 hectares with a few exceeding two hectares [7]. In the midst of limited land for the cultivation of both maize and *Jatropha curcas*, it is surmised that the integration of these in an alley cropping could help provide an appropriate output from these plants.

Agroforestry is credited with improving the utilization of space by improving recycling of nutrients and organic matter. This translates into improved soil chemical, physical and biological characteristics with a reduction in the use of chemical fertilizers and improved infiltration of water. There is higher aggregate biomass production from an agroforestry mixture than from monoculture. Microclimate extremes are reduced as is soil erosion. Agroforestry thus provides a more favourable environment for sustained cropping, the creation of habitat diversity and provides a more continuous flow of more products over time [8]. A promising agroforestry technology for the humid and sub-humid tropics, which has been developed during the past decades, is alley cropping.

Alley cropping also known as hedgerow intercropping has been the subject of intensive research at the International Institute for Tropical Agriculture (IITA) in Nigeria [9, 10]. The concept of alley cropping was formalized at IITA where the term was defined as the growing of crops, usually food crops, in alleys formed by trees or woody shrubs that are established mainly to hasten soil fertility restoration and enhance productivity [11]. Currently it entails growing food crops between hedgerows of planted shrubs and trees, preferable leguminous species. It is a management-intensive system that can lead to increased crop yields and productivity of the land. This study was therefore designed to determine the appropriate hedgerow spacing in order to maximize grain yield of maize cv. obatampa in a *Jatropha curcas* hedgerow intercropping system.

2. Objectives for the Study

The specific objectives of the study were therefore to:
1. Determine the growth and yield of maize cv. Obatampa in alleys of *Jatropha curcas* grown on ferric acrisol.
2. Assess the effect of *Jatropha curcas* on soil chemical properties.
3. To evaluate the economics of using *Jatropha curcas* in an alley cropping system.

3. Materials and Method

3.1. Description of the Study Area

The experiment was laid out at Ayakumaso which is about 3 km from Sunyani. It lies between latitude 7°55′N and7°35′N and longitude 2°00′W and 2°30′W (SMA, 1998). The area has a tropical climate, with high temperatures averaging 23.9°C. Its mean monthly temperature varies between 23°C and 33°C with the lowest in August and highest in March and April respectively. It has a double maxima rainfall pattern. Rainfall ranges, from an average of 1000 mm to 1500 mm. The major rainy season occurs from April to end of July whilst September to October is the minor wet season. It has relative humidity of about 70%. The vegetation type is the dry semi-deciduous forest. The soil texture at the study site is silt loam and classified as ferric acrisols.

3.2. Design of Experiment

The experiment was laid using the Randomized Complete Block Design. Three hedgerow spacing treatments of 2m x 1m, 3m x 1m, 4m x 1m of *Jatropha curcas* and a control (No hedgerow) were used. These were replicated three times. Plot sizes for each treatment were 12 m x 5 m. The 2 m x 1 m, 3 m x 1 m, and 4 m x 1 m spacing had 6, 4, and 3 hedgerows respectively. The alleys of *Jatropha curcas* were established on the 1st of May, 2008. The test crop used in the alleys was Maize (*Zea mays*, var Obatampa) and sown at a spacing of 100 cm x 40 cm giving a population of 25,000 plants ha-1. The planting date for the maize was 21st July, 2008 and repeated on the 25th July, 2009.

3.3. Soil Analysis

300 g of soil samples were collected at a depth of 0 – 30 cm prior to the establishment of the experiment and taken to the Soil Research Institute (SRI) for analysis. The samples were thoroughly mixed before the analysis. At the end of the experiment, composite soil samples were again taken from

each treatment. The soil samples were air-dried and analyzed for soil pH, Organic C, Total N, Organic matter, Ca, Mg, K, Na, T.E.B., Exchangeable acid and Base saturation.

3.4. Measurement of Test Crop

Vegetative traits such as plant height, plant diameter, number of leaves and stover weight were determined by randomly selecting 10 maize plants from each plot. The stover weight was determined after drying at 72 hours at 65°C. Plant height was measured with the measuring tape and diameter was measured using the vernier caliper. At harvest, 10 maize plants from each plot were taken to determine yield and yield components.

3.5. Data Analysis

All data recorded were analyzed using the GEN-STAT package. The Analysis of Variance (ANOVA) was generated to determine if there were any significant differences between the treatments. The Fishers Least Significant Difference (LSD) was then used to separate the means between the treatments at 5% probability level.

3.6. Limitations of the Study

Limitations encountered during the study were:
1) Data was collected for two years and therefore can only predict what is likely to happen in the third and subsequent years.
2) The research did not take into consideration planting distances greater than 4 m between rows because the objective of the research is to make optimum use of land.

4. Results

4.1. Effect of Jatropha Curcas Spacing on Growth Parameters of Maize

The effect of *Jatropha curcas* spacing on the growth of maize is shown in Table 1. Plant height, diameter, number of leaves, number of nodes per plant were not affected by *Jatropha curcas* spacing in the first year. However, in the second year, significant differences ($P < 0.05$) in height and diameter of maize were realized (Table 1). Spacing of 2 m x 1 m differed with the control, 3 m x 1m and 4 m x 1m treatments. An increase in plant height of 17.19%, 22.39% and 23.38% were obtained for 3 m x 1 m, 4 m x 1 m and control respectively with respect to 2 m x 1 m in the second year. The control (No hedgerow) had the highest plant height (2.72 m and 2.48 m), diameter (20. 35 mm and 19.09 mm), number of leaves (11 and 10) and number of nodes (12.33 and 12.33) for the years 2008 and 2009 respectively (Table 1). Diameter growth in the second year was highest in 4 m x 1 m but it did not differ significantly from the control and 3 m x 1 m treatment. It increased by 19.7%, 16.9% and 18.5% for control, 3 m x 1 m and 4 m x 1 m respectively with respect to 2 m x 1 m. Stover weight was however significantly different ($P < 0.05$) in the first and second year. In the first year, significant difference was realized between 2 m x 1 m and the other treatments. No significant effect was attained between the control (No hedgerow), 3 m x 1 m and 4 m x 1 m treatments. Similar results were obtained in the second year where the control and 4 m x 1 m treatments did not differ significantly (Table 3). However, these differed significantly from 2 m x 1 m and 3 m x 1 m treatments (Table 3).

Table 1. *Effect of J. curcas spacing on growth parameters of Maize.*

Treatments (Hedgerow spacing)	Height (m)		Diameter at 1st node (mm)		Number of leaves		Number of nodes per plant	
	2008	2009	2008	2009	2008	2009	2008	2009
Control	2.72	2.48	20.35	19.09	11.00	10.00	12.33	12.33
2m x 1m	2.43	2.01	21.03	16.36	10.12	9.67	11.53	11.33
3m x 1m	2.51	2.36	21.22	19.12	11.00	10.33	11.63	11.39
4m x 1m	2.55	2.46	21.55	19.38	11.33	10.00	12.33	12.03
S.E.	0.87	0.16	0.39	1.07	0.36	0.21	0.28	0.54
LSD (0.05)	NS	0.32	NS	2.70	NS	NS	NS	NS

*NS means Not Significant

4.2. Effect of Jatropha Curcas Spacing on Yield and Yield Components of Maize

Table 2. *Effect of J. curcas spacing on yield and yield components of maize.*

Treatments	100 seed weight		Number rows/cob		Number seed/row	
	2008	2009	2008	2009	2008	2009
Control	43.00	25.17	14.67	14.22	35.00	30.56
2m x 1m	41.80	21.15	14.00	14.17	30.08	26.33
3m x 1m	41.88	23.73	13.67	14.58	31.62	27.67
4m x 1m	42.77	24.67	15.00	15.00	33.67	28.00
S.E.	3.42	0.97	0.63	0.25	1.40	3.60
LSD (0.05)	NS	2.36	NS	NS	NS	NS

*NS means Not Significant

The results showed significant differences in the yield of maize at different spacing of *Jatropha curcas*. Maize yields ranged between 2.05 tons ha-1 and 4.47 tons ha-1 in the first year and 1.56 and 2.99 tons ha-1 in the second year. The differences in yield were significant in both years; however, the differences ($P < 0.05$) for 3 m x 1m and 4 m x 1 m treatments in both years were not significant (Table 3). There were no significant differences in 100 seed weight, number of rows/cob, number of seeds/row, weight of ear and weight of seed/cob with respect to *Jatropha curcas* spacing in the first year (Table 2 and 3). In the second year, however, significant differences ($P < 0.05$) were observed for 100 seed weight, weight of ear, weight of seed/cob. 100 seed weight was highest (25.17 g) at No hedgerow (control) but did not differ significantly from the spacing at 3 m x 1 m (23.73 g) and 4 m x 1 m (24.67 g) (Table 2). Significant differences were found between 100 seed weight at 2 m x 1m spacing and all other spacing. Weight of ear and weight of seed/cob showed the same trend for the second year. The maximum weight of ear of 228.1 g was recorded for the control while

the lowest was 198.7 g for 2 m x 1 m. Weight of seed/cob was also highest (125.8 g) at No hedgerow (control) and lowest (99.30 g) at 2m x 1 m in the second year. In both cases, the highest results obtained from the control plot did not differ significantly ($P < 0.05$) from 3 m x 1 m and 4 m x 1 m treatment (Table 3). Generally, yield and yield components of maize were lower in the second year.

Table 3. Effect of J. curcas spacing on yield and yield components of maize.

Treatment	Weight of ear (g)		Weight of seed/cob (g)		Stover weight (tons/ha)		Grain Yield (tons/ha)	
	2008	2009	2008	2009	2008	2009	2008	2009
Control	267.2	228.1	202.8	125.8	11.99	8.95	4.47	2.99
2m x 1m	239.3	198.7	133.5	99.30	8.72	4.38	2.05	1.58
3m x 1m	242.3	217.1	164.3	112.5	11.77	4.88	2.82	2.00
4m x 1m	245.2	218.5	178.9	119.5	11.86	7.49	3.80	2.10
S.E.	34.5	2.83	25.9	3.76	0.37	1.14	0.51	0.11
LSD(0.05)	NS	9.80	NS	13.03	1.28	3.95	1.77	0.38

*NS means Not Significant

4.3. Effect of Jatropha Curcas Spacing and Cultivation of Maize on Soil Chemical Properties

Data on soil chemical properties were analysed for year 2008 and 2009 after harvesting. The soil chemical properties were not significantly ($P < 0.05$) affected by *Jatropha curcas* spacing and maize cultivation. Also, soil samples taken before the experiment, the control treatment (No hedge), 2m x 1 m, 3 m x 1 m and 4 m x 1 m treatments did not differ

significantly (Table 5 and Table 6). However, the results showed that before the experiment was carried out, the site had high proportion of total nitrogen, organic matter and exchangeable potassium when compared to the ranking in Table 4. ECEC was however moderate. After two years of cultivating *Jatropha curcas* with maize a similar trend in ranking was observed.

Table 4. Soil nutrients (mineral content) and its ranking.

Nutrient	Rank/Grade
Phosphorus P (ppm), (Blay -1)	
< 10.0	Low
10.0 – 20.0	Moderate
> 20.0	High
Potassium, K (ppm); Exchangeable K (cmol) (+)/Kg	
< 50;< 0.2	Low
50 – 100 ;0.2 – 0.4	Moderate
> 100;> 0.4	High
Calcium, Ca (ppm) / Mg = 0.25 Ca	
< 5.0	Low
5.0 – 10.0	Moderate
> 10.0	High
ECEC (cmol) (+)/ Kg	
< 10.0	Low
10.0 – 20.0	Moderate
> 20.0	High
Organic matter (%)	
< 1.5	Low
1.6 – 3.0	Moderate
> 3.0	High
Nitrogen (%)	
< 0.1	Low
0.1 – 0.2	Moderate
> 0.2	High

Table 5. Chemical properties of the soil after harvesting in 2009.

Treatment	PH	Org. C.	Total N (%)	Org. Matter	Exchangeable cations me/100g			
					Ca	Mg	K	Na
Before the experiment	5.43	2.75	0.24	4.74	10.80	3.61	0.75	0.13
Control	5.87	2.69	0.25	4.65	12.20	3.61	0.62	0.12
2m x 1m	5.68	2.99	0.26	5.15	8.73	8.39	0.83	0.17
3m x 1m	5.95	3.24	0.28	5.58	12.06	7.53	0.90	0.15
4m x 1m	5.75	2.76	0.24	4.76	9.66	4.19	0.79	0.15
C.V. (%)	4.98	9.00	9.14	8.99	29.94	14.78	8.05	10.35
S.E.	0.29	0.26	0.023	0.45	3.19	0.825	0.064	0.015
LSD(0.05)	NS	NS	NS	NS	NS	NS	NS	NS

*NS means Not Significant

Table 6. Chemical properties of the soil after harvesting in 2009.

Treatments	T.E.B.	Exch. Acid (Al + H)	E.C.E.C. me/100g	Base saturation (%)	Available – Bray's	
					P ppm	K ppm
Before the experiment	15.29	0.28	15.59	98.71	5.67	177.32
Control	16.55	0.15	16.53	98.68	4.59	167.40
2m x 1m	18.12	0.24	18.36	98.93	8.29	184.13
3m x 1m	20.64	0.13	20.80	99.37	8.60	217.61
4m x 1m	14.79	0.18	14.77	98.81	5.94	160.70
C.V. (%)	15.73	44.45	15.33	0.59	3.97	42.44
S.E.	2.71	0.08	2.67	0.58	0.64	4.45
LSD (0.05)	NS	NS	NS	NS	NS	NS

4.4. Land Equivalent Ratio of Cultivating Jatropha Curcas with Maize

One way to assess the benefits of growing two or more crops together or intercropping is to measure productivity using the Land Equivalent Ratio (LER). It was proposed to help judge the relative performance of a component of a crop combination compared to sole stands of that species. LER is the sum of the relative yields of the components species. That is:

$$LER = C_i/C_s + T_i/T_s$$

Where C_i = Crop yield under intercropping
C_s = crop yield under sole crop
T_i = Tree yield under intercrop
T_s = Tree yield under sole system

Using the LER as a measure of both beneficial and negative interaction between the crops, all the treatments were beneficial over that of the control (Table 7). The highest LER of 1.62 and 1.20 was attained at 4 m x 1 m for the year 2008 and 2009 respectively. Lowest LER of 1 was obtained at the control treatment for both years (Table 7).

Table 7. Land equivalent ratio of the various treatments for year 2008 and 2009.

Treatment	Crop	Partial LER	Total LER 2008	Crop	Partial LER	Total LER 2009
Control	Maize	4.47/4.47 = 1	1	Maize	2.99/2.99 = 1	1
2m x 1m	Maize	2.05/4.47 = 0.45	1.45	Maize	1.58/2.99 = 0.53	1.14
	Jatropha	0.5/0.5= 1		Jatropha	1.22/2.00 = 0.61	
3m x 1m	Maize	2.85/4.47 = 0.6	1.54	Maize	2.00/2.99 = 0.67	1.18
	Jatropha	0.47/0.5 = 1		Jatropha	1.03/2.00 = 0.51	
4m x 1m	Maize	3.80/4.47 = 0.8	1.62	Maize	2.10/2.99 = 0.70	1.20
	Jatropha	0.41/0.5 = 0.9		Jatropha	1.00/2.00 = 0.50	

*LER greater than 1 shows intercropping is advantageous, LER < 1 show disadvantage, LER = 1 show no effect.

5. Discussion

5.1. Effect of Jatropha Curcas Hedgerow Spacing on Growth of Zea Mays L.

Plant heights, number of leaves, number of nodes per plant were not significantly affected by *Jatropha curcas* spacing in the first year. In the second year, however differences were observed in plant height and diameter. Maximum plant height (2.48 m) in the control plants was highest and lowest in at the

2 m x 1m (2.01 m). Diameter of maize stalks was highest at the 4 m x 1 m spacing (19.38 mm) and lowest at 2 m x 1 m (16.36 mm). The results suggest that the dense population of *Jatropha curcas* at the 2 m x 1 m spacing may have accounted for reduced plant height and diameter of maize. The reduced height and diameter may be due to limiting supply of water and nutrients from the soil and other environmental resources at dense population of *Jatropha curcas*. Since the maize component is smaller, its root will be confined to soil horizons that are also available to the roots of

Jatropha curcas, but *Jatropha curcas* can exploit soil volume beyond the reach of maize. Therefore, the effects of nutrients and water competition will be more severe for maize culminating in reduced height and diameter. These findings are in conformity with results of [12] and [13] who observed reduced plant height at high plant population. A contrary finding to this study has also been reported. It is reported that plant height and internodes length increased with increasing plant population because of competition for light [14]. [15] also recorded highest plant height in dense population.

5.2. Effect of Jatropha Curcas Hedgerow Spacing on Yield and Yield Component of Zea Mays

Generally no significant differences were observed in the first year. Differences in yield and yield components of maize started emerging in the second year even though yield was generally lower than in the first year. The lower yields observed in the second year at the closer spacing of 2 m x 1 m could be attributed to competition for nutrients, space and water. Yield reductions involving one or all component in intercropping have been attributed to interspecific competition for nutrients, moisture and/or space [14, 16]. [17] and [18] concluded that the main reasons for the comparatively poor crop performance under alley cropping treatments were root competition and shading. It is possible that in the second year *Jatropha curcas* competed with maize for nutrients and water thus causing a reduction in yield and yield parameters at closer spacing

(2 m x 1 m). [18] also noted that reduced crop yields, due to root competition between hedgerows and crops in the alleys, were detected at 11 months after hedgerow establishment, and that competition increased with age of the hedgerows as measured by steadily declining crop yields close to the hedgerows. This is corroborated by this study since, there were no significant differences in the first year but subsequently occurred in the second year. The ITTA study by [19] showed that maize and cowpeas yields were generally lower under alley cropping than when grown as sole crops. This can be confirmed in this study where yields and yield components where generally higher in the control treatment (No hedgerow) than the hedgerow intercropping for both first and second years. It is significant to note that the control did not differ significantly from wider spacing of 3 m x 1 m and 4 m x 1 m. [20] observed that though 2 m hedgerow spacing gave higher biomass, the yield of maize was reduced in this hedgerow spacing compared to the 4 m hedgerow spacing. Also [21] showed that competition for light was a more critical factor than root competition for intercropped maize between teak trees. Low yields from maize rows adjacent to *Leucaena leucocephala* hedgerow was attributed to shade [10].

5.3. Effect of Jatropha Curcas Spacing and Cultivation Of Maize on Soil Chemical Properties

The chemical properties of the soil did not decline significantly over the two-year period of establishment of

Jatropha curcas hedgerows and the cultivation of maize. This implies that *Jatropha curcas* can be integrated into our land use system without a significant deterioration of soil chemical properties in the short term. Soil under alley cropping was higher in organic matter and nutrient content than soil without trees [22]. An experiment to compared the effect of *Cassia siamea*, *Gliricidia sepium* and *Flemingia macrophylla* in alley cropping trial found that soil organic matter and nutrient status were maintained at higher levels with *Cassia siamea* (which surprisingly, is not a N2-fixing species) [23]. This study showed a higher organic matter content for the *Jatropha curcas* hedgerow treatments over the non alleyed treatments even though these were not significantly different. This increase although not significant could be due to the short duration of the study and the addition of leaf litter into the soil from the hedgerows. Over a period of six years, the relative rates of decline in the status of nitrogen, pH and exchangeable bases of the soil were much less under alley cropping than under non-alley cropped [19]. This was attributed to the nutrient cycling capability of *Leucaena leucocephala* hedgerow, as there was evidence of a slight increase in soil pH and exchangeable bases during the third and fourth years after the establishment of the hedgerows. Could the slight increase in exchangeable bases in the *Jatropha curcas* hedgerows be attributed to nutrient cycling capability of *Jatropha curcas*? In this study, it was observed that there was an extension of *Jatropha curcas* roots laterally. This may provide an avenue for the intercept nutrients and recycling to the topsoil.

5.4. Land Equivalent Ratio of Cultivating Jatropha Curcas with Maize

The highest Land Equivalent Ratio (LER) attained at the hedgerow width of 4 m shows the beneficial effect of intercropping *Jatropha curcas* and maize at this treatment. The results for the year 2008 (LER of 1.62) means that an area planted as monoculture (Maize) would require 16.2% more land to produce the same yield as the same area planted to *Jatropha curcas* and maize combination. According to [24] intercropping was highly advantageous for pepper and Cardamon. Pepper intercropped with grevillea produced 3.9 times more than in monoculture and Cardamon intercropped with Grevillea and Pepper yielded 2.3 times more than in monoculture [24]. The high LER showed a very clear benefit from intercropping *Jatropha curcas* and maize at 4 m x 1 m hedgerow spacing. It reported that, on average, mixtures are 12% more productive than pure stands, based on 202 direct observations, or 13% more productive, based on 604 estimates using yield-density relationships [25]. It can be stated that a *Jatropha curcas* and maize mixtures would be beneficial than planting maize as a monocrop.

6. Conclusion

Hedgerow intercropping of *Jatropha curcas* with maize could prove useful if a spacing of 4 m x 1 m is adopted. It should be noted that closer spacing of *Jatropha curcas* could create competition with the associated crop resulting in reduced yields. Even though alley width of 4 m spacing did not give the highest yield compared to the control, its highest land equivalent ratio implies it could be the most appropriate spacing. Soil chemical properties were not affected within the two year of cultivating *Jatropha curcas* with maize. It implies that its use in alley cropping would not result in any deterioration in soil chemical status within a short term. In the midst of declining land area for the cultivation of *Jatropha curcas* as a biofuel crop, its use in alley cropping at the appropriate hedgerow spacing could be exploited to produce the crop.

Acknowledgements

I am very grateful to the Agroforestry Practices to Enhance Resource-Poor Livelihood project (APERL), sponsored by CIDA for the financial assistance received to carry out this study as part of my PhD Study.

References

[1]   Shekhawat, Benniwal, Gour, V.K., Kumar, N. and Sharma N. (2009) Promoting Farming for the future. Jatropha World.

[2]   United States Census Bureau. (2011). Economic census. Summary statistics by 2011. US Census Bureau. USA.

[3]   FAO (2003). Forestry outlook study for Africa: Sub regional Report West Africa. Food and Agriculture Organization of limited Nations. Rome. Italy.

[4]   World Bank (2004). Agricultural Sector Investment Program. World Bank Report. 2004.

[5]   Abalu, G. and Hassan R., (1999). Agricultural productivity and natural resources use in South Africa. Food Policy 23 (6): 477- 490.

[6]   Scoones, I. (1998). Sustainable rural livelihoods: A framework for analysis. Working paper 72. Institute of Development Studies, Brighton, UK.

[7]   OCAR (Agricultural and Rural Development Department) (2002). Appraisal Report on Community Forestry Management Project. April, 2002.

[8]   Cameron, D.M., Gutteridge, R.C. and Rance, S.J. (1991). Sustaining multiple production systems. Forest and Fodder trees in multiple use systems in the tropics. Tropical Grasslands 25:165.

[9]   Kang, B.T., Wilson, G.F. and Nangju, D. (1981a). Leucaena (*Leucaena leucocephala.* Lam.De wit.) pruning's as nitrogen source for maize (*Zea mays L.*) fertilizer Research 2 (4): 279.

[10] Kang, B.T., Wilson, G.F. and Sipkens, L. (1981b). Alleycropping maize (*Zea mays L.*) and Leucaena (*Leucaena leucocephala* Lam.) in Southern Nigeria: Plant and Soil 63: 165.

[11] Wilson, G.F. and Kang, B.T. (1981). Developing stable and productive biological cropping systems for the humid tropics. In Biological Husbandry-A scientific approach to organic farming. Edited by B. stonehouse. Butterworth, London.

[12] Genter, C.F and H.M. Camper. (1973). Component Plant Part Development on Maize as Affected by Hybrids and Population Density. Agron. J. 65:669.

[13] Dimchovoski, P. (1978). The effect of the Size and Shape of the Area Allocated to Maize Plants on the Productivity of Maize Hybrid Knezsha. 2L. 602 Resteriew dni Nausle:12: 56-64. Field Crop Absts. 29(5): 3654, (1976). In : Zamir. S. I.et al. 1999. Effect of Plant Spacing on Yield and Yield Components of Maize. International Journal of Agriculture and Biology. Vol 1. No3:152-153.

[14] Enyi, B.A.C., (1973). Growth rate of Three Cassava varieties Under Varying Populations Densities. J. Agric. 81:15 - 28.

[15] Sharma, T.R and I.M Adamu. (1984). The effect of Plant Population on the Yield and Yield Contributing Characters in Maize (Zea mays L.). Zeitschrif-fur-Acker and Pflanzenbau, 153:315-318. In: Zamir S.I. et al. 1999. Effect of Plant Spacing on Yield and Yield Components of Maize. International Journal of Agriculture and Biology. Vol.1. No.3:152-153.

[16] Okpara, D.A. and C.P.C O Maliko, (1995). Productivity of Yambean (*Sphenostylis stenocarpa*)/ Yam (*Dioscoria sp.*) Intercropping. Indian J. Agric. Sci. 65:880-882.Reyes T., R. Quiroz, O. Luukkanen and F. de Mendibura. (2009). Spice crops agroforestry systems in the East Usambara Mountains, Tanzania: growth analysis. Agroforestry Syst. DOI 10. 1007/s10457-009-9210-5.

[17] Szott, L.T. (1987). Improving the Productivity of Shifting Cultivation in the Amazon Basin of Peru Through the use of Leguminous Vegetation. Ph.D. Dissertation, North Carolina State University, Raleigh, NC, USA. In:Nair, P.K. An Introduction to Agroforestry, Kluwer Academic Publishers. US Census Bureau (2002). World population: 1950 – 2050. US Census Bureau International Data base 10 – 2002.

[18] Fernandes, E.C.M. (1990). Alley copping on an Acid soil in the Peruvian Amazon : Mulch fertilizer and Hedgerow root pruning Effect PhD Dissertation. North Carolina State University, Raleigh, NC, USA.

[19] Lal, R. (1989). Agroforestry Systems and soil surface management of a tropical altisol. Parts I-VI, Agroforestry Systems 8 (1) : 1-6.

[20] Lawson, T. L. and Kang B. T. (1990). Yield of maize and cowpea in an alley cropping system in relation to available light. Agriculture and Forestry Meteorology 52: 347 – 357.

[21] Verinumbe, I. Okali, D. U. U. (1985). The influence of coppiced Teak (*Tectona grandis,*) regrowth and roots on intercropped maize. Agroforestry Systems 3: 381 – 386.

[22] Atta-Krah, A. N. (1990). Alley farming with leucaena: effect of short grazed fallows on soil fertility and crop yields. Experimental Agriculture 26:1.

[23] Yamoah, C.F., Agboola, A.A., and Mulongoy, K. (1986). Decomposition, nitrogen release and weed control by prunings of selected alley cropping shrubs. Agroforestry systems 4: 239 – 246.

[24] Reyes T., R. Quiroz, O. Luukkanen and F. de Mendibura. 2009. Spice crops agroforestry systems in the East Usambara Mountains, Tanzania: growth analysis. Agroforestry Syst. DOI 10. 1007/s10457-009-9210-5.

[25] Jollife, P. A. (1997) Are mixed populations of plant species more productive than pure stands? Nordic ecological society. Oikos 80(3): 595 – 602. DOI: 10.2307/3546635. In: Reyes T., R. Quiroz, O. Luukkanen and F. de Mendibura. 2009. Spice crops agroforestry systems in the East Usambara Mountains, Tanzania: growth analysis. Agroforestry Syst. DOI 10. 1007/s10457-009-9210-5.

Structure and Regeneration Status of Menagesha Amba Mariam Forest in Central Highlands of Shewa, Ethiopia

Abiyou Tilahun[1, *], Teshome Soromessa[2], Ensermu Kelbessa[2]

[1]Department of Biology, College of Natural Science, Debre Berhan University, Debre Berhan, Ethiopia
[2]Department of Plant Biology and Biodiversity Management, Science Faculty, Addis Ababa University, Addis Ababa, Ethiopia

Email address:

abiytila22@gmail.com (A. Tilahun)

Abstract: This study was conducted in Menagesha Amba Mariam Forest, a dry evergreen afromontane forest in central highlands of Ethiopia. The aim of the study was to determine vegetation structure, and regeneration status of the forest. Sixty-nine sample plots (20 m x 20 m) were laid following altitudinal gradient and each quadrat has a 125 m altitudinal drop. Herbaceous species were collected from five (1 m x 1 m) sub-plots laid at four corners each and one at the centre of the large quadrat. All plant species found in each plot were recorded, collected, pressed and identified following Flora of Ethiopia and Eritrea. Diameter at Breast Height (DBH) and height were measured for trees and shrubs having DBH > 2.5 cm. The analysis of vegetation revealed that the forest possesses the highest number of DBH, height and density of species at the lower classes. Vertical stratification revealed that most of the species in the Menagesha Amba Mariam Forest were found in the lower storey. The total density of tree stems per hectare and basal area of trees with DBH >2.5 cm were 4,362.08 and 84.17 m^2 ha^{-1} respectively. The total density of tree species greater than 2 cm and 10 cm DBH were found to be 860.56 which is greater than those with DBH >20 cm (197.46). Thus, the regeneration prevalence of small individuals (seedlings and saplings) was at good condition. Menagesha Amba Mariam Forest, which is one of the remnant dry evergreen afromontane forests in central Ethiopia, is under high degree of anthropogenic impact, which needs further attention as it is quite close to the nearby towns.

Keywords: Altitudinal Gradient, Density, Dry Evergreen Afro Montane Forest, Regeneration, Structure

1. Introduction

Ethiopia has diverse flora and fauna in tropical Africa due to its great geographical diversity, vegetation types, soil types and diverse climatic conditions led to the emergence of habitats that are suitable for the evolution and survival of various plant and animal species (Gebre Egziabher, 1991). The vegetation of the country is very heterogeneous and has a rich endemic element. In Ethiopia, forest cover has been declining rapidly. Most of the remaining forests of the country are confined to south and south-western parts of the country (Bekele, 1993).

Loss of forest cover and biodiversity owing to human-induced activities is a growing concern in many parts of the world including our country (Demissew, 1980). Deforestation is one of the main causes of the prevailing land degradation in Ethiopia. Some parts of northern Ethiopia that currently are bare and experience severe land degradation once had a good vegetation cover. The current rate of deforestation and loss of fertile topsoil results in massive environmental degradation

(Bekele, 1993). Many of the studies focusing on forests or vegetation of specific regions in Ethiopia (Hedberg, 1951 & 1957; Mooney, 1963; Gilbert, 1970; Coetzee, 1978; Friis et al., 1982; Sharew, 1982; Woldu, 1985; Demissew, 1988; Uhlig, 1988; Woldu et al., 1989; Uhlig & Uhlig, 1990; Woldu & Backeus, 1991; Haugen, 1992; Tadesse , 1992; Bekele, 1993 and 1994; Miehe & Miehe, 1994; Yeshitela and Bekele, 2003; Shibru and Balcha, 2004; Soromessa et. al., 2004) have been carried out. Moreover, the vegetation resources of Ethiopia, including forests, woodlands and bush lands, have been studied by several scholars (Pichi-Sermolli, 1957; von Breitenbach, 1963; Westphal, 1975; Chaffey, 1979; Gebre Egziabher, 1986, 1988; Friis, 1986; Friis and Tadese, 1990; Soromessa and Demissew, 2002) who have employed different methods of vegetation classification. Almost all the aforementioned studies have warned about the intractable loss of this natural resource (Mekuyie,2014). However, the current tree planting trends that have been started elsewhere will be a commitment to overcome deforestation of forest resources that will again have a significant impact on climate change issues.

Even though Ethiopia is rich in biodiversity with high endemism and most of the forests provide socio-economic benefits and ecological function for longtime (Zewdie, 2007), most of its biodiversity is being now threatened, endangered and some are also locally extinct. This is due to habitat destruction and fragmentation, over exploitation of wildlife and their habitat beyond the limit of regeneration (Teketay, 2001).

Menagesha Amba Mariam (MAM) Forest is also one of these exposed forests, and there was no research carried out in the forest previously. Therefore, in order to implement appropriate forest management measures that could minimize forest losses, adequate information on factors affecting natural forest and the rate at which they cause depletion have to be obtained. Hence, the present situation of such fragile ecosystem grabs the attention and interest of researchers. Therefore, the present study is broadening its scope to assess the structural distribution together with regeneration status of the forest. Hence, this study was initiated to be conducted on the forest with the major objective of investigating structure and regeneration status of Menagesha Amba Mariam Forest in central highlands of Shewa, Ethiopia.

2. Materials and Methods

2.1. Study Area Description

The present study was conducted in Welmera Wereda, Oromia National Regional State, central high lands of Ethiopia (Fig. 1). The study forest is located at about 30 km west of Addis Ababa, and has total area around 84.354 hectare. The forest is known to have gradient of altitude and is situated approximately between $9^{0}01'$- $09^{0}03'$ N and $38^{0}35'$ - $38^{0}36'$ E. The altitudinal range of the study area varies from 2574 - 2948 m above sea level.

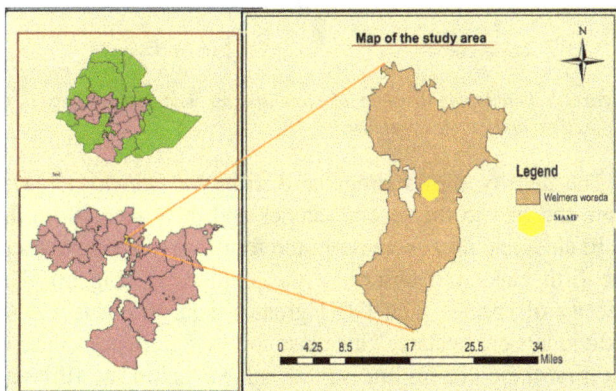

Figure 1. Location map of the study forest.

2.2. Reconnaissance Survey, Sampling Design and Sampling

Initially reconnaissance survey was conducted in October, 2008 in order to identify the possible sampling sites and number of transect lines to be laid across the forest and the altitudinal range of the forest area was determined.

Systematic sampling was used for the study. Sampling sites from the forest were arranged octagonally by eight line transects from the peak of the mountain to all directions covering the whole range of altitudes. Eight transects were laid at 200 m interval at the peak, 550 m at the middle of the mountain and 1.5 km at the bottom. The transect lines radiate from the top of the mountain to eight directions and each of them contains different number of plots depend on the length of transect. This is because the study area has a shape like Frustum of a cone. Quadrats of 20 m x 20 m (400 m) were placed at 125 m altitudinal drop between each quadrat and five sub-plots (1 m x 1 m) within each corner and one at the centre of the main plot for herbaceous plants were used to gather vegetation data (van der Maarel, 1979). Transects were placed on the ground using the Magellan NAV5000 Pro GPS navigation system. Sixty-nine Sample plots (2.76 ha) of each 400 m^2 were laid at every 125 m drop in altitude in the study site.

Altitude and geographical coordinates were measured for each sample plot using 'Pretel' digital altimeter, and Magellan NAV5000 Pro GPS respectively. Then a complete list of herbs, shrubs, lianas, epiphytes, and trees were made in each plot. Plant speciemens were collected, pressed, dried,identified, and checked at the National Herbarium of the Addis Ababa University using specimens in the Herbarium and published volumes of Flora of Ethiopia and Eritrea.

2.3. Structural Data Analysis

All individuals of trees and shrubs with a Diameter at Breast Height (DBH) greater than 2.5 cm, and height greater than 1 m were measured for DBH. Individuals were counted as seedlings (height $\leq$ 1 m and DBH $\leq$ 2 cm) and saplings (height > 1 m and DBH $\leq$ 2 cm).

2.3.1. Density

Density is defined as the number of plants of a certain species per unit area. Tree density was computed by converting the count from the total quadrats in the hectare basis (Mueller-Dombois and Ellenberg, 1974).

2.3.2. Basal Area (BA)

Basal area calculations were made on the diameter measurements of the stem with DBH of >2.5 cm and above. It is expressed in square centimeter/ hectare (cm^2/ha) (Hutchings, 1986; Mueller-Dombois and Ellenberg, 1974). There is direct relationship between DBH and basal area. Basal area (BA) = π (d / 2)^2, where d =DBH. $\Rightarrow$ / C, where C = circumference, d =diameter

2.3.3. Frequency

Frequency is defined as the proportion of sample quadrats in which individuals of a species are recorded. It is obtained by using quadrats and expressed as the number of quadrats occupied by a given species per number thrown or, more often, as a percentage (Goldsmith et al., 1986). Frequency measure indicates the uniformity of the distribution of the species in the study area, which again tells about the habitat preference of the species (Silvertown and Doust, 1993; cited in Eshete *et al.,* 2005). It gives an approximate indication of the homogeneity of the stand under consideration (Kent and Coker, 1992). The higher the frequency, the more important the plant is in the community (Denu, 2007). A better idea of the importance of species with the frequency can be obtained by comparing the frequency of occurrence of the entire tree species present. High frequency value shows that the plant is widely distributed in the study area but abundance does not always indicate the importance of a plant community.

2.3.4. Importance Value Index (IVI)

The Important Value Index (IVI) permits a comparison of species in a given forest and depicts the sociological structure of a population in its totality in the community. It often reflects the extent of the dominance, occurrence and abundance of a given species in relation to other associated species in an area (Kent and Coker, 1992). Importance Value Index (IVI) = RD + RF + RDO, Where RD is Relative density, RF is Relative frequency and RDO as Relative Dominance.

2.4. Diversity Indices

2.4.1. Shannon Wiener Diversity Index

Shannon-Wiener (Shannon and Wiener, 1949) index is the most applicable index of diversity (Greig-Smith, 1983). Like Simpson's index, Shannon's index accounts for both abundance and evenness of species present.

2.4.2. Simpson's Index (D)

Simpson's index measures the probability that two individuals randomly selected from a sample will belong to the same species. It is an index of dominance and hence inversely related to evenness and richness. It is often expressed as diversity index. Simpson's index can be calculated by the formula $D = (1-\lambda)$, where $\lambda = pi^2$. The value of D ranges between 0 and 1. With this index, 0 represents infinite diversity and 1, no diversity. The similarity index used was Sorensen's similarity coefficient (SI) = 2a /2a + b + c ; where "a" is the number of common species, "b" is number of species in vegetation of one site and "c" is number of species in vegetation of another site (Kent and Coker, 1992). Jaccard's coefficient (S_J) was also used for assessing the similarity among plant species clusters in terms of species composition. It is calculated by the formula (S_J = a/ a+b+c) where "S_J" is Jaccard's similarity coefficient, "a" is number of common species to the compared clusters , "b" is number of species in one cluster and "c" is number of species in the other cluster.

3. Results and Discussions

3.1. Analysis of Vegetation Structure

3.1.1. Density of Tree Species

Analysis on density distributions by diameter classes for tree species showed different patterns. Such patterns of species population structure can indicate variation in population dynamics. To observe regeneration of species in the study area, 32 plant species were selected based on their mean cover abundance value.

The total density of mature species with DBH > 2.5 cm in the forest was 1058.02 stems ha^{-1}. This was classified into seven density classes: 1) ≤ 1, 2) 1.01- 5, 3) 5.01-10, 4) 10.01- 20, 5) 20.01-35, 6) 35.01-50 and 7) > 50 stems ha^{-1}. The density of each species was calculated and compared as the number of individuals per hectare with DBH greater than 2 cm, 10 cm and 20 cm (Table 1). The density of trees with DBH greater than 2 cm is 705.1 individuals / ha.

Table 1. *Distribution of trees in different DBH classes.*

DBH (cm)	Density ha^{-1}	Percentage
< 10	705.1	66.60
>10 (a)	155.46	14.70
>20 (b)	197.46	18.70
Ratio of a to b	0.8	

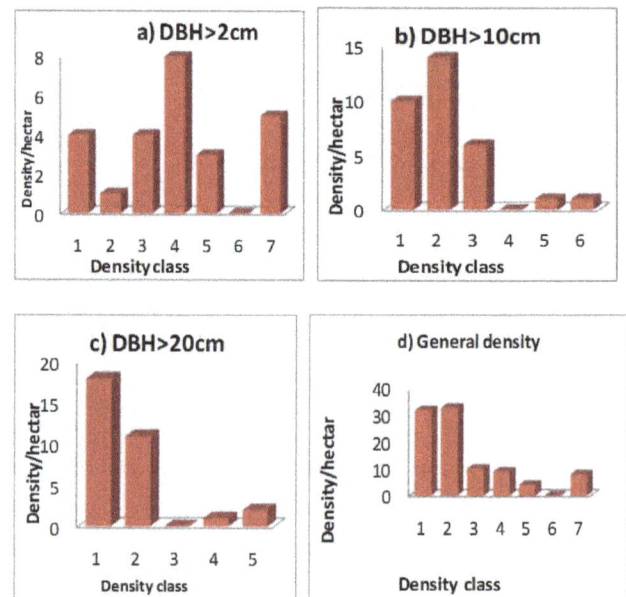

Figure 2. *Population structure of plant species with DBH >2 cm, >10 cm, >20 cm and their general density.*

The density shows irregular distribution which increases from the first to the second classes and then decrease in the third class and then increase in the fourth; it also decreases in the fifth class and absent in the sixth class (Fig.2a). The density of species with DBH greater than 10 cm is 155.46 individuals per hectare. This class shows significant values in the first three classes and very small values in the fifth and seventh classes and totally absent in the fourth and sixth classes (Fig.2b). The density of woody species at DBH greater than 20 cm is 197.46 individuals per hectare. The first two classes contain maximum values of this class while the fourth and seventh classes show very small values and classes: three, five and six have no representatives (Fig. 2c).

The overall density distribution of the forest shows high value in the first two classes and small values in the rest classes except class six, which has no representatives in all the density classes (Fig. 2d). As Table 2 shows, the largest proportion of the species density at DBH > 2 cm is contributed by *Olea europaea* (17.7%) followed by *Juniperus procera* (15.8%) *Olinia rochetiana* (11.3%) and *Erica arborea* (9.73%) which constitute the 1st, 2nd, 3rd and 4th largest proportions. The density of species (Table 3) at the DBH class > 10 cm alone is 155.46 individuals per hectare.

At this DBH class, the largest proportion of species density is contributed by *Juniperus procera* (34.00 %), *Olea europaea* subsp. *cuspidata* (20.5%) *and Myrica salicifolia* (5.1%). Tree species with DBH class > 20 cm has 197.46 individuals ha[-1]. Among the tree species *Juniperus procera, Olea europaea* subsp. *cuspidata* and *Scolopea theifolia* covers 50.8 %, 25.5 %, and 9 % of the species density respectively.

Table 2. *Density ha[-1] of seedling, sapling and mature tree species in Menagesha Amba Mariam Forest.*

No	Botanical name	Habit	Seedling Density ha[-1]	Sapling Density ha[-1]	Tree D ha[-1]	Sum of densities
1	*Acacia abyssinica*	T	14.13	27.9	68.12	110.15
2	*Apodytes dimidiata*	T	0	0	3.62	3.62
3	*Arundinaria alpina*	S	1.5	11.23	15.22	27.95
4	*Bersama abyssinica*	T	30.1	19.6	65.94	115.64
5	*Buddleja polystachya*	T	5.44	0.72	20.65	26.81
6	*Carissa spinarum*	S	22.5	8	189.86	220.36
7	*Croton macrostachyus*	T	1.1	2.54	7.97	11.61
8	*Podocarpus falcatus*	T	36.23	32.25	92.39	160.87
9	*Dombeya tórrida*	T	6.9	14.5	26.45	47.85
10	*Dovyalis abyssinica*	T	15.6	23.2	78.62	117.42
11	*Ekebergia capensis*	T	0.36	4.4	8.33	13.09
12	*Erica arborea*	T	68.5	117	260.51	446.01
13	*Ficus sur*	T	0.36	145.1	6.52	151.98
14	*Hagenia abyssinica*	T	0.72	3.3	17.39	21.41
15	*Hypericum revolutum*	T	8.33	23.6	35.14	67.07
16	*Juniperus procera*	T	19.2	81.5	365.22	465.92
17	*Maesa lanceolata*	T	1.1	6.2	7.97	15.27
18	*Maytenus arbutifolia*	T	4.7	19.2	37.32	61.22
19	*Maytenus obscura*	T	3.62	2.17	12.68	18.47
20	*Millettia ferruginia*	T	0	1.1	2.17	3.27
21	*Myrica salicifolia*	T	8.7	8	34.06	50.76
22	*Nuxia congesta*	T	6.2	48.6	87.32	142.12
23	*Olea europaea* subsp. *cuspidata*	T	31.9	146.4	385.14	563.44
24	*Olinia rochetiana*	T	79.7	133.3	297.46	510.46
25	*Osyris quadripartita*	T	8.7	21.7	112.32	142.72
26	*Pittosporum viridiflorum*	T	11.6	56.9	81.16	149.66
27	*Scolopia theifolia*	T	5.44	18.5	46.01	69.95
28	*Prunus africanus*	T	13.8	40.6	72.1	126.5
29	*Rhamnus staddo*	T	1.8	0.72	6.52	9.04
30	*Rhus glutinosa*	T	5.44	22.1	29.35	56.89
31	*Rhus vulgaris*	T	30	60.9	128.26	219.16
32	*Sideroxylon oxyacanthum*	T	8.1	64.9	142.39	215.39
	Total		451.77	1166.13	2744.18	4362.08

The ratio of the density of individuals greater than 10 cm to those greater than 20 cm is taken as a measure of the distribution of the size classes (Grub *et al.*, 1963). This ratio is 0.8 for the Menagesha Amba Mariam Forest, indicating only slight variability between the proportion of small-sized and large-sized individuals.

According to Grubb *et al*. (1963), the ratio of 'density at DBH class >10 cm' to 'density at DBH class >20 cm' can be used as a measure of the distribution of the different size classes. The dominance of small sized (DBH>2-9.9 cm) individuals in the forest is largely due to the high density of *Olea europaea* subsp. *cuspidata* (124.28 stems/ha), *Juniperus procera* (111.23 stem/ha), *Olinia rochetiana* (79.71 stems ha $^{-1}$) and *Erica arborea* (68.5 stems/ha) (Table 2).

3.1.2. Diameter at Breast Height (DBH)

The distribution of trees in different DBH classes is given in Figure 3. The number of stems in DBH class less than 10 cm (classes 1 and 2) is 708.7/ha (67.18%), 188.04/ha (17.8%) in 10-25 cm, 89.13/ha (8.42%) in 25-40 cm and 69.93/ha (6.6%) in >40 cm. The overall density distribution of individuals in the various size classes showed decline towards the higher DBH classes (i.e., as DBH size classes increase the number of individuals decrease). The highest densities of species were recorded in the lowest DBH size classes. The distribution of trees in DBH class from lower to higher showed a decreasing trend, but the percentage DBH of trees in DBH class >140 cm was contributed by *Juniperus procera*, *Olea europaea* subsp. *cuspidata* and few individuals of *Hagenia abyssinica*. The total DBH of trees in lower classes is much higher when compared to the DBH of the intermediate and higher classes. Distribution of all individuals in different size classes showed relatively an inverted J-shape distribution for both DBH and height classes (Fig. 3). This pattern indicates that the majority of the species had the highest number of individuals in lower DBH and height classes. This pattern implies that the forest vegetation has good reproduction and recruitment potential.

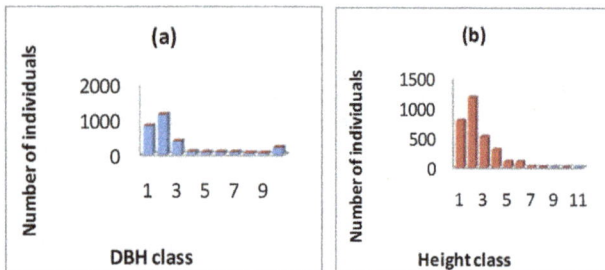

Figure 3. General DBH and Height class versus number of individuals in MAM Forest.

Analysis of density distribution by diameter classes of woody species resulted in different patterns (Figure 4a-j). The first pattern shows high value in the 5.1-10 cm diameter class and almost uniformly represented in the rest classes. This pattern is detected for Juniperus procera (Fig.4a). Figure 4c, d, and g also showed strong peaks up to 10 cm and followed by abrupt decline from 10.1-20 cm and then totally absent in the higher classes (>20 cm). This pattern was represented by Erica arborea, Osyris quadripartita, and Rhus vulgaris. Population patterns (Figure 4e & j) indicating

selective removal of higher-class individuals which shows good reproduction but, discontinuous recruitment. Representative species for this pattern are Prunus africana and Acacia abyssinica.

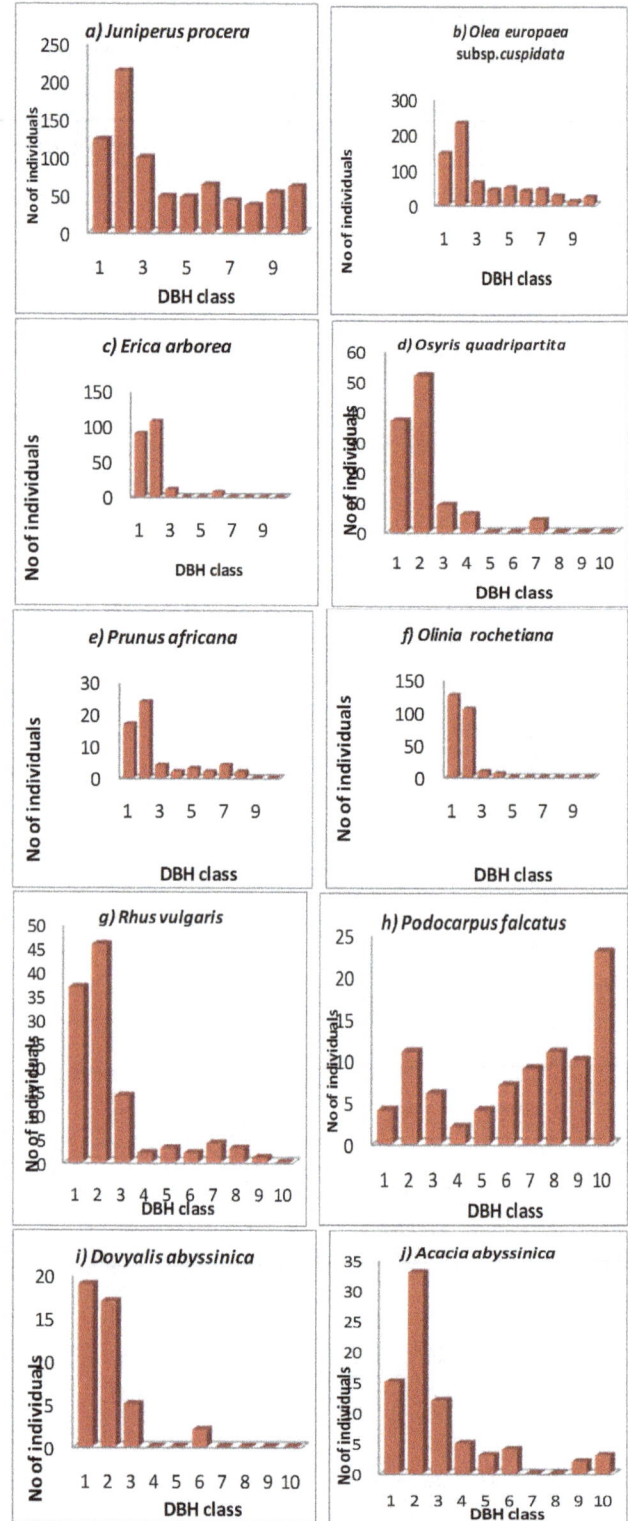

Figure 4. (a-j). Top dominant tree species population structure of the Forest.

Figure 4b shows high value up to 10 cm (class1 & 2) and uniformly decline up to the ninth class and also increases in the tenth class. This pattern relatively shows an inverted J-shape and exemplified by *Olea europaea* subsp. *cuspidata*. The other recognizable pattern (Figure 4h) shows high value in the higher DBH classes and small value in DBH classes<5 cm, and 15-25 cm and then highest value in the last class (>45 cm). This pattern revealed selective cutting and removal of medium sized individuals have taken place. In this case, the juveniles are not well represented and show poor regeneration. The representative example is *Podocarpus falcatus*.

3.1.3. Tree Height

The trees in the study area could be conveniently divided into eleven height classes. The overall height class distribution of individuals in Menagesha Amba Mariam Forest showed higher number of individuals in the lower size and a gradual decrease towards the middle and upper size trees indicating continuous representation of individuals in all height classes. Trees in height class I and II, together make 65.5%/ha. Trees in the height classes III and IV together are found to be 24.5%. The old trees are found in the height class VII and VIII and their percentage distribution is 9.52%/ha. Height classes I and II are represented by *Olea europaea* subsp. *cuspidata, Juniperus procera, Erica arborea, Sideroxylon oxyacanthum* with 15.78%, 15, 6%, 9.91% and 8.8% respectively while height class VIII is dominated by *Juniperus procera* (72..7%) and *Apodytes dimidiata* (27.27%). The upper canopy is dominated by these five tree species in the area. *Olea europaea* subsp. *cuspidata, Juniperus procera,* and *Prunus africana* are the emergent trees and grow above all the canopy trees except *Podocarpus falcatus* and *Hagenia abyssinica* at some inaccessible and well protected parts of the forest. The analysis based on relative density, diameter and height classes carried out for tree species of the forest resulted in different patterns. Generally, four patterns of population structure were analyzed. Each structural pattern reveals different population dynamics (Fig. 5 a-j).

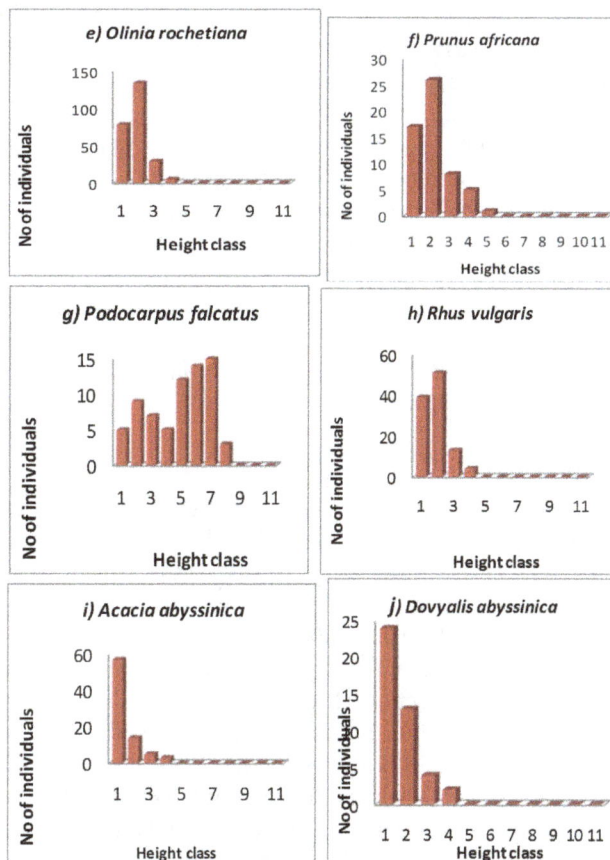

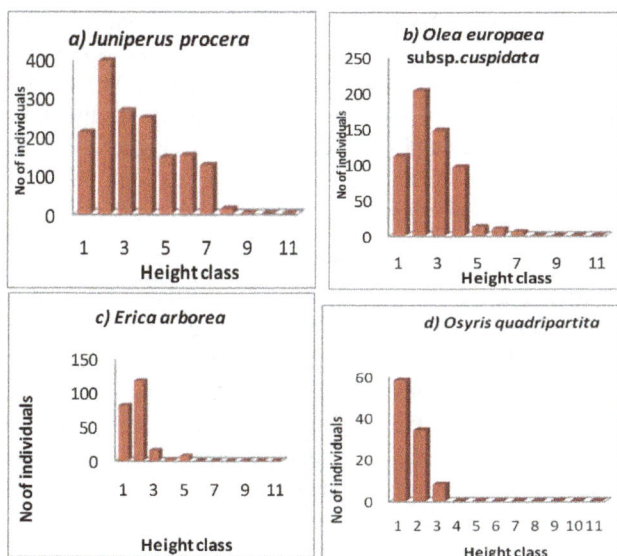

Figure 5. (a-j). *Top dominant tree species population structure of MAMForest.*

The first pattern was formed by species with highest density in the second-class, medium value in the first, third and fourth classes, small values in classes five six and seven and no value in the rest classes. This pattern shows better reproduction but a bad recruitment potential in the forest. This pattern was observed height class of Olea europaea subsp. cuspidata and Juniperus procera (Fig. 5a & b).

The second pattern (Fig. 5c, e, f & h) was indicated by species well represented in the lower height classes and absent in the higher classes which are species with no reproduction and only very few large and old individuals. This pattern was frequent in most shrubs and some trees. This indicates that there is an indiscriminate exploitation of large individuals of this species. Species with such a pattern could become endangered in the future, because individuals are being harvested before reaching reproductive ages, and this could result in the future decline of the species population because these reflect good reproduction but, bad recruitment. The height class of Erica arborea, Olinia rochetiana, Prunus africana and Rhus vulgaris shows this type of pattern.

The third pattern (Fig. 5j) indicates a normal distribution of species with reversed J-shape. Maximum values occurred in the first class and then reduce gradually up to the fourth class. This pattern represents good reproduction status and regeneration potential. *Osyris quadripartita* is the representative of this pattern.

The fourth pattern was indicated by low density in the lower height classes and high density in the higher height classes and no value in the last class. The species under this pattern have big individuals that are less competent to reproduce and hence represent poor reproduction, regeneration and thus a declining population. The pattern shows relatively J-shape curve and represented by height class of *Podocarpus falcatus* (Fig. 5g).

Variations in the population structure of plant species may be attributed to environmental factors that can interrupt in regeneration, differences in the regeneration behavior of the species, human intervention, and change in climate, natural and artificial disturbances.

3.2. Vertical Structure

According to this scheme, three simplified vertical structures are distinguished in tropical forests. These are upper, middle and lower storeys. Based on the above scheme the forest vegetation was classified into three strata. The upper includes where individual tree height exceeds 2/3 of the top height while the middle storey includes species having height between 1/3 and 2/3 of the top height and the lower stratum (storey) is less than 1/3 of the top height(Lamprecht, 1989). Based on the result obtained from the study the top height is 30 m, which indicates that the upper storey is greater than 20 m, the middle storey is between 10 & 20 m and the lower storey is less than10 m. The result obtained from this forest study indicated that the highest species density, 192 species (86.5%) was found in the lower storey. Most of the apparent gaps were filled up with under storey species canopy. The lower storey is mainly covered with herbs, shrubs and small trees (Figure 6). Twenty-five species of plants (11.5 %) reach to the middle storey. Species like Erica arborea, Olinia rochetiana, Croton macrostachyus, Hagenia abyssinica, Buddleja polystachya, Bersama abyssinica, Nuxia congesta, Osyris quadripartita, Ficus sur, Acacia abyssinica, Maytenus arbutifolia, Maytenus obscura, Dovyalis abyssinica, Pittosporum viridiflorum, Prunus africana, Rhus vulgaris, Rhus glutinosa, Sideroxylon oxyacanthum, Myrica salicifolia, Acacia mearnsii, Dombeya torrida, Arundinaria alpina, Scolopia theifolia, Pinus patula, and Eucalyptus globulus have representatives in the middle storey. The Upper storey also contains Only 5 emergent tree species (2.3%) of the total individuals in the forest that include Podocarpus falcatus, Juniperus procera, Olea europaea subsp. cuspidata, Apodytes dimidiata and Millettia ferruginea. There are many species, which could not attain the upper and the middle storey by their nature. All species that have representative in the upper also appeared in the middle and lower storey. Most of these trees have started dying-back from their tips and degenerating and some of them are completely absent.

3.2.1. Basal Area (BA)

The basal area of Menagesha Amba Mariam Forest was 84.17m²/ha. The highest (44.38%) and the lowest (0.01%) BA ha⁻¹ was contributed by *Juniperus procera* and *Maesa*

lanceolata respectively. It is important to note here that species with the highest basal area do not necessarily have the highest density, indicating size difference between species (Bekele, 1994; Shibru and Balcha, 2004; Denu, 2007). It was reported that BA provides a better measure of the relative importance of the species than simple stem count (Cain and Castro, 1959; cited in Bekele, 1994). Thus, species with the largest contribution to BA can be considered as the most important species in the forest. Consequently, the most important tree species in Menagesha Amba Mariam Forest are *Acacia abyssinica, Hagenia abyssinica, Olea europaea* subsp. *cuspidata, Juniperus procera, Podocarpus falcatus* and *Rhus vulgaris.*

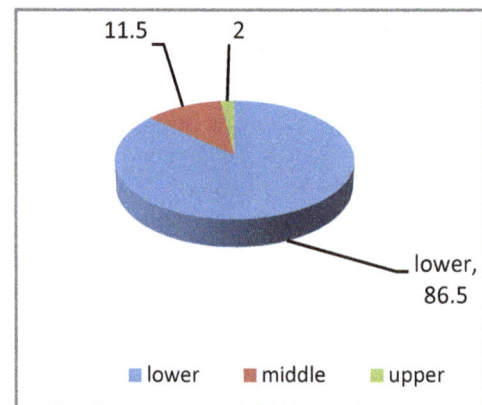

Figure 6. *Percentage density of trees in lower, middle and upper storey.*

3.2.2. Frequency

The trees and shrubs were classified into five frequency classes on the bases of their percentage frequency values: 1= 0-20, 2= 21-40, 3= 41-60, 4= 61-80 and 5= > 81. Two tree species (*Olea europaea* subsp. *cuspidata, Juniperus procera*) are most frequently occurred (in 68 and 66 quadrats out of 69) respectively. The species with more than 50% distribution were *Carissa spinarum, Dovyalis abyssinica, Erica arborea, Nuxia congesta, Rhus vulgaris, Olinia rochetiana, Osyris quadripartita* and *Pittosporum viridiflorum.* The species with the least occurrence are *Aruninaria alpina, Apodytes dimidiata, Ekebergia capensis, Millettia ferruginea, Rhamnus staddo, Hagenia abyssinica* and *Ficus sur.* This study revealed that there is high frequency in the lower frequency classes and low frequency in the higher frequency classes. This indicates that the vegetation has high heterogeneity and low homogeneity.

3.2.3. Importance Value Index (IVI)

Importance Value Index combines data from three parameters, which include RF, RD and RDO (Kent and Coker, 1992). IVI is useful to compare the ecological significance of species (Lamprecht, 1989). It was also stated that species with the greatest importance value are the leading dominant species of specified vegetation (Shibru and Balcha, 2004). Percentages of species in the IVI classes were 32.5%, 9.8%, 28.6%, 27.6% and 1.5% for classes 1, 2, 3, 4, and 5 respectively (Table 4).

Table 3. *IVI classes, values, sum of species belonging to each class and their percentage value.*

IVI class &value	No of species	Sum of IVI	Percentage (%)
5(<1)	6	4.22	1.5
4(1-10)	17	82.93	27.6
3(10.1-20)	6	85.73	28.6
2(20.1-30)	1	29.52	9.8
1(>30)	2	97.6	32.5

The highest IVI value (88.7%) was in classes 1, 3 and 4, while the remaining (11.3%) at classes 2 and 5 (Figure 7). The five most dominant tree species of Menagesha Amba Mariam Forest occupied 54.5% of the total important value index (Table 3). The dominant species were *Erica arborea, Olinia rochetiana, Podocarpus falcatus, Olea europaea* subsp. *cuspidata* and *Juniperus procera*. Figure 7 shows that much of IVI was attributed by a few species. These tree individuals were tolerant species that resist high pressure of disturbance, natural and environmental factors, and the effect of local communities.

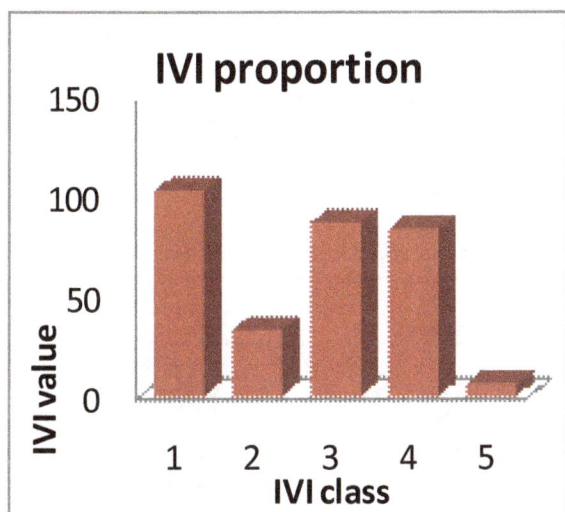

Figure 7. *The IVI proportions of most frequent and dominant trees and shrubs in the study area, IVI class: 1=>30, 2=20.1-30, 3=10.1-20, 4=1-10, 5=<1.*

In terms of abundance, distribution and basal area, the contribution of *Olea europaea* subsp. *cuspidata* and *Juniperus procera* were the highest of all tree species and accounted for 32.53%. These two species were the most frequent (16.9%) and they had the highest BA (59.7 m^2 ha^{-1}) out of the total BA (84.17 m^2 ha^{-1}). *Millettia ferruginea* and *Ekebergia capensis* had the lowest relative IVI values and were found to be the least dominant species among the species in the study site. *Juniperus procera* was 150 times more important than *Millettia ferruginea*. Similarly, *Olea europaea* subsp. *cuspidata* was 75 times more important than *Ekebergia capensis* in Forest. Priority for conservation of these species must be given based on their IVI values (i.e. the first priority for species with highest IVI value and the last priority of conservation for species with least IVI values) (Table 3). The result in Table 4 shows three species

(*Juniperus procera, Olea europaea* subsp. *cuspidata* and *Podocarpus falcatus* are grouped in priority class one that require immediate conservation and protection while species like *Millettia ferruginea, Ekebergia. capensis, Rhamnus staddo, Apodytes dimidiata , Maeasa lanceolata* and *Arundinaria alpina* in the last priority class and they need the last priority of conservation. The rest of the species are in the intermediate priority classes (2, 3 and 4), indicating that they need intermediate conservation programme.

3.3. Regeneration Status of Some Woody Species

The density and composition of seedlings and saplings indicate the status of regeneration. The total density of seedling, sapling and trees were 451.77 ha^{-1}, 1166.13 ha^{-1} and 2744.18 ha^{-1}. The ratio of seedlings to saplings is 0.4 and saplings to trees (0.43) and seedlings to mature trees is 0.17. The above ratio indicates that the distribution of seedling populations less than that of sapling and that of saplings is less than mature individuals (i.e. density of seedling<saplings<mature trees). The distribution of seedlings, saplings and mature trees shows three distinct patterns.

First distribution pattern contains least density of seedlings, intermediate in saplings and highest in the tree levels. This pattern was exhibited by Hagenia abyssinica, Olea europaea subsp. cuspidata, Juniperus procera, Osyris quadripartita, Pittosporum viridiflorum, Scolopia theifolia, Prunus africana, Dovyalis abyssinica, Ekebergia capensis, Rhus glutinosa, Rhus vulgaris and Sideroxylon oxyacanthum (Fig.8a).

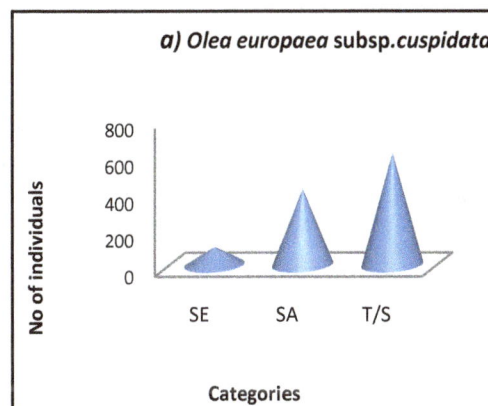

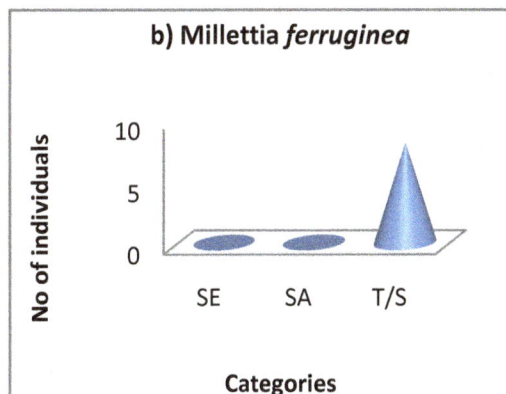

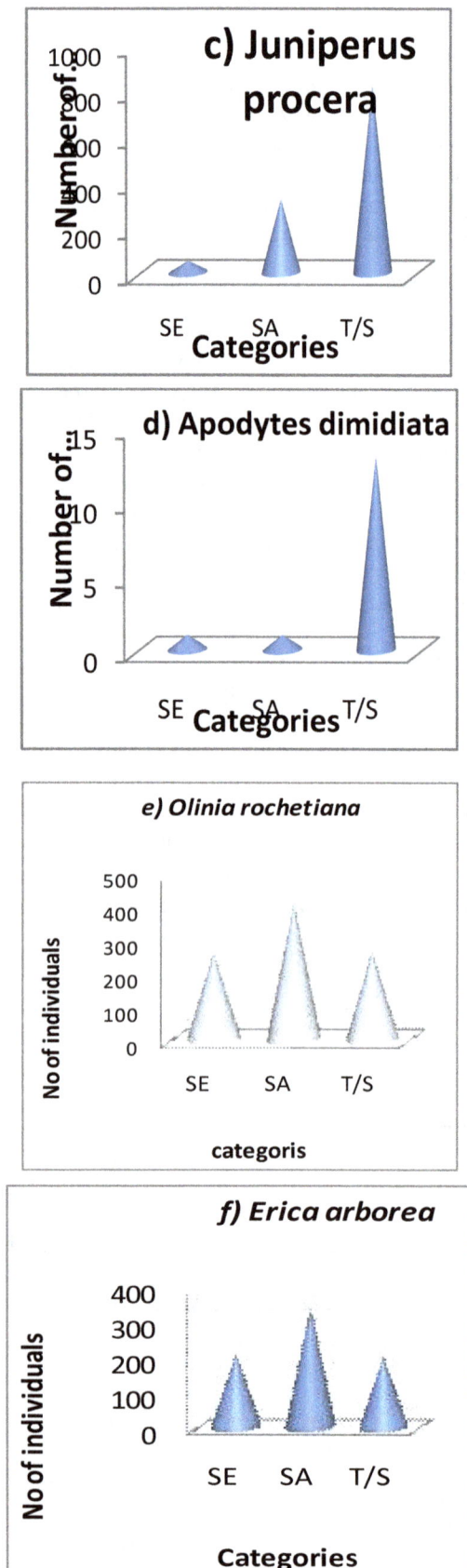

Figure 8. (a-f). *Seedlings, saplings and tree/shrub distribution of some selected species occurring in MAMForest.*

The second distribution pattern represents the highest population in sapling stages and medium in seedlings and trees. The pattern is represented by *Olinia rochetiana, Rhus vulgaris* and *Erica arborea* (Fig.8e and f). Few seedlings per species may be due to disturbances like seed predators, grazers, canopy cover for seedling recruitment, nature of seeds dormancy breakage in relation to environmental factors, physical factors, pathogens, and others those affecting the nature of propagation and reproduction.

Pattern three shows complete absence of seedlings and saplings. This includes Apodytes dimidiata and Millettia ferruginea and has very rare mature individuals in the forest (Fig.8b). The result from regeneration analysis of species revealed the complete absence or small amount distribution of seedlings and saplings shows insufficient recruitment hampered regeneration for the majority of species. This may be due to sensitiveness of seedlings and saplings to the disturbance such as seed predators, grazers and browsers, lack of safe site for seedling recruitment, litter accumulation, pathogens and environmental variables.

To set priority for regeneration analysis species in the study site, the species were classified in to three groups. Group 1, those species that were totally absent in regeneration category, Group 2, species whose density was greater than zero and less than 50 individuals ha[-1] and Group 3, species having individuals greater than 50 individuals ha[-1] (Table 4).

Table 4. *List of species under regeneration status group.*

Regeneration status		
Group 1	Group 2	Group 3
	Arundinaria alpina	*Acacia abyssinica*
	Buddleja polystachya	*Bersama abyssinica*
	Croton macrostachyus	*Carissa spinarum*
	Ekebergia capensis	*Podocarpus falcatus*
	Ficus sur	*Dombeya torrida*
	Hagenia abyssinica	*Dovyalis abyssinica*
	Maesa lanceolata	*Erica arborea*
	Maytenus obscura	*Hypericum revolutum*
Apodytes	*Myrica salicifolia*	*Juniperus procera*
dimidiata	*Rhamnus staddo*	*Maytenus arbutifolia*
Millettia		*Nuxia congesta*
ferruginea		*Olea europaea* subsp.
		cuspidata Olinia rochetiana
		Osyris quadripartita
		Pittosporum viridiflorum
		Scolopia theifolia
		Prunus africana
		Rhus glutinosa
		Rhus vulgaris
		Sideroxylon oxyacanthum

Species classified as Group 1 and Group 2 is recommended to be given the highest priority for conservation purpose. The result shows certain gaps between floristic composition and structure of matured stands and the regeneration. Some of the matured trees lacked seedlings and /or saplings. This suggests that their regeneration from seedling and sapling is reduced and these species may disappear in the future. Abundance of seedlings and saplings are indicators of the establishment of young individuals. The regeneration potential of plant species could depend on factors like soil seed bank, physical factors and anthropogenic activities.

4. Conclusion

Result of the present study revealed that most tree species were in poor regeneration and recruitment level. The total density of tree stems per hectare was 4362.08 indicating that the vegetation of the Forest has densely populated and dominant trees like *Juniperus procera, Olea europaea* subsp. *cuspidata, Olinia rochetiana* and *Rhus vulgaris*. The species population structure showed different dynamics. Most species have high population in the lower DBH and Height classes. Few species occur in all DBH and Height classes showing variation in population size. The forest is characterized by high density of trees in the lower class than in the higher. The total Basal Area of trees whose DBH >2.5 cm was 84.17 m^2 ha^{-1}. This value is supposed to be high in basal area coverage in dry evergreen montane forest. IVI of 59.14% was attributed by *Juniperus procera, Olea europaea* subsp. *cuspidata, Podocarpus falcatus, Olinia rochetiana, Erica arborea* and *Rhus vulgaris*. These species were also important in ecological significance. Species classified as regeneration classes as Group 1 (with no regeneration like *Apodytes dimidiata* and Group 2 (>zero & <50) individuals ha^{-1} as *Millettia ferruginea, Hagenia abyssinica, Ficus sur, Ekebergia capensis, Maytenus obscura, Croton macrostachyus, Maesa lanceolata, Buddleja polystachya, Rhamnus staddo* and *Myrica salicifolia*) are recommended to be given high priority for conservation purpose.

References

[1]   Bekele, T. (1993) Vegetation Ecology of Remnant Afromontane Forests on the Central Plateau of Shewa, Ethiopia. *Acta Phytogeogr. Suec.* 79:1-59.

[2]   Bekele, T. (1994) Comprehensive Summaries of Uppsala Dissertations from the Faculty of Science and Technology: Studies on Remnant Afromontane Forest on the Central Plateau of Shewa, Ethiopia. Acata Universities Upsaliensis, Uppsala.

[3]   Chaffey, D.R. (1979) Southwest Ethiopia Forest Inventory Project, a Reconnaissance Inventory of Forest in Southwest Ethiopia. Land Resources Development Centre, Tolworth Tower Surbition Survey, England.

[4]   Coetzee, J.A. (1978) Phytogeogarphical aspects of the montane forests of the chain of mountains on the eastern side of Africa. *Erdwiss Forsch.* 11: 482-494.

[5]   Demissew, S. (1980) A study on the structure of a montane forest. The Menagesha-Suba State Forest. Unpublished M.Sc Thesis, Addis Ababa University, Addis Ababa.

[6]   Demissew, S. (1988) The Floristic composition of the Menagesha State Forest and the Need to Conserve Such Forest in Ethiopia. *Mount. Res. Devt.* 8: 243-247.

[7]   Denu, D. (2007) Floristic composition and ecological study of Bibita forest (Gurda Farda), southwest Ethiopia, Unpublished MSc Thesis, Addis Ababa University, Addis Ababa.

[8]   Eshete, A., Teketay, D. and Hulten, H. H. (2005) The Socio-Economic Importance and Status of Populations of *Boswellia papyrifera* (Del.) Hochst in Northern Ethiopia: The Case of North Gondar Zone. *Forests Trees and Livelihoods,* 15: 55-74.

[9]   Friis, I. (1986) The forest Vegetation of Ethiopia. Acta Univ. *Ups. Symb. Bot. Ups* XXVI: 31-47.

[10] Friis, I. (1992) Forest and Forest trees of North East Tropical Africa. New Bulletin, Additional series, XV: 1-396.

[11] Friis, I. and Tadesse, M. (1990) The evergreen forests of tropical Northeast Africa. *Mitt. Inst. Allg. Bot. Hamburg.* 23a: 249-263.

[12] Friis, I., Rasmussen, F.N. and Vollesen, K. (1982) Studies in the flora and vegetation of Southwest Ethiopia. *Opera Botanica* 63: 8-70.

[13] Gebre Egziabher, T.B. (1986) Ethiopian vegetation – past, present and future. *SINET: Ethiop. J. Sci.* 9: 1-13.

[14] Gebre Egziabher, T.B. (1988) Vegetation and environment of the mountains of Ethiopia: implications for utilisation and conservation. *Mount. Res. Dev.* 8: 211-216.

[15] Gebre Egziabher, T.B. (1991) Diversity of Ethiopian flora. In: Plant Genetic Resources of Ethiopia, pp. 75-81, (J.M.M Engels, J.G. Hawkes & Melaku Worede, eds.) Cambridge University Press, Cambridge.

[16] Gilbert, E.F. (1970) Mount Wachacha: A botanical Commentary. *Walia* 2: 3-12.

[17] Goldsmith, F. B., Harrison, C. M. and Morton, A. J. (1986) Description and analysis of vegetation. In: Methods in Plant Ecology, (Moore, P.D. and Chapman, S. B. eds). Black Well Scientific Publications, Oxford. Pp. 437- 515.

[18] Greig-Smith, P. (1983) Quantitative Plant Ecology (3rd ed.). Butterworths, London.

[19] Grub, P.J., Lloyd, J. R., Pennington, J. D. & Whitmore, J.C. (1963). A comparison of montane and lowland rain forest in Ecuador. I. The forest structure, physiognomy and floristics.

[20] Haugen, T. (1992) Woody vegetation of Borana, South Ethiopia, a study on the main vegetation types of the area. *SINET: Ethiopi. J. Sci.* 15: 117-130.

[21] Hedberg, O. (1951) Vegetation belts of the East African Mountains. *Sven. Bot. Tidskr.* 45:140-204.

[22] Hedberg, O. (1957) Afroalpine vascular plants, a taxonomic revision. *Symbolae Botanicae Upsaliensis* 15 (1): 1-411.

[23] Hutchings, M. J. (1986) Plant population biology. In: Methods in Plant Ecology, (Moore, P. D. and Chapman, S. B. eds). Black Well Scientific Publications, Oxford. pp.377- 435.J. Ecol. 51: 567-610.

[24] Kent, M and Coker, P. (1992) Vegetation Description and Analysis. A practical approach. Bolhaven Printing Press, London. John Wiley and Sons. Inc. New York. pp 547.

[25] Lamprecht, H., (1989) Silviculture in the tropics   Tropical forest ecosystem and their tree species, Possibilities and methods for their long-term utilization. T2 Verlagsgesselschaft GmbH, Postatch 1164, D101 RoBdort, Republic of Germany Pp 290-296.

[26] Miehe, G. and Miehe, S. (1994). Ericaceous Forests and Heath lands in the Bale Mountains of South Ethiopia: Ecology and man's Impact. Stiftung Walderhaltung in Africa, Hamburg.

[27] Mooney, H.F. (1963). An account of two journeys to the Araenna Mountains in Bale province (Southeast Ethiopia), 1958 and 1959-1960. *Proc. Linn. Soci.* 172: 127-147.

[28] Mueller-Dombois, D. and Ellenberg, H. (1974). Aims and Methods of Vegetation Ecology.

[29] Pichi-Sermolli, R.E.G. (1957) Una carta geobotanica dell' Africa Orientale (Eritrea, Etiopia, Somalia). *Webbia* 13: 15-132.

[30] Shannon, C.E. and Wiener, W. (1949). The Mathematical Theory of Communication. University of Illinois, Chicago, USA.

[31] Sharew, H. (1982) An Ecological Study of a Forest in Jemjem in Sidamo. MSc Thesis, Addis Ababa University, Addis Ababa. Unpubli.

[32] Shibru, S. and Balcha, B. (2004) Composition, structure and regeneration status of woody species in Dindin Natural Forest, Southeast Ethiopia; An implication for conservation. *Ethiopian Journal of Biological Science* 3:15-35.

[33] Soromessa, T. and Demissew, S. (2002) Some uses of plants by the Benna, Tsemay and Zeyise people, southern Ethiopia. *EJNR* 4 (1): 107-122.

[34] Soromessa, T., Teketay, D. and Demissew, S. (2004) Ecological study of the vegetation in Gamo Gofa zone, southern Ethiopia. *Tropical Ecology* 45: 209-221.

[35] Tadesse, M. (1992) A survey of the evergreen forests of Ethiopia. *NAPRECA Monograph Series No.* 2: 1-18.

[36] Teketay, D. (2001) Deforestation, Wood Famine, and Environmental Degradation in Ethiopia's Highland Ecosystems: Urgent Need for Action. Forest Stewardship Council (FSC Africa), Kusami, Ghana. *Northeast African Studies* 8: 53-76.

[37] Uhlig, S.K. (1988) Mountain forests and upper tree limit on the southeastern plateau of Ethiopia. *Mount. Res. Dev.* 8: 227-234.

[38] Uhlig, S.K. and Uhlig, K. (1990) The floristic composition of a natural montane forest in southeastern Ethiopia. *Feddes Repert.* 101: 227-234.

[39] Van der Maarel, E. (1979). Transformation of cover abundance values in phytosociology and its effects on community. *Vegetatio* 39: 97-114.

[40] Von Breitenbach, G. (1961). Forests and woodlands of Ethiopia, a geobotanical contribution to the knowledge of the principal plant communities of Ethiopia, with special regards to forestry. *Ethiop. For. Rev.* 1: 5-16.

[41] Von Breitenbach, G. (1963). Indigenous Trees of Ethiopia. Ethiopian Forestry Association, Addis Ababa.

[42] Westphal, E. (1975). Agriculture in Ethiopia. Addis Ababa/Wageningen (College of Agriculture, Haileselassie I University and the center for agricultural publication and documentation. Wageningen Agricultural University).

[43] Woldu, Z. (1985) Variation in grassland vegetation on the central plateau of shewa, Ethiopia in relation to edaphic factors and grazing conditions. Doctoral thesis, Uppsala University Dissertations Botanicae, 84, J.Cramer,Vaduz.

[44] Woldu, Z. and Backeus, I. (1991) The Shrubland Vegetation in Western Shewa, Ethiopia and its Possible Recovery. *J. Veg. Sci.* 2: 173-180.

[45] Woldu, Z., Feoli, E. and Nigatu, L. (1989) Partitioning an elevation gradient of vegetation from southeastern Ethiopia by probability methods. *Vegetation* 18:189-198.

[46] Yeshitela, K. and Bekele, T. (2003)The woody species composition and structure of Masha-Aneracha forest, Southwestern Ethiopia. *Ethiop. J. of boil. Sci..* 2 (1):32-48.

[47] Zewdie, A. (2007) Comparative Floristic study on Menagesha-Suba State Forest on Years 1980 and 2006, Unpublished M.Sc. Thesis, Addis Ababa University, Addis Ababa.

[48] Muluken Mekuyie Fenta. Human-Wildlife Conflicts: Case Study in Wondo Genet District, Southern Ethiopia. Agriculture, Forestry and Fisheries. Vol. 3, No. 5, 2014, pp. 352-362. doi: 10.11648/j.aff.20140305.14

[49] Siboniso M. Mavuso, Absalom M. Manyatsi, Bruce R. T. Vilane. Climate Change Impacts, Adaptation and Coping Strategies at Malindza, a Rural Semi-Arid Area in Swaziland. American Journal of Agriculture and Forestry. Vol. 3, No. 3, 2015, pp. 86-92. doi: 10.11648/j.ajaf.20150303.14

8

Survey of Nematode Destroying Fungi from Selected Vegetable Growing Areas in Kenya

Wachira P. M.[1, *], **Muindi J. N.**[2], **Okoth S. A.**[1]

[1]School of biological Sciences, University of Nairobi, Nairobi, Kenya
[2]Faculty of Science, Department of Biology, Catholic University of East Africa, Nairobi, Kenya

Email address:
pwachira@uonbi.ac.ke (Wachira P. M.)

Abstract: Plant parasitic nematodes cause up-to 5% yield losses to a wide range of economic crops. In Kenya vegetables yield loss attributed to plant parasitic nematodes is estimated to 80%. Over the years, nematode control has heavily on the use of chemical nematicides which unfortunately leads to biological magnification and elimination of the beneficial microorganisms in the soil. This has triggered a growing interest in search of alternate management strategies. The objective of this study was, therefore, to document nematode destroying fungi in selected major vegetable growing areas in Kenya as a step towards developing self-sustaining system for management of plant parasitic nematodes. Soil samples were collected from five vegetable production zones in Kenya which were Kinare, Kabete, Athi-river, Machakos and Kibwezi and transported to the laboratory for extraction of the nematode destroying fungi. Soil sprinkle technique as described by Jaffee *et al.,* (1996) was used to isolate the fungi from the soil while identification was done using identification keys described by Delgado *et al.,*(2001). From the study a total of 171 fungi isolates were identified as nematode destroying fungi. The highest population was recorded at Kabete area recording 33.9% of the total record, followed by Machakos, Kibwezi, Athi-river and the least in Kin are with 24.6, 22.2, 11.7 and 7. 6% of the total population in that order. *Arthrobotrys* was the most frequent genera with a mean occurrence of 7.3 followed by *Monacrosporium* with 6 and *Stylophage* with 5.2. *A.dactyloides* was significantly (P=0.002) affected by the agro-ecological zones with the highest occurrence being recorded in Kabete and the least in Athi-river. The highest diversity index and species richness of nematode destroying fungi was recorded in Kibwezi while the least was recorded in Athi-river. The genera *Arthrobotrys* had the highest number of trapped nematodes with a total population of 57, followed by *Monacrosporium* and least was *Stylopage* with 45 and 36 respectively, within a period of 104 hours. From the study, it is evident that agricultural practices affect the occurrence and diversity of nematode destroying fungi and *Arthrobotrys* can be developed as a bio-control agent for management of plant parasitic nematodes.

Keywords: *Arthrobotrys*, Biological Control, Plant Parasitic Nematodes

1. Introduction

Horticultural crops both for local consumption and export are important in Kenya. One-tenth of vegetables in Kenya are grown for export. They are recognized for their health and nutritional benefits; provide cash income and employment for close-to two million people in Kenya (Dobson, *et al.,* 2004). Production of vegetables in Kenya, especially for the expanding domestic market, is still limited by major pest and disease problems (Dobson, *et al.,* 2004). Plant parasitic nematodes have been identified as a major production constrains affecting vegetable production, reducing its yield quality and quantity (Nchore *et al.,* 2010). They are responsible to up to 80%, on vegetables production

(Kaskavalci, 2007).Vegetable production in Kenya is characterized by high chemical inputs for pest and soil fertility management (Mutsotso *et al.,* 2005). These practices have been associated with increases in soil born disease and decline in beneficial soil microorganisms (Wachira *et al,* 2008). Specifically vegetable damage by Root-Knot nematodes in Kenya has been reported with infected plants being rendered unacceptable for international markets (Nchore *et* al., 2010). The root knot nematodes increase wounding of the root system providing points of ingress for the pathogen. They may also modify the tissue becoming more suitable for bacteria colonization (Hayward, 1991). Globally it is estimated that US$500 million is spent on root-knot nematode control (Keren-Zur *et al.,* 2000, Pinkerton *et al.* 2000.) where range of strategies are used to control them. These include the use of nematicides, organic manure amendment and resistant cultivars. Overall, although nematicides are effective in managing root-knot nematodes and other plant parasitic nematodes, they are expensive and environmental pollutants when not applied at the right time, in the right way and in the right dosage increasing the cost of production and reducing the profit for the farmers (Republic of Kenya, Taita District Development Strategies 2002-2006).Their use is also curtailed by their threat to groundwater, soil biodiversity as well as long waiting periods between use, harvesting and marketing of crops (Bridge, 1996).

Alternatively, soil beneficial microorganisms could be used to reduce the effect of plant parasitic nematodes hence, reducing the application of chemical nematicides to the soil. The beneficial microorganisms are non-polluting and thus environmentally safe and acceptable. Usually they are species specific to target pest and therefore no chances affecting non target species unlike the chemicals which are broad spectra in action. Nematode destroying fungi are such beneficial microorganism that can be used to control plant parasitic nematodes. They are micro fungi that are natural enemies of nematodes. They naturally capture, kill and digest nematodes in the soil (Rodrigues *et al.*, 2001, Nordbring-Hertz *et al.*, 2002). They comprise three main groups of fungi, the nematode trapping, the endoparasitic fungi and the egg- and cyst-parasitic fungi (Nordbring-Hertz *et al.*, 2002; Masoomeh, *et al.*, 2004,). After trapping the nematodes, they penetrate the cuticle and invade the entire body cavity and then digest them completely. This group of fungi has drawn much attention due to their potential for development as biological control agents of plant parasitic nematodes (Jansson and Persson, 2000; Sanyal, 2000; Masoomeh, *et al.*, 2004). About 70% of fungi genera and 160 species are associated with nematodes but only a few of them can be used as biological control agents of nematodes (Elshafie *et.al.* 2006). This study was therefore aimed at documenting the occurrence and diversity of nematode destroying fungi and testing their as bio-control against plant parasitic nematodes.

2. Materials and Methods

Soil samples were collected from five different vegetable growing areas in Kenya. These areas were Kinare, Kabete, Athi-river Machakosand Kibwezi in the order of altitude and temperature. Kinare was a high altitude area and the coldest among the zones and Kibwezi being the lowest and hottest. The vegetable gardens in each zone were mainly dominated by spinach, kales, tomatoes, cabbage and pepper among other vegetables. From each of the study areas five farms under intensive vegetable production were randomly selected for this study. From each of the farm, five different vegetable gardens were sampled. From each vegetable garden, five soil samples were collected and mixed in a bucket to make one composite sample. One kilogram of soil was then re-sampled from the composite sample in the bucket, put in plastic bags,

labeled and placed in a cool box. Soil sampling was done using a soil auger which was sterilized using ethanol after every sampling point to avoid cross contamination. All the samples were later transported to the laboratory for isolation of nematode destroying fungi.

Isolation of nematodes destroying fungi was done using the soil sprinkle technique as described by Jaffee *et al.*, 1996 where tap water agar (TWA) was prepared by dissolving 20 g agar in one liter of tap water. The medium was autoclaved and cooled before use after amending it with 0.1 gl^{-1} of streptomycin sulfate under a laminar flow. One gram of soil sample was sprinkled on the medium in the petri dish and a suspension of *Meloidogyne* species of approximately 1000 nematodes was added into the petri dishes as baits (Christina *et al.*, 1999).The plates were then incubated at room temperature and observed daily from the third week up to sixth week under a dissecting microscope. The examination was focused on trapped nematodes, trapping organs and conidia of the nematodes destroying fungi (Wachira *et al.*, 2008).

Taxonomic classification of the nematode destroying fungi was done using the slide culture technique where slides were observed under a microscope while identification of the genus was done using identification keys described by Delgado *et al.*, (2001).After identification of nematode destroying fungi; pure cultures of the three most frequently isolated fungi isolates were made for efficacy experiment. A mycelia block (5 mm) was inoculated into PDA media in a petri dish and allowed to grow for five days before approximately 50 plant parasitic nematodes were added. The efficacy of the fungi isolates was monitored for a period of 3-6 weeks. Trapped nematodes were counted for five days, after 3 weeks of incubation. All the data in this study was analyzed through analysis of variance (Kindt & Coe 2005)

3. Results

171 fungi isolates were identified as nematode destroying fungi. They were grouped into three genera and five taxa. The three genera were *Arthrobotrys*, *Monacrosporium* and *Stylopage*. *Arthrobotry*s was the frequently encountered fungi genera. This generawas represented by, *A. oligospora, A. dactyloides* and *A. longispora,* the genera *Monocrosporium* was represented by *M. cionopagium* while genera *Stylopage* was represented by *Stylopage grandis. A.oligospora* had the highest frequency of occurrence followed by *A.dactyloides, M. cionopagium, S. grandis* and the least was *A. longispora* with. occurrence frequencies of 46.20,45.61,5.85,1.17 and 1.17% in that decreasing order (Fig I).

The fungi isolates were recovered in all the vegetable production zones. Except for the *A. dactyloides*, all the isolates were not significantly (P> 0.05) affected by the agro-ecological zones. The highest occurrence of *A.dactyloides* was 40 being recorded in Kabete while the least was 4 being recorded in Athi-river This species was also recorded in Machakos, Kibwezi and Kinare with records of 19, 10 and 5, respectively, in that decreasing order. Among all the isolates, only *A.oligospora* and *A.dactyloides* occurred in all agro-ecological zones. *M.cionopagium* occurred in all zones except in Kinare while *S.grandis* was present in both Kibwezi and Kinare with *A. longispora* being recorded in Kibwezi only (Table 1).

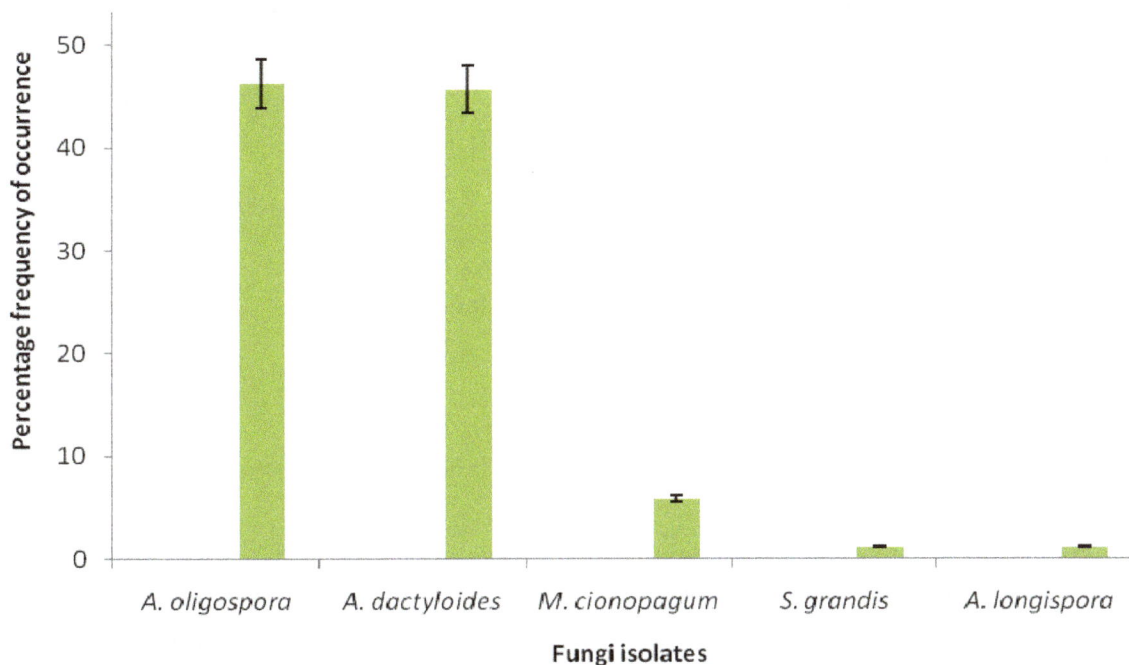

Figure 1. *Percentage occurrence of nematode destroying fungi in vegetable production areas in Kenya.*

Table 1. *Occurrence of nematode destroying fungi in major agro-ecological zones in Kenya.*

Zone	A. dactyloides	A. oligospora	A. longispora	M. cionopagium	S. grandis	Total
Kabete	20	17	0	1	0	58
Machakos	19	22	0	1	0	42
Kinare	5	7	0	0	1	13
Kibwezi	10	20	2	5	1	38
Athi-river	4	13	0	3	0	20
P.value	0.002	0.395	0.062	0.165	0.062	

The highest number of nematode destroying fungi was recorded in Kabete (Table 1), followed by Machakos, Kiwezi, Athi-river and finally Kinare with total mean abundance of 11.6, 7.6, 7.4, 4.0, and 2.6 in that decreasing order (Table 2). Kibwezi recorded the highest diversity index with a mean of 0.930, then Machakos with 0.637 while Kabete recorded the least diversity index mean of 0.411. Mean richness and abundance varied between the vegetable production zones. The highest mean species richness was recorded in Kibwezi while the least was recorded at Athi-river. All the agro ecological zones differed significantly ($P = 9.587 \times 10^{-4}$) in terms of species abundance. Kabete had the highest species abundance with a mean of 11.6 and least was Kinare with a species mean abundance of 2.6 (Table 2)

Table 2. *Mean shannon, species richness and abundance of nematode destroying fungi in different vegetable growing areas in Kenya.*

Zone	N	Mean Shannon	Mean richness	Mean abundance
Kibwezi	5	0.930	3.4	7.6
Machakos	5	0.637	2.2	7.4
Athi-river	5	0.483	1.6	4.0
Kinare	5	0.482	1.8	2.6
Kabete	5	0.411	2.2	11.6
P value				9.587×10^{-4}

Detection of nematode destroying fungi increased with increase in number of the soil samples taken. It was evident that all possible isolates of nematode destroying fungi were recorded in this study from the samples collected. Collecting and processing additional samples would not significantly increase the number of isolates (Fig 2).

There was a significant (P=0.003) difference on efficacy between the three most frequent nematode destroying fungi species. *Arthrobotryrs oligospora* was the most efficient nematode destroying fungi with a mean of 7.3, followed by *Monacrosporium* and the least was *Stylopage* with mean records of 5.9 and 5.1 respectively.

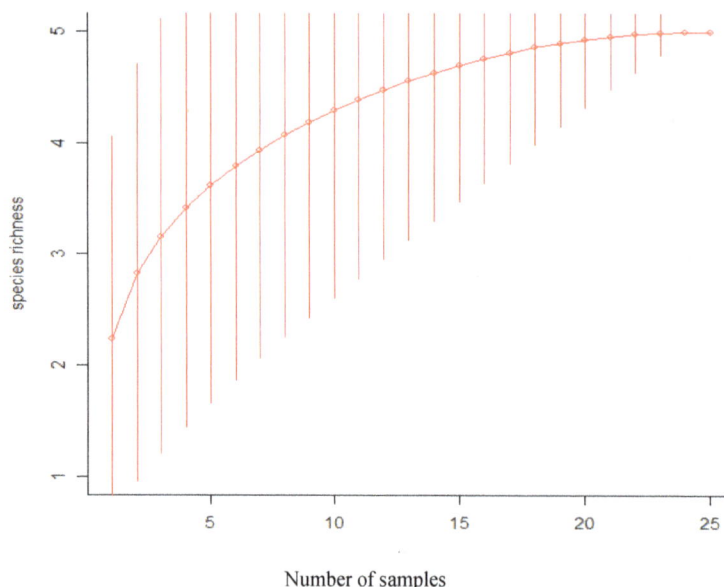

Figure 2. *A total species cumulative curve for nematode destroying fungi in the vegetable production zones in Kenya.*

4. Discussion

Nematode destroying fungi were isolated from all the targeted vegetable production zones. They occurred in different frequencies and diversity. The study has demonstrated the diverse occurrence of nematode destroying fungi in nature and especially in vegetable production zones. These findings agree with previous reports on nematode destroying fungi that they are widespread in all habitat but with different densities and diversities (Birgit *et al.*, 2002, Wachira *et al.*, 2008).

Arthrobotrys oligospora was the most abundant species of nematode destroying fungi in the study area. Other studies on nematode destroying fungi have made the same observation. It has never been clear why this is the most frequently encountered fungi. Wachira *et al.*, (2008) had suggested that farming practices like weeding could be the cause of this high occurrence. It has also been suggested that it is due to its ability to exist both as a saprophyte and a plant parasitic nematode feeder (Sobita and Anamika, 2011). Due to this high occurrence, this fungus is attracting a lot of other interesting studies (Niu and Zhang, 2011).

It was expected that the highest number of fungi would be isolated from areas with low temperature (Kinare). However this was not the case. Kinare had the least number of fungi. This was attributed to the high use of chemical fertilizers and pesticides since all the vegetables were all aimed for market. In a study on long-term effects of manures and fertilizers on soil productivity and quality, it was reported that fertilized soils had lower contents of organic matter and numbers of microfauna than manured soils (Edmeades, 2003).The highest number of nematode destroying fungi was recovered from Kabete. Soils in this area had been collected from the University of Nairobi farm where animal manure is frequently applied. This may explain the high number of nematode destroying fungi since they have been associated with increase of beneficial microorganisms in the soil (Wachira and Okoth, 2009). From this study, there was high fungal population in areas where manure was applied and low fungal population being recorded in areas where chemical fertilizer was applied.

Temperature is an important factor regulating microbial activity and shaping the soil microbial community. It determines moisture levels in the soil which is key to fungal germination and growth. High temperatures lead to low soil moistures which is attributed to low fungal spore germination. A study by Haugen and Smith (1992) reported that at high temperatures there is low germination of fungi spores leading to low fungi populations and while in low temperatures there is high fungal germination leading to high fungal population. This was reported in reverse in our study. Although Machakos and Kibwezi had high temperatures, high population of nematode destroying fungi was reported. This could be attributed to irrigation activities which ensured moist conditions throughout the growth season. This soil moisture, coupled with high temperature enhanced the fungal germination since fungal spore germinates better in moist and warm conditions.

The efficacy test showed that genus *Arthrobotrys* was the most effective in trapping plant parasitic nematodes. Previous studies on fungi belonging to these genera have consistently showed that it was able to trap 90% of all the nematodes in petri dish and in liquid culture in 16-40 hours (Rajeswari and Sivakumar, 1999).Therefore, due to its' high occurrence, this fungi can be developed for the management of plant parasitic nematode and its potential should be investigated. The investigation should concentrate on suitable carriers for the fungi and also mode of application. This would reduce over reliance on chemical nematicides and also develop a self regulation system in the soil that controls the soil born pests.

5. Conclusion

Additional evidence has been provided from this study that nematode destroying fungi are naturally occurring in agricultural habitats. It is also evident that agricultural activities targeting high crop production, like application of chemical fertilizers and pesticides are directly affecting the soil biodiversity. The results of this study can be used for further research to establish the potential of nematode destroying fungi in regulation of plant parasitic nematodes.

Acknowledgement

The University of Nairobi is acknowledged for providing laboratory equipment and space. We also wish to thank the small scale farmers in the five different vegetable production zones for their cooperation during this study and for providing free access into their farms.

References

[1]   Bridge, J. 1996.Nematode management in sustainable and subsistence agriculture. Plant Parasitic Nematodes in Temperate48 F. JAITEH et al. Agriculture. CABI International WallingfordUK. pp. 203-209.

[2]   Christina, P., Stefan, O and J.Hans. 1999. Growth of Arthrobotryssuperba from a birch woodresource base into soil determined by radioactive tracing.Thorvaldsensvej 40, DK-1871.

[3]   Delgado, A. E., Pinero, A. J and L., M.Urolaneta.2001. Invitro preparatory activity of nematophagous fungi from Costa Rica with potential use for controlling sheep and goat parasitic.

[4]   Dobson, H.M., Matthews, G.A., Olembo, S., Baleguel, P and T., L.Wiles.2004. Application challenges for small-scale African farmers: a training initiative in Cameroon. International Advances in Pesticide Application. Aspects of Applied Biology 71:385–392.

[5]   Edmeades, D. C.2003.The long-term effects of manures and fertilizers on soil productivity and quality: a review Nutrient Cycling in Agroecosystems66: 165–180.

[6]   Elshafie AE, Al-Mueini R, Al- Bahry, Akindi A, Mohmoud I, Al- Rawahi S. 2006. Diversity and trapping efficiency of nematophagous fungi from Oman. Phytopathology Mediterrenian. 45: 266 - 270

[7]   Farrell .F.C., Jaffee.B.A and D.,R. Strong.2006.The nematode trapping fungus Arthrobotrysoligospora in soil of the Bodega marine reserve: distribution and dependence on nematode-parasitized moth larvae. Soil Biology and Biochemistry 38:1422-1429.

[8]   Haugen, L. M and S., E. Smith.1992.The effect of high temperature and fallow period on infection of mung bean and cashew roots by the vesicular-arbuscular mycorrhizal fungus Glomus:Plant and Soil 145: 71 -80.

[9]   Hayward, A.C.1991.Biology and epidemiology of bacterial wilt caused by Pseudomonas solanacearum. Annual Review Phytopathology 29: 65-87.

[10] Jaffee, B.A., Srong, D.R and A., E.Milton.1996.Nematode trapping fungi of natural shrubland: Tests for food chain involvement. Mycologia, 88:554-564.

[11] Jansson, H.B., Persson, C and R. Odeslus.2000. Growth and capture activities of nematode destroying fungi in soil visualized by low temperature scanning electron microscopy. Mycologia 92:10 – 15.

[12] Kaskavalci, G. 2007.Effect of soil solarization and organic amendment treatments for controlling Meloidogyne incognita in tomato cultivars in Western Anatolia. Turkish Agricultural Forum31: 159-167.

[13] Keren-Zur, M., Antonov, J., Bercovitz, A.,Feldman, A., Keram, G., Morov and N. Rebhum. 2000. Baccillusfirmusformulation for the safe control of root knot nematodes. The BCPC Conference.Pests and Disease, Brighton, UK.pp.307-311.

[14] Kindt, R and R. Coe, 2005. Tree diversity analysis. A manual and software for common statistical methods for ecological and biodiversity studies. Nairobi: World Agro-forestry Center (ICRAF).

[15] Masoomeh, S.G., Mehdi, R.A., Sharokh, R.B., Ali, E.R.Z and E. Majid.2004. Screening of soil and sheep faecal samples for predacious fungi:Isolation and characterization of the nematode – trapping fungus Arthrobotrys oligospora. Iranian Biomedical. 8: 135 – 142.

[16] Nchore, S.B., Waceke, J.W and KariukiG., M.2010. Incidence and prevalence of root-knot nematode Meloidogyne species in selected indigenous leafy vegetables in Kisii and Trans-Mara Counties of Kenya. In: Transforming Agriculture for improved livelihoods through Agricultural Product Value Chains. 12th KARI Biennial Scientific Conference,November 8-12, 2010,Nairobi,Kenya.KARI,pp. 675-681.

[17] Nordbring-Hertz, B., H.B. Jansson and A. Tunlid .2002.Nematophagous Fungi: Encyclopedia of Life Sciences. Macmillan Publishers Ltd., London.

[18] Pinkerton JN, Ivors KL, Miller ML and Moore LW, 2000.Effect of solarization and cover crops on populations of selected soil borne plant pathogens in western Oregon. Plant Diseases 84: 952-960.

[19] Rajeswari.S and C., V.Sivakumar.1999.Occurrence of nematophagous fungi (Hyphhomycetes and their predacious ability in Tamil Nadu).Journal of Biological control 13:107-110.

[20] Rodrigues, M.L.A., Castro, A.A., Oliveira, C.R., Anjos, D.H.S., Bittencourt, V.R.E.P and J., V.Aranjo.2001. Trapping capabilities of Arthrobotryssp and Monacrosporium thaumasium on Cyathostoma, larvae. Veterinary Parasitology10: 51 – 54.

[21] Sanyal, P.K.2000. Screening for Indian isolates of predacious fungi for use in biological control against nematode parasites of ruminants. Veterinary Research Communications 24: 55–62.

[22] Sobita Simon and Anamika.2011. Management of Root Knot Disease in Rice Caused by Meloidogyne graminicola through Nematophagous Fungi. Journal of Agricultural Science 3: 122 – 127.

[23] Wachira, P.M and Okoth, S.A. 2009. Use of nematode destroying fungi as indicators of land disturbance in TaitaTaveta, Kenya Tropical and Subtropical Agroecosystems, 11: 313 - 321

[24] Wachira, P.M., Kimenju, J.W., Okoth, S., Mibey, R.K and J. Mungatu.2008. Effect of land use on occurrence and diversity of nematode destroying fungi in TaitaTaveta, Kenya. Asian journey of Plant Sciences 7: 447-453.

9

Key Informant Perceptions on the Invasive *Ipomoea* Plant Species in Kajiado County, South Eastern Kenya

Kidake K. Bosco[1,*], **Manyeki K. John**[1], **Kirwa C. Everlyne**[1], **Ngetich Robert**[1], **Nenkari Halima**[2], **Mnene N. William**[1]

[1]Arid and Range Lands Research Institute-Kiboko, Kenya Agricultural and Livestock Research Organization, Makindu, Kenya
[2]Agricultural Sector Development Support Program, Ministry of Agriculture, Livestock and Fisheries, Kajiado County, Kajiado, Kenya

Email address:
bkkidake@gmail.com (K. K. Bosco)

Abstract: Invasion of rangelands by undesirable plant species is one of the challenges facing rangeland productivity and to an extension livestock production in East Africa. They have affected communities in different ways in areas where they grow. Focus group discussions and interviews were held in two sites in pastoral and agro-pastoral regions of Kajiado County to get perceptions of farmers, livestock keepers and other stakeholders concerning the invasive plant species Ipomoea. This was accompanied by visits and field excursions to areas heavily infested by the invader species. The interviewed key informants agreed that the plant has more detrimental effects to the environment, ecologically and to the economy of the region. There is need for urgent interventions involving all stakeholders to curb the spread of the species, which is currently at an unprecedented rate. These include efforts by relevant institutions such as Government, Non-Governmental institutions through mobilization, training and capacity building and demonstrations in order to reverse the trend. Any trainings should however include aspects of recovery of invaded and degraded land primarily through pasture improvement and other interventions as this will enhance the utilization of these areas for increased livestock productivity and reverse degradation.

Keywords: Invasion, Ipomoea, Rangelands, Semi-Arid Lands

1. Introduction

Invasive plant species are hazards that have shown negative environmental and socio-economic impacts in East African drylands (Obiri, 2011). They have led to degradation of the environment leading to serious impacts on local communities' resource base. Invasive species remain one of the most understudied in developing countries (Pysek et al., 2008). In semi-arid rangelands, based on their impact and effect on grazing areas and natural pastures, invasive species cause massive losses in livestock production.

The livestock sector contributes to about 42% of the agricultural Gross Domestic Product (GDP) of Kenya. In the arid and semi-arid lands (ASALs), characterized by extensive rangeland grazing systems, livestock accounts for about 90% of employment and 95% of households derive their incomes from this subsector (Nyariki et al., 2005). Changing production systems have resulted in different degrees of degradation and productive capacity. Overgrazing has particularly impacted negatively on vegetation resources and biodiversity in general (Nyariki et al., 2009). In the rangelands, the major feed resource base is natural pastures found in communal grazing areas and individual farms as well.

Invasion by shrubs has been cited as one of the major causes of range deterioration in the Southern rangelands (Macharia, 2004). One of the species, which is a problem invasive species in natural and established pastures in Kajiado County, is Ipomoea spp. The species has been described as one of the most undesirable forage species for grazing livestock (Lusigi et al., 1984). Ipomoea spp. is a creeping annual herb, widespread in the semi-arid districts of Southern Kenya, which colonizes and spreads rapidly immediately after the onset of the rainy season (Mganga et al., 2010a). The species is mainly found in disturbed or degraded sites. The plant exhibits most characteristics common to invasive species, which include capacity for rapid growth and so expansion, capacity to disperse and reproduce widely or by nurturing fewer progeny but with great efficiency (Emerton and Howard, 2008). The species also capable of effective competition with local species – for food, space, light and water.

Key informants are an important source of information and knowledge. The focus of this study was to better understand perceptions from farmers, pastoralists and key informants on the invasive weeds mainly in pasturelands and croplands in Mashuru and Kajiado Central divisions of the semi-arid Kajiado County.

2. Materials and Methods

2.1. Study site

The study was conducted in Kajiado County, of the Southern rangelands of Kenya. The climate of the region is arid to semi-arid falling between zone IV and V with very little potential for rain fed cropping. The mean annual rainfall ranges from 300-800 mm with a bimodal pattern experienced, the long rains are received in March to May while the short rains last from October to December. The vegetation varies from open grasslands, bushland to wooded grasslands with different distinct associations. Soils found in the region are predominantly shallow red sandy soils resulting from different soil formation processes and geological associations (de Leeuw et al., 1991). The predominant land use and management system in the region is free range grazing with the main livestock species kept being cattle and small stock. In most locations of Kajiado central, grazing land is used communally and individually, while livestock is managed by individual families.

2.2. Methodology

Two cluster sites were purposively selected based on the extent of the *Ipomoea* spp. invasion, - Kajiado central (Olbelbel and Sajiloni) and in Mashuru (Mashuru and Nkatu). The two sites were settled on after a consultation with government and non-government departments who identified these areas in the county as highly infested with *Ipomoea* spp. A survey questionnaire tool was then developed to capture the farmers' perceptions on the most problematic weeds and the challenge they pose in pasture and crop fields. The team adopted a participatory methodology in carrying out the study where key informant persons in administration and ministries in charge of State Department of Agriculture, Livestock and Fisheries (MoALF) in Kajiado were consulted and followed by intensive one-on-one interviews. In addition, focus group discussions (FGD) were conducted to triangulate the information obtained from key informants. Two FGDs and four key informant interviews were conducted. Visits to some of the remote areas that are highly infested were done with guidance of the local community.

3. Results and Discussion

3.1. Origin of the Weed

According to the informants, the causative agent of the invasive species was reported as overgrazing and climate change. Unique climatic episodes cited was the 1997/98 *el nino* rain events, believed to have brought about the species in agreement with what was reported by Macharia P. N. (2004). In as much as frequent droughts and rainfall events also cause changes in vegetation attributes, the plant is reported to have been in the region as early as the 1960s (Kedera and Kuria, 2003); and had simply increased (Macharia and Ekaya, 2005). In East Africa, origins and patterns of introduction of invasive species are however not known (Gichua et al., 2013).

3.2. Extent and Spread

Visual estimation by authors of this work place the spread of the plant in different heavily infested localities in Kajiado at 60% - 80% of the pasturelands. This is corroborated by results from the FGDs from which participants estimated the degree of invasion in pasture fields at 65% and 50% respectively in Kajiado Central and Mashuru areas of the county. The degree of invasion was however less for crop fields (10% for both sites) – mainly due to the low number of persons practicing crop production.

Continued degradation due to poor management, has however encouraged the establishment and spread of *Ipomoea* species. Invasive species can enter and establish in a disturbed or degraded habitat more easily than into a system that is stable (Emerton and Howard, 2008). Invasive plants are known to have characteristics of non-native species favouring local conditions and being more vigorous than native species. These include attributes such as unpalatability, formation of thickets, production of spines and thorns, allelopathic effects, toxicity to animals and fire tolerance (Mworia, 2011). More so, some of the species are able to produce many seeds that have long dormancy which tolerate the dry seasons. These attributes are in agreement with what was mentioned by the participants in reference to *Ipomoea* spp. Its ability to out-compete other existing grass and forage species in Kajiado County was also mentioned and noted by the farmers. This is through competition for nutrients and water from soil due to its aggressive growth nature.

Table 1. *Physiognomic features and environmental conditions for Ipomoea establishment.*

Features	Kajiado Central	Mashuru
Terrain	Lowland levels (flat land)	Flat and gently sloping areas
Soil	Black cotton and Red soil	Red soil
Rainfall	Normal rainfall	Little rainfall
Others	Highly overgrazed lands	Highly overgrazed lands
	Drastic change in climate leading to land degradation hence – opportunistic weeds	

3.3. Physiognomic and Environmental Features Preferred by the Plant Species

The conditions required for *Ipomoea* establishment are summarized in Table 1.

The community indicated that the plant prefers red soils although it can also be found in black cotton soils. Mostly *Ipomoea* spp. weeds are found in lowlands although they can also be found in the fairly sloppy lands. The onset of rains normally herald a resprouting of the seeds that are dispersed over the season upon drying.

3.4. Control of Invasive Ipomoea Species

Past and present attempts of technologies and innovations to control *Ipomoea* spp. invasion, source of the information, mode of dissemination and effectiveness were also assessed in the area. These are outlined in Table 2.

Table 2. Past control strategies of Ipomoea spp in Kajiado County.

Innovation/ technology	Source	Mode of dissemination	Effectiveness*
Manually uprooting using small equipment	Own initiative	Learn from neighbours	Fairly effective
Hand removal of young seedlings	Own initiative[1] and ASAL programme[2]	Seminars	Effective
Application of soda Ash	ASAL programme[2]	Training and Demonstration	Not effective
Spraying using chemicals	ASAL programme[2]	Training and Demonstration	Not effective

*Very effective, effective, fairly effective, not effective
[1]Both done in Olbelbel and Sajiloni
[2]Happened only in Olbelbel

Only hand removal and manual uprooting were found to be effective in both sites while the others were not effective at all. The manual removal has specifically been applied in localized areas such as individual grazing fields or around compounds. It is also an advantageous method due to its selective nature and avoidance of non-target species. A repeated follow-up control however is generally required, as well as subsequent rehabilitation measures, because disturbed ground and soil erosion in cleared areas may encourage re-invasion (Boy and Witt, 2013). Two examples of fenced sites, previously invaded, had been reclaimed through manual removal followed by pasture establishment and management. This represents a success story in the region. The benefits of hand removal are high in the short term, although for this to be sustained in the long term.

Other methods that have been tried in the same region include chemical control and other integrated approaches, in which soda ash and wood ash were applied to freshly cut stems stumps of the plant in an attempt to dry them up. However, success of managing the weed was not realized during these initiatives, tested by a local non-governmental

organization (NGO). Instead, according to one participant, there was an incidence of increase of the species in one area where the plants were cut and soda ash applied to the fresh stumps.

3.5. Control Constraints

Table 3 shows the perceptions on controlling the weed species by the community.

Table 3. Summary of community perception constraints associated in controlling Ipomoea species.

Attribute	Kajiado	Mashuru
High financial* cost	√	√
High technical skill required	×	×
Low benefit of controlling *Ipomoea*	×	×
Labour intensive	√	√

*On average an acre would cost KES 4,000 ($40) on manual uprooting in the two sites
√ = Yes, × = No

The high cost and the labour intensiveness being the key reason to failure of control and eradication of the weed. Perceptions on constraints associated with controlling *Ipomoea* spp. were similar in the two sites. Manual removal of the weed, being the most appropriate and beneficial option for many, was ranked as expensive and labour intensive initiative. An excursion visit to some farmers who have heavily invested in uprooting of the weed revealed information that the initial costs of control are very high. One pastoralist, whose 20-acre farm was heavily infested with the weed, spent Ksh. 36, 000 (about $360) to uproot some of the plants at part of the farm. Based on farmer estimates, up to Ksh. 4000 ($40) had been spent to control the weed per acre, which is high and unaffordable to many farmers and livestock keepers. However, those interviewed mentioned a subsequent reduction in control costs in the subsequent season. This is because there are only a few new and young plants that are easy to uproot having grown or sprouted.

Table 4. Summary of community perceptions on the positive and negative effects of Ipomoea spp.

Custer site	Positive	Negative
Kajiado	Sources of honey production Concoction of its leaf sap and water used to control fleas	Reduction in pasture* Bitter honey that also cause dizziness During flowering it causes flu Poisonous when eaten by animals Restricted access and movement
Mashuru	Sources of honey production Beautiful flowers – aesthetic value	Reduction in pastures* Chickens get sick when they feed on flower Bitter honey which also causes dizziness

*noted as the most important effect of the invader species in all the sites.

3.6. Positive and Negative Effects

Positive and negative consequences of *Ipomoea* spp. were

also enumerated by the community. They are summarized in the Table 4.

The notable effect of invasive *Ipomoea* spp. was the reduction in pastures in the grazing lands at both sites. The weed poses the greatest challenges to pasture and eventually livestock production in semi-arid rangelands of Kenya. The plant has managed to colonize most of the grazing regions in the sites visited. Fresh and heavy biomass of *Ipomoea kituiensis* has been reported to suppress the growth and development of the grasses underneath it resulting in high incidences of grass seedling mortality. This is through deprivation of sunlight necessary for normal photosynthetic function. The net effect is the poor establishment of grasses with much of the denuded areas remaining bare after the end of the rains (Mganga et al., 2010b). *Ipomoea hildebrandtii, has* been reported to depress native grass biomass production in addition to changes in site hydrologic and nutrient dynamics patterns in Kajiado County (Mworia et al., 2008).

During the dry season, out of desperation, livestock may sometimes feed to some invasive species resulting in serious health hazards and even death if ingested. Though no cases for cattle mortality were reported by the participants, there is a possibility that livestock may feed on the species since it remains green even during the dry season. All these lead to reduced carrying capacity in the long run and economic losses for the community.

Other effects of the invasive *Ipomoea* species noted elsewhere include toxicity to grazing wild animals. *Ipomoea carnea* species seeds have been reported as a threat to wild herbivores when ingested (Lahkar et al., 2011). Kajiado County is home to many ungulates roaming the community conservancies and plains. Impacts on wild herbivores are likely to be similar to those reported for livestock. In general, all informants agreed that the species causes more negative effects than positive hence no benefits are accrued from it. However, ecologically, invasive species have the potential for carbon sequestration depending on plant traits. Such traits include wood and lignin concentration, fire resilience and tolerance, resprouting, deep rooting and herbivore defence traits (Mworia, 2011). *Ipomoea* spp found in the region possess some of these traits.

Table 5. Ranking of possible options and strategies in controlling the spread of Ipomoea spp.

Listing in order of importance	Kajiado Central	Mashuru
1	Use herbicide to eliminate *Ipomoea*	Use Herbicide to eliminate *Ipomoea*
2	Holistic community advocacy to remove *Ipomoea*	Financial aid for labour costs to remove the plant
3	Train in pasture improvement and rehabilitation of degraded land	Train in pasture improvement and rehabilitation of degraded land
4	Financial aid from different partners or government	N/a
5	Sensitisation on management of rangelands	N/a

3.7. Possible Control Mechanisms and Approaches

The farmers revealed key research areas and possible strategies of controlling the invasive *Ipomoea* spp. weeds. These are summarized in Table 5 in order of importance for the two sites.

Research on appropriate herbicides for a complete elimination of Ipomoea weeds topped the lists. All possible solutions are to be accompanied by extensive training on pasture improvement that is lacking in the two sites.

Due to the costs and limited success of the control of invasive species, many government institutions across the globe have abandoned the management of these species hence their continuous spread and establishment (Borokini and Babalola, 2012). The participants suggested the need for financial aid for holistic elimination of this weed, possibly from the County and Central Governments, in initiatives such as money-for-work programs. These would go a long way in controlling invasive species as well as employment creation by involving the youth.

The best method of management of invasion however remains the prevention of establishment and spread (Borokini and Babalola, 2012). For semi-arid grazing lands in the Southern rangelands, rehabilitation of degraded grazing lands using natural pastures such as Cenchrus ciliaris, Enteropogon macrostachyus and Eragrostis superba is beneficial due to the competitive advantage these species possess in suppressing weeds such as Ipomoea kituiensis (Mganga et al., 2010b). Cenchrus ciliaris in particular is primarily recommended because of its allelopathic properties and deep root system hence the necessity to include the species in any reseeding initiative (Mganga, 2009). Application of organic fertilizers hastens the process of recovering such lands. Having many species in an area, that is, high biodiversity also helps in control and prevention of invasion (Obiri, 2011). In this case, rehabilitating degraded areas with many different grass species mixtures is recommended as well as other plant species. This would subsequently result in invasion resistance.

4. Conclusions and Recommendations

Short-term solutions that have been applied in the management and control of the spread and impact of the species are not adequate. There is need for an integrated approach involving other initiatives such as rangeland rehabilitation through reseeding to restore severely degraded areas – an opportunity to prevent the problem of invasion from occurring. Training and advocacy on pasture improvement and suitable range management practices to conform to the changing conditions remain key. The trend in many parts of Kajiado leans towards land privatization hence fenced demonstrations for a village; responsible model famer groups and/or individuals are ideal initiatives of control of invasive *Ipomoea* species and other weeds. Of course, economic and ecological analysis of the process of eradicating or controlling invasive species indicates that the most cost-effective and least environmentally damaging method is spread prevention. In addition, more research should also focus on the potential benefits and use of these invasive species in semi-arid regions.

References

[1] Borokini, T. I. and F. D. Babalola. (2012). Management of invasive plant species in Nigeria through economic exploitation: lessons from other countries. *Management of Biological Invasions, 3*(1), 45-55.

[2] Boy, G. and A. Witt. (2013). *Invasive Alien Plants and their Management in Africa.* Nairobi: UNEP/GEF Removing Barriers to Invasive Plant Management Project.

[3] de Leeuw, P. N. de, Grandin, B. E. and S. Bekure. (1991). Introduction to the Kenyan rangelands and Kajiado district. In S. Bekure, P. N. de Leeuw, B. E. Grandin, and P. J. Neate, *Maasai Herding: An analysis of the livestock production system of the Maasai pastoralists in Eastern Kajiado District, Kenya* (ILCA systems Study 4 ed., p. 172). ILCA (International Livestock Centre for Africa).

[4] Emerton, L., and G. Howard (2008). *A Toolkit for the Economic Analysis of Invasive Species.* Nairobi: Global Invasive Species Programme.

[5] Gichua, M., Njoroge, G., Shitanda , D., and D. Ward. (2013). Invasive Species in East Arica: Current status for Informed policy decisions and management. *JAGST, 15*(1), 45-55.

[6] Kedera, C. and B. Kuria. (2003). Invasive alien species in Kenya: status and management. IPCC Secretariat. Identifiation of risks and management of invasive alien species using the IPPC framework. Proceedings of the workshop on invasive alien species and the International Plant Protection Convention, Braunschweig, Germany, 22-26 September 2003. pp. 199-204.

[7] Lahkar, B. P., Talukdar, B. K. and P. Sarma. (2011). Invasive species in grassland habitat: an ecological threat to the greater one-horned rhino (Rhinoceros unicornis). *Pachyderm*, 33-39.

[8] Lusigi, W. J., Nkurunziza, E. R. and S. Masheti. (1984). Forage Preferences of Livestock in the Arid Lands of Northern Kenya. *Journal of Range Management, 37 (6)*, 542-548.

[9] Macharia, P. N. (2004). Community based interventions as a strategy to combat desertification in the arid and semi arid rangelands of Kajiado district, kenya. *Environmental Monitoring and Assessment, 99*, 141-147.

[10] Macharia, P. N. and W. N. Ekaya. (2005). The Impact of Rangeland Condition and Trend to the Grazing Resources of a Semi-arid Environment in Kenya. *Journal of Human Ecology, 17*(2), 143-147.

[11] Mganga, K. Z. (2009). *Impact of grass reseeding technology on rehabilitation of the degraded rangelands: a case study of Kibwezi district, Kenya.* Nairobi: MSc Thesis, University of Nairobi,.

[12] Mganga, K. Z., Musimba, N. K., Nyariki, D. M., Nyangito, M. M., Mwang'ombe, A. W., Ekaya, W. N. and W. M. Muiru. (2010a). The challenges posed by Ipomoea kituensis and the grass-weed interaction in a reseeded semi-arid environemnt in Kenya. *International Journal of Current Research, 11*, 001-005.

[13] Mganga, K. Z., Nyangito, M. M., Musimba, K. N., Nyariki, M. D., Mwangombe, A. W., Ekaya, W. N., Muiru W. M., Clavel, D., Francis, J., von Kaufmann, R. and J Verhagen. (2010b). The challenges of rehabilitating denuded patches of a semi-arid environment in Kenya. *African Journal of Environmental Science and Technology, Vol 4 (7)*, 430-436 .

[14] Mworia, J. K. (2011). Invasive Plant Species and Biomass Production in Savannas. In I. Atazadeh, *Biomass and Remote Sensing of Biomass.* InTech Open Access Publishers.

[15] Mworia, J. K., Kinyamario, J. I., and E. A. John. (2008). Impact of the invader *Ipomoea hildebrandtii* on grass biomass, nitrogen mineralization and determinants of its seedling establishment in Kajiado, Kenya. *African Journal of Range and Forage Science, 25*, 11-16.

[16] Nyariki, D. M., Makau, B. F., Ekaya, W. N. and J. M. Githuma. (2005). *Guidelines for Emergency Livestock Off-take .* Nairobi: Arid Lands Resource Management Project. Office of the President; Agricultural Research Foundation (AGREF).

[17] Nyariki, D. M., Mwan'gombe, A. W. and D. M. Thompson. (2009). Land-Use Change and Livestock Production Challenges in an Integrated System: The Masai-Mara Ecosystem, Kenya. *Journal of Human Ecology, 26*(3), 163-173.

[18] Obiri, J. F. (2011). Invasive plant species and their disaster-effects in dry tropical forests and rangelands of Kenya and Tanzania. *Journal of Disaster Risk Studies, 3*(2), 417-428.

[19] Pyšek, P., Richardson, D. M., Pergl, J., Jarošík, V., Sixtová, Z. and E. Weber. (2008). Geographical and taxonomical biases in invasion ecology. *Trends in Ecology and Evolution 23(5)*, 237-244.

10

Germination of *Allanblackia floribunda* Seeds: The Effect of Soak Duration in Fluridone on Germination and Seedling Growth

Faith Ileleji, Elsie I. Hamadina[*], Joseph A. Orluchukwu

Crop and Soil Science Department, Faculty of Agriculture, University of Port Harcourt, Port Harcourt, Nigeria

Email address:

elsieile@yahoo.com (E. I. Hamadina)

Abstract: *Allanblackia* seeds contain about 72% white fat (mostly of oleic and stearic acid), with high medicinal and industrial value, but the plant has not been domesticated, so seeds are only gotten from the wild. The demand for seeds of the wild *Allanblackia floribunda* exceeds supply and efforts to domesticate the plant to increase seed availability have been unsuccessful due to long seed dormancy periods. Soaking scarified seeds from immature fruits in water or fluridone shortened dormancy to less than 3 months, but the effects of different soak durations on germination, dormancy and seedling vigor, are not fully understood. This study aimed to determine the effects of three soak durations (1, 6, and 12 h) in water or fluridone (10 μM or 30 μM) on seed germination and seedling growth. The study was a 3 x 3 factorial experiment arranged in a Completely Randomized Design, using scarified immature seeds. At 16 weeks after treatment, seedlings were transplanted into polypots filled with topsoil and assessed weekly for the effects of treatments on seedling growth (leaf number and plant height). The % germination increased as the duration of soak in water or 10 μM fluridone increased from 1 to 12 h. The tendency to obtain 75-100% germination was higher when seeds were soaked in 10 μM fluridone (i.e., for 6 or 12 h) as compared to soaking in water for up to 12 h. The shortest duration (72 d, i.e., approx. 2.5 months) to achieve >75% germination was when *Allanblackia* seeds were soaked for 6 h in 10 μM fluridone. Soaking the seeds for 12 h in 10 μM fluridone resulted in 100% germination in 3 months. At transplanting, seedlings from fluridone treated seeds were taller, and had more leaves than those from seeds soaked in water for 1 or 6 h, but the reverse was observed in seedlings from seeds soaked for 12 h in water. However, these effects diminished after 2 weeks, when the seedlings have become established. This study has shown a promising method for achieving rapid and efficient germination of *Allanblackia* seeds with no noticeable adverse effects on seedling establishment.

Keywords: *Allanblackia floribunda*, Rapid Germination, Seed Dormancy, Fluridone

1. Introduction

Allanblackia trees produce large berry-like fruits that contain seeds (30 to 50) of high food, cosmetic and pharmaceutical importance (Foma and Abdala, 1985; Bonanome and Grundy, 1988). *Allanblackia* seeds are mainly obtained from collections of wild mature fruits that drop off the tree. The seeds contain about 72% white fat (Foma and Abdala, 1985) that remains solid at ambient temperatures and consists mostly of oleic and stearic acid (Eckey, 1954; Hilditch, 1958, *cf* http://www.fao.org/docrep/X5043E0d.htm). Oleic and stearic acids are reported to lower plasma cholesterol levels (Bonanome and Grundy, 1988). Also, the fat from *Allanblackia* seeds has a relatively high melting

point (35°C), which makes it more valuable than other fats because it can be used without the need to transform its consistency for margarines and similar products (Nkengfack *et al.*, 2002). These characteristics make *Allanblackia* seeds extremely valuable to food and cosmetic industries.

Currently, the demand for *Allanblackia* seeds is much higher than its supply: only 20% of the demand is obtainable from the wild (Leakey *et al.*, 2000; Adubofuor *et. al.*, 2013). To produce large number of *Allanblackia* trees, the seeds must be domesticated (Munjuga *et al.*, 2008). Unfortunately, efforts to domesticate *Allanblackia* through seed propagation is hindered by the lack of a method to induce spontaneous and high rate of seed germination (Atangana *et al.*, 2006; Ofori *et al.*, 2011). *Allanblackia* seeds could remain dormant for as long as 2 years, and only one in ten seeds eventually germinates (Ofori *et al.*, 2011).

The need to increase the rate of *Allanblackia* seed germination, and possible methods to stimulate early germination, is the drive behind research aimed to understand the contol of *Allanblackia* seed dormancy. In a previous study, the duration of dormancy, from treatment of seeds obtained from mature fruits to germination, was reduced to 4-10 months, and germination was increased to 40% by: (1) buring the fruits in sand for 2-3 months to soften the seed coat followed by scarification and storage in black polybags, or (2) scarifing the seeds followed by storage in black polybags (Peprah *et al.*, 2008). Seed rot is however high under these methods. A novel method of germinating *Allanblackia* using scarified seeds from immature fruits was recently proposed (Ileleji and Hamadina, 2015), which led to 75% germination in 72 days (2.5 months) by soaking scarified seeds from immature fruits for 6 h in 10 µM fluridone or water, as compared to 50% (in 100 d) for seeds from mature fruits. The response of the seeds soaked in fluridone and water suggests that young scarified seeds still exhibit dormancy. We propose that this may be caused by the presence of growth inhibitor(s), perhaps abscisic acid, and that the effect of the inhibitor is related to the duration of soak in water or ABA inhibitor. Fluridone is a plant growth regulator that is known to induce germination in many plant parts that exhibit dormancy by competitively inhibiting C_{40} carotenoid, which is the precursor of abscisic acid (ABA) biosynthesis (Fong *et al.*, 1983; Zeevaart and Creelman, 1988). Also, ABA can leach out of plant tissues soaked in water (Villiers and Wareing, 1960; loveys and Van Dijk, 1988). However, the most effective soak duration in fluridone or water on germination of scarified seeds from immature fruits is not known. Also, information on the effects of such approach on seedling establishment and growth is hard to come by. The main aim of this study is to determine the effect of soak duration (1, 6, and 12 h) in water or fluridone (10 or 30 µM) on seed germination and seedling growth of *Allanblackia*.

2. Materials and Methods

2.1. Collection of Immature Allanblackia Fruits

Allanblackia floribunda fruits were collected in the month of August from Rivers State Sustainable Development (RSSDA) Allanblackia Project site at Igbu Idonka, Etche, Rivers State of Nigeria. In this study, all the fruits were plucked by detaching from the tree either at the point of attachment of the fruit stalk (pedicel) to the stem or along the stalk. *Allanblackia floribunda* was choosen because it is the dorminant *Allanblackia* species found in Nigeria.

2.2. Fruit and Seed Assessment for Maturity

The plucked fruits were assessed for maturity using the indices detailed by Bewley and Black (1994), even when the

plucked fruits were expected to be immature. The rationale for collecting immature fruits in this study relates to the hypothesis that dormancy may be avoided by collecting seeds before the desiccation and hardening of the seed coat, and that recalcitrant seeds, when collected before the hardening of the seed coat are germinable and may require no pretreatment to germinate (Bewley and Black, 1994).

2.3. Chemicals Used

The chemicals and reagents used in this study were fluridone (1-methyl-3-phenyl-5-[3-trifluoromethy1= (phenyl)] -4-(1*H*)-pyridinone), Dimethyl sulphuroxide (DMSO), nitric acid, potassium hydroxide and sodium hypochlorite. 1N KOH (potassium hydroxide) was used as base stabilizer and HNO_3 (Nitric acid) as acid stabilizer, and sodium hypochlorate as disinfectant. Fluridone was purchased from Chem Services Inc., West Chester, USA. Fluridone was chosen because it is known to inhibit ABA biosynthesis and induce germination. Also, if dormancy in immature *Allanblackia* seeds resulted from ABA concentration then, inhibiting ABA biosynthesis should induce early germination.

2.4. Fruit and Seed Preparation

Allanblackia floribunda fruits were wrapped in plantain leaves for two weeks to soften them prior to seed extraction. The seeds were extracted from the fruits, cleaned, weighed, scarified and weighed again. Scarification was done because seeds tend to germinate faster when the hard seed coat is removed or thinned down.

2.5. Experimental Treatments

Scarified seeds from immature fruits were soaked in one of three fluridone (FLU) concentrations (0 µM, 10 µM, and 30 µM fluridone solutions) for one, six or twelve hours. Treated seeds were then stored/incubated in disinfected transparent plastic containers wrapped in black polyethylene bags. The seeds were observed weekly for germination. Thus, there were nine treatment combinations: 0 µM FLU 1h soak duration (control), 0 µM FLU 6h soak duration, 0 µM FLU 12h soak duration, 10 µM FLU 1h soak duration, 10 µM FLU 6h soak duration, 10 µM FLU 12h soak duration and 30 µM FLU 1h soak duration, 30 µM FLU 6h soak duration, 30 µM FLU 12h soak duration. A one-hour soak duration was considered the maximum duration it could take to wash and surface-sterilizes seeds in a routine seed preparation activity.

2.6. Seed Incubation and Incubation Conditions

Scarified, treated seeds were placed in sterile transparent plastic containers. The containers were kept moist by spraying some distilled water on the inner part of the cover. Each container was then wrapped in a black polythene bag and placed in a warm screen house.

The sides of the screen house were covered with white polyethylene film to about 3/4 of the height of the screen house while the ¼ of the screen house was covered with white net for ventilation. Temperature and relative humidity, inside the seed storage containers, and screen house were monitored using a 433MHZ cable free Oregon scientific sensor model BTHR968.

2.7. Data Collection

Seeds were observed weekly for shoot bud emergence until 16 WAT. To study the effect of treatments on seedling growth, all germinated seeds were transplanted at 16 weeks after treatment when seedling height was ≥1.8 cm. One seedling was transplanted into a polypot filled with 3 kg top soil. The height of each plant was measured weekly for eight weeks. The number of leaves per plant was counted weekly.

2.8. Experimental Design and Data Analysis

This experiment was a 3 x 3 factorial experiment arranged as a Completely Randomized Design; with three fluridone concentrations and three soak durations. Data was analyses using GenStat Release 10.3DE, Copyright 2011, VSN International Ltd. Rate of germination was determined using percentages. Data on duration from treatment date to germination was analyzed using survival data analysis. Survival data analysis is a statistical tool for analyzing data where the outcome variable is time until the occurrence of an event, which in this study was shoot emergence. This program is most suitable in the analysis of data on duration from a defined time to the occurrences of a named event, and where some elements in the study survive (i.e., where some seeds do not germinate) longer than the period of the study (Collet, 1994; Hoon, 2008). Thus for such seeds, the duration to germination is unknown but at least as long as the duration of the study and so, it is censored and used in the model. Censoring is important in survival analysis, representing a particular type of missing data. Also, because the distribution of this type of data is usually not a normal distribution, the regular Analysis of Variance (ANOVA) cannot be used in this analysis.

Plant height data was analyzed using two-way analysis of variance. Means were separated using standard error of difference (SED). Leaf number data was square root transformed and then analyzed using General Treatment Structure (no blocking) in analysis of variance (ANOVA).

3. Results

3.1. Fruit and Seed Assessment for Maturity

Fruit assessment at collection suggests that the fruits were immature; the fruit was greenish with no split or opening on the fruit's surface, the pedicels/stalk was still attached to the fruits, the fruit-pulp was hydrated, hard and difficult to cut through and the seeds were tightly held to the pulp. The seeds on the other hand were germinable; they appeared well filled, embryo was present, they sunk in floating tests for germinability, the endosperm was slightly hard and the seed coat was easy to cut through. Average seed fresh and dry weights were 6.5 g and 0.18 g respectively and percentage moisture content was up to 94%.

3.2. Percentage Shoot Emergence

Shoot emergence (germination) was first noticed at 6 WAT. By 7 WAT, shoot emergence was observed in many treatments and shoots were up to 2 mm. Soaking seeds for 1 h in water (Control) led to 60% shoot emergence by 11 WAT and >70% by 14 WAT (Figure 1). Also, soaking the seeds in water for 6 hours led to 60% by 8 WAT and above 70% by 14 WAT. For the 12 h soak in water treatment, percentage shoot emergence was nearly 100% by the 14 WAT. In the fluridone treatments, soaking seeds in 10 µM FLU for 1h, led to more than 45% shoot emergence by 14 WAT. Increasing the soak duration in 10 µM FLU to 6 h increased the percentage germination (80% by 10 WAT). Soaking the seeds for 12 h in 10 µM FLU led to almost 100% germination in 14 WAT. Seeds soaked in fluridone concentration of 30 FLU 1hrs, 30 FLU 6hrs, and 30 FLU 12 h lead to less than 40% by 9 WAT, 14 WAT and 11 WAT respectively (Figure 1).

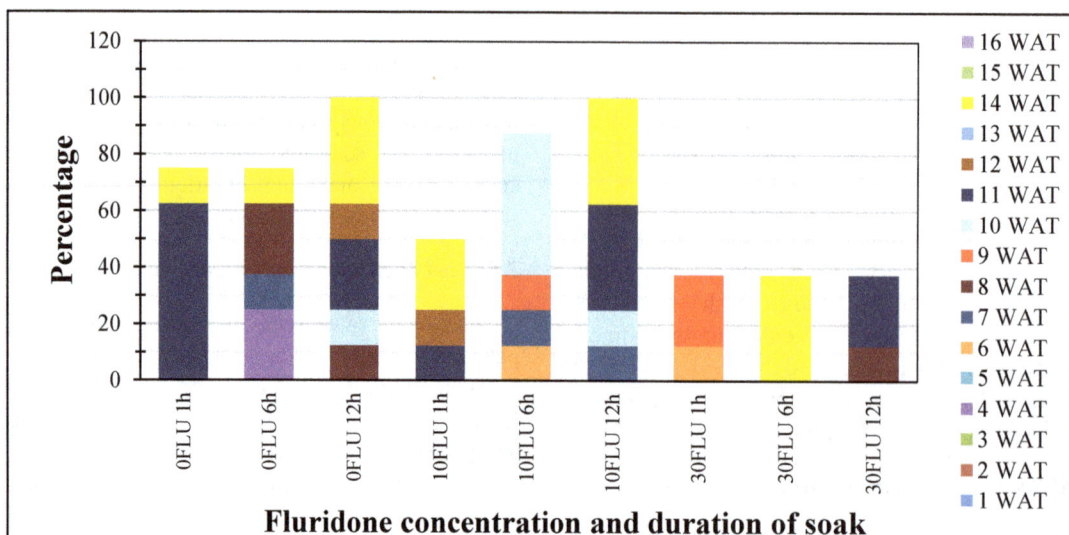

Figure 1. *Effect of fluridone concentration and soak duration on rate and percentage shoot emergence*

Thus, prolonging the soak duration in water to 12 h increased the percentage seed germination. Similarly, percentage seed germination increased with increase in soak duration in 10 µM fluridone. However, percentage shoot emergence was inhibited by soaking seeds in 30 µM fluridone compared to the control or 10 µM fluridone treatment, and prolonging the duration of soak in 30 µM fluridone for 12 h did not significant change rate or percentage seed germination.

3.3. Effects of Treatments on Time Taken to Germinate

The Kaplan Meier log cumulative hazard (germination) curve shows that the treatment curves were significantly different at P< 0.01 (Figure 2). A maximum of 50% germination occurred in 79 days when scarified *Allanblackia* seeds collected from immature fruits were soaked for 1 h in water, *i.e.,* 0 µM FLU 1h (control). In contrast, the duration to 50% germination was significantly hastened by 20 d by soaking seeds for 6 h in water. However the treatment did not lead to 70% or greater germination by the end of the study. Increasing the duration of soak in water to 12 h led to greater than 50 and 75% germination in 100 days.

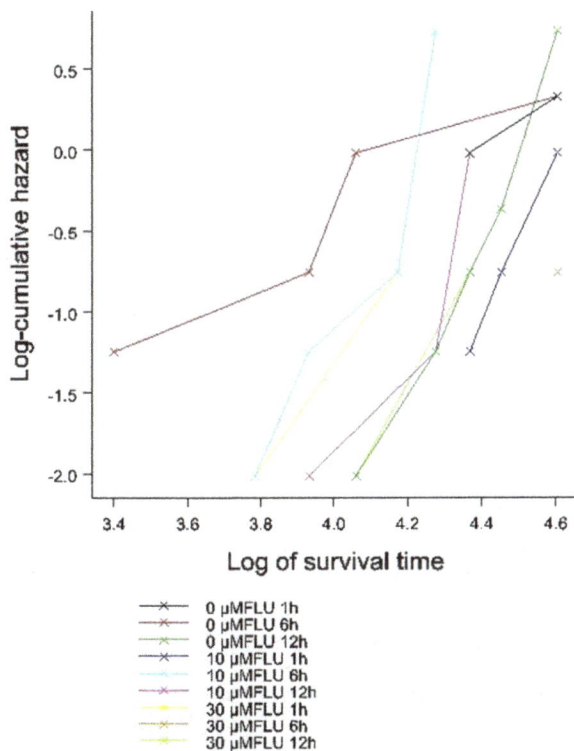

Figure 2. Kaplan-Meier log cumulative hazard (germination) curve for the effect of treatment on time taken to germinate for Allanblackia seeds. Log rank test of equality; P<0.01; test statistic=21.264.

In the 10 µM fluridone treatments, 50% germination was observed in 72 and 79 d when seeds were soaked in 10 µM FLU for 6 or 12 h respectively. Greater than 75% germination was obtained in 72-100 d when seeds were soaked in 10 µM FLU for 6 or 12 h respectively. These treatments clearly stimulated germination in more seeds and

shortened the duration to germination compared to the control. However, the effects of these treatments were comparable to that of a prolonged (12 h) soak in water.

In the 30 µM fluridone treatments, neither 50 nor 75% germination was obtained by the end of the study. Prolonging the soak duration did not hasten the duration to germination. Only 25% germination was possible with these treatments.

3.4. Effect of Treatments on Seedling Growth

3.4.1. Effect on Number of Leaves Per Seedling
At transplanting and throughout the study period, the number of leaves per plant did not differ significantly (P<0.05) across the treatments. The mean number of leaves (square root transformed) was 1.101, 1.330, 0.208, 0.202, and 0.230 at 0, 2, 4, 6 and 8 weeks after transplanting.

3.4.2. Effect on Seedling Height
At transplanting, seedlings were taller under most of the fluridone treatments than in the control except where the seedlings developed from seeds that were soaked for 12 h in either 10 or 30 µM FLU (Table 1). The tallest seedlings were observed under the 30 µM FLU 6 h treatment. At 2, 4, 6 and 8 weeks after transplanting, there were no significant (P<0.05) differences in seedling height across treatments.

Table 1. Seedling plant height at transplant as affected by treatments.

Fluridone (µM)	Duration of Soak (h)		
	1	6	12
0	1.90	2.83	3.30
10	3.27	2.67	2.00
30	2.43	3.90	2.23
S.E.D	0.638		

4. Discussion

This study seeks to determine the effect of soak duration (1, 6, and 12 h) in water or fluridone (10 µM and 30 µM) on germination of scarified *Allanblackia* seeds collected from immature fruits, and on subsequent seedling growth. In this study, percentage germination increased as the duration of soak in water or 10 µM fluridone increased from 1 to 12 h. This effect suggests the presence of an endogenous growth inhibitor, possibly abscisic acid, that is soluble in water and whose activity is inhibited by fluridone (an inhibitor of Abscisic acid biosynthesis). Abscisic acid (ABA) is long known to leach in water allowing dormant seeds to germinate (Villiers and Wareing, 1960). Also, fluridone has severally been shown to inhibit ABA biosynthesis/ content in seeds and tubers (Zeevaart and Creelman, 1988; Le Page-Degivry et al., 1990; Mulwa and Nwanza, 2006). Although information on the changes in ABA content of *Allanblackia* seeds during seed development and through germination is rare , many studies have shown that its content reduces as dormant seeds/ tubers progress towards germination (Zeevaart and Creelman 1988; Suttle, 1995). We suggest therefore that prolonged soak in water may be necessary to significantly leach out ABA while prolonged soak in fluridone may be necessary to prevent the wearing off of a fluridone effect over the long period before germination. Wearing off effect of fluridone has been reported in prolong *invitro* (Hamadina *et al.*, 2010) and open field pot studies (Hamadina and Eze, 2013) on yam dormancy. In the 30 μM fluridone treatments, however, increasing the duration of soak did not increase percentage germination nor did it induce early seed germination. The tendency of 30 μM fluridone to slow down the *Allanblackia* seed germination has been reported (Ileleji and Hamadina, 2015).

The likelihood of obtaining 75-100% germination is higher when germinable, scarified, seeds from immature *Allanblackia* fruits are soaked in 10 μM fluridone (i.e., for 6 or 12 h) than in water (i.e., for 12 h). Soaking *Allanblackia* seeds for 6 h in 10 μM fluridone is recognized for its role in bringing about over 75% (indeed 80%) germination over the shortest period of time (72 d, *i.e.,* approx.. 2.5 months). Soaking seeds for 12 h 10 μM fluridone or in water is recognized for bring about maximum percentage (100%) germination though over a long period of 100 d, i.e., approx. 3.5 months). Although the use of water for germinating seeds is cheaper and safer, this results shows that the use of a recognized inhibitor of ABA biosynthesis is a more effective method of stimulating germination than the use of a method that leaches out already biosynthesized growth inhibitor. Further studies are however required to confirm this.

The effect of fluridone treatments on the seedling growth parameters reported in this study suggests that fluridone, particularly at 10 μM, supports shoot growth rather than inhibit it but this effect may be transient; declining as the seedlings becomes more autotrophic. Although fluridone have been shown to cause reduction in rice mesocotyle growth (Watanabe *et al.*, 2001), it has also been shown to increase micro tuber number and support shoot growth in yam (Hamadina *et al.*, 2010) and increase germination and seedlings growth of Mungbean (Thind *et al.,* 1997). The transient tendency of fluridone observed in this study may prove useful in *Allanblackia* germination effort but this need to be verified. Thus, this study confirms that *Allanblankia* seeds from immature fruits harvested in the month of August have very high germination rates (70-100%) that are dependent on soak duration in water or fluridone. The outcome of this study has high consequence for large scale seedling propagation when compared to a germination rate of <10% obtainable over 6 to 24 months when seeds from mature fruits are propagated or the 6-40% germination obtainable over 10 months when seeds from mature fruits are scarified and incubated in black polythen bags.

5. Conclusion

In conclusion, this study has shown a promising method for achieving rapid and efficient germination of *Allanblackia* seeds with no noticeable adverse effects on seedling establishment. The outcome of this study has high potential for large scale seedling propagation using seeds from mature fruits are scarified and incubated in black polythen bags.

Acknowledgement

The authors wish to acknowledge assistance of the Rivers State Sustainable Development Agency (RSSDA) for providing the fruits used in this study.

References

[1] Adubofuor, J., W. Sefah and J. H. Oldham. 2013. Nutrient composition of *Allanblackia paviflora* seed kernels and oil compared with some plant fats and oils and application of the oil in soap preparation. J. Cereals and Oil seeds, 4(1): 1-9.

[2] Atangana, A. R., Z. Tchoundjeu, E.K. Asaah, A.J. Simons, and D.P. Khasa. 2006. Domestication of *Allanblackia floribunda*: Amenability to vegetative propagation. Forest Ecology and Management 237: 246-251. doi: 10.1016/j.foreco.2006.09.081.

[3] Bewly J. D., and Black, M. 1994. Seeds – Physiology of development and germination. Plenum Press, New York, 445 pp.

[4] Bonanome, A., and S.M. Grundy. 1988. Effect of dietary stearic acid on plasma cholesterol and lipoprotein levels. New England Journal of Medicine 318(19): 1244-1248.

[5] Collet, D. (1994). Modelling Survival Data in Medical Research. Chapman and Hall, London.

[6] Eckey, E.W., Vegetable Fats and Oils, Reinhold Publishing Corp., 1954. pp 695-6

[7] Foma, M., and T. Abdala. 1985. Kernel oils of seven species of Zaire. Journal of the American Oil Chemists Society 62(5): 910-911.

[8] Fong F, Koehler DE, Smith JD. 1983. Fluridone induction of vivipary during maize seed development. In: Kruger JE, LaBerg DE, eds. Third International Symposium on Pre-Harvest Sprouting in Cereals. Boulder, Co. USA: Westview Press, 188-95.

[9] Hamadina, E.I., and G. Eze. 2013. Pre Tuber Application of Fluridone: Effect on Vegetative Growth and Seed Tuber Dormancy in Yam (D. alata). American Journal of Experimental Agriculture, Vol. 4(4): 415-426, 2014

[10] Hamadina, E.I.; Craufurd, P.Q.; Battey, N.H.; Asiedu, R. 2010. In vitro micro-tuber initiation and dormancy in yam. Annals of Applied Biology vol. 157 issue 2 September 2010. p. 203-212

[11] Hilditch, Chemical Contribution of Natural Fats. pp 264-5 TANG. Tr. Bull., 1958, Part I, p.13

[12] Hoon, T.S. (2008). Using Kaplan Meier and Cox regression in survival analysis: An example. ESTEEM, Vol.4, pp.3-14.

[13] Ileleji, F. O. and E. I., Hamadina. 2015. Improving Seed Germination in *Allanblackia floribunda*: Effect of Seed Age and Fluridone. Nigerian Journal of Agriculture, Food and Environment, Vol. 11(2):24-32

[14] Le Page-Degivry M-T, Barthe P, Garello G. 1990. Involvement of endogenous abscisic acid in onset and release of *Helianthus annuus* embryo dormancy. Plant Physiology 92, 1164-8.

[15] Leakey, R.R.B., J-M.Fondoun, A. Atangana, and Z. Tchoundjeu. 2000. Quantitative descriptors of variation in the fruits and seeds of *Irvingia gabonensis*. Agroforestry Systems 50: 47-58.

[16] Loveys, B. R. and van Dijk, HM. 1988. Improved extraction of abscisic acid from plant tissues. Austr. J, Plant Physiol. 15:421-427.

[17] Mulwa R.M.S., Nwanza L.M. (2006) Biotechnology approaches to developing herbicide tolerance/selectivity in crops. African journal of Biotechnology,5,396-404

[18] Munjuga, M., Ofori, D., Sawe, C., Asaah, E., Anegbeh, P., Peprah, T., Mpanda, M., Mwaura L., Mtui, E., Sirito, C., Atangana, A., Henneh, S., Tchoundjeu, Z., Jamnadass, R., Simons, A.J. (2008). Allanblackia propagation protocol. World Agroforestry Centre (ICRAF), Nairobi, Kenya, ISBN 978-92-9059-231-0.

[19] Nkengfack, A.E., G.A. Azebaze, J.C. Vardamides, Z.T. Fomum, and F.R. van Heerden. 2002. A prenylatedxanthone from *Allanblackia floribunda*. Phytochemistry60: 381-384.

[20] Ofori, D.O., Peprah, A.T., Cobbinah, J.R., Atchwerebour, H.A., Osabutey, F., Tchoundjeu, Z., Simons, A.J., Jamnadass, R. 2011. Germination requirements of *Allanblackia parviflora* seeds and early growth of seedlings. New Forest, 41:337-348 DOI: 10.1007/s11056-011-9252-1.

[21] Ofori, D.O., Peprah, A.T., Cobbinah, J.R., Atchwerebour, H.A., Osabutey, F., Tchoundjeu, Z., Simons, A.J., Jamnadass, R. (2011). Germination requirements of *Allanblackia parviflora*

[22] Peprah T, Moses Munjuga, Daniel Ofori, Corodius Sawe, Ebenezar Asaah,Paul Anegbeh, Mathew Mpanda, Lucy Mwaura, Eustack Mtui, Chrispine Sirito, Alain Atangana, Samuel Henneh Zac Tchoundjeu, Ramni Jamnadass and Tony Simons 2008. Allanblackia propagation protocol.

seeds and early growth of seedlings. New Forest, 41:337-348 DOI: 10.1007/s11056-011-9252-1.

[23] Suttle, J. C. (1995). Postharvest changes in ABA levels and ABA metabolism in relation to dormancy in potato tubers. Physiologia Plantarum, 95, 233-240.

[24] Thind, S. K., Chanpreet and Miridula. 1997. Effect of fluridone on free sugar level in heat stressed mungbean seedlings. Plant Growth Regulation 22(1):19-22

[25] Villiers, T.A. and Wareing, P.F. 1960. Interaction of growth inhibitor and natural germination stimulator in the dormancy of *Fraxinus excelsior* L. I, 185, 112-114.

[26] Watanabe, H., Takahashi, K., and Saigusa, M. 2001. Morphological and anatomical effects of abscisic acid (ABA) and fluridone (FLU) on the growth of rice mesocotyls. Plant Growth Regulation 34(3):273-275

[27] Zeevaart, J.A.D., and Creelman, R.A. (1988).Metabolism and physiology of abscisic acid.Annu. Rev. Plant Physiol. Plant Mol. Biol. 39, 439–473.

11

Features of Ontogeny of Wheat Hybrid of Type Dwarf II

Ruzanna Robert Sadoyan

Scientific Center of Agriculture, Ministry of Agriculture, Echmiadzin, Republic of Armenia

Email address:

ruzannasad@mail.ru

Abstract: *The* genes of hybrid depression widely spread in the genus Triticum lead to inviability of hybrid plants and prevent the successful implementation of breeding programs. At the same time the phenomenon of hybrid depression serves as a basic model for the study of profound changes in hybrid plants resulting of expression of various genes. Research of the intensity and orientation of these changes and regularities of ontogenetic development of wheat is necessary to evaluate the viability of hybrids. *We* have investigated the influence of complementary genes of hybrid dwarfism on the root system, intensity of photosynthesis and the activity of catalase. It was shown that the interaction of complementary dominant genes of hybrid dwarfism has multilateral impact on the ontogenetic development of wheat hybrid Dwarf II. Depression in above ground and underground plant organs was manifested. Photosynthesis in Dwarf II hybrids proceeded more intensively than in the parental forms, but the catalase activity was interrupted in leaves and roots. Notable decrease of the volume and total absorbency of the root surface was detected.

Keywords: Wheat, Hybrid Depression, Morphological Parameters, Photosynthesis Intensity, Catalase Activity

1. Introduction

The evolution of wheat in some regions of the world have undergone negative dominant mutations, leading to the appearance of depressive or lethal hybrid plants, thereby preventing the success of breeding programs. The divergence of some members of the complementary systems of depression has resulted in their biotypical and eco-geographical localization. Genetic factors in specific combinations have leaded to the interruption of vital activity of regulatory mechanisms, widespread in the genus Triticum. They have diverse manifestations (hybrid necrosis, chlorosis, and hybrid dwarfism), complementary nature and differentiation, depending on species and biotypical and eco-geographical localization [1, 2, 3]. The research of these factors is of great importance, because the knowledge of the changes in physiological traits, associated with genetic gains in yield potential, presents a basic key for understanding the yield-limiting factors and future breeding strategies [4, 5, 6].

One of the effective approaches in the research of genetic changes in physiological traits is the investigation of three different independent complementary genetic systems (hybrid necrosis, red chlorosis and white spotted chlorosis) leading to the identical physiological changes: disruption of

assimilation apparatus, decay of pigments and lower level of viability. Each form of depression has special morphological character, specific development and definitive manifestation. For hybrid dwarfness the intense synthesis and accumulation of chlorophyll is typical. Selection dwarfs wheat cultivars have greater photosynthetic capacities than the taller semi-dwarfs, they averaged 20 % higher maximum net photosynthetic rates compared to the taller semi-dwarfs. The higher rates of photosynthesis are occurring only at anthesis with slightly greater carboxylation efficiencies and significantly increased chlorophyll concentration per unit leaf area [7]. Ellis and Leech [8] stated that the chloroplast population of developing leaf tissues was increased. Depression in hybrids was manifested also on the level of catalase activity in the leaves and roots. The role of dwarf genotypes was specially discussed in the roots formation process, variation in root depth and water extraction [9, 10]. According to Wasson *et al.* [11] the direct selection of yield leads to the development of varieties with optimal root system. Manschadi *et al.* [9] stated the importance of root system characteristics to soil exploration and below-ground resource acquisition that are strongly related to plant adaptation to sub-optimal conditions.

According to the hypothesis of Hermsen [12], the phenomen of "hybrid dwarfness" is assumed to be determined by the complementary interaction of three genes D_1, D_2 and D_3, differing in dominance and in contribution to the dwarf phenotype. Depending on the genotype of the parental forms in various stages of development, the hybrids express the features of dwarfness with different levels of viability. Three dwarf types are described [12]. *Type 1-dwarfs* are dwarf during their whole life cycle and generally do not produce seeds. *Type 2-dwarfs* are typically dwarf. They have dark green leaves, shortened and thickened stems with productive semisterile ears. In the process of development the hybrid plants are lagging from one of the late maturing parent for 15-20 days. *Type 3-dwarfs* in the phase of full tillering are forming abundant shoots and light green leaves. At the beginning of phase of earing the hybrid plants are actively growing and almost reaching the level of parental forms. At the end of the growing season the *type 3-dwarfs* are characterized by abundance of thin stems, small ears and grains.

The objective of our research was the dwarf hybrids physiological traits investigation, as one of approaches for the analysis of gene-to-phenotype relationship.

2. Materials and Methods

Our research was conducted on the basis of the Armenian Scientific Center of Agriculture of Ministry of Agriculture, located in Echmiadzin at Ararat region. The region is characterized by dry and sharply continental climate and the cultivation of agricultural crops is conducted under irrigation.

F_1 hybrids of wheat were obtained by artificial pollination of the Amby ♀ and Delfi ♂ parental forms. Neutered ears were kept in the parchment isolators, then pollinated after 3-5 days. The first hybrid generation (F_1) seeds and parental lines were sown in 5 kg pots in triplicate

each.

We have determined the changes of some physiological characters of hybrids, compared with parental forms. The intensity of photosynthesis with application of wet combustion technique with minor modifications was used [13]. A disk of tissue from the treated leaf was placed in a small test tube containing 5 ml of chromic acid which was sealed in a 7.5 cm wide-mouthed McCartney bottle (using a screw cap with a 2,5 mm thick neoprene seal) containing 2 ml of 0.2SN NaOH. The bottles containing leaf disks from the daily samples were autoclaved at 97 kPa for 20 min, and left overnight at room temperature. Complete oxidation of the tissues occured under these conditions, and the released CO_2 was absorbed by the NaOH in the bottles. Loss of water from the NaOH solution by absorption in the chromic acid reduced the volume of the NaOH to about 1.7 rnl on average.

The total and working absorbing surfaces of the root system were colorimetrically determined by application of methylene blue according to the method of Sabinin and Kolosov [14].

Estimation of catalase activity, expressed as μM H_2O_2 hydrolyzed mg^{-1} protein min^{-1} in the plant tissue samples was determined according to the method of Luck [15]. The assay is based on the estimation of residual hydrogen peroxide (H_2O_2) by oxidation with potassium permanganate ($KMnO_4$).

Statistical data processing was performed by Student's t-test.

3. Results and Discussion

Completed ontogeny with delayed progress of development for hybrid Dwarf II was detected. The levels of development and morphological parameters of dwarf hybrid completely differed from parental forms.

The morphological differences between hybrid and parental forms are presented in Table 1.

Table 1. *The morphological parameters of hybrid Dwarf II and parental forms.*

Samples	Height of plants, cm	Productive tillering psc/plant	Length of ear, cm	Grain number in ear, psc
Amby ♀	104,90±1,55	4,20 ±0,25	9,10±0,32	37,0±1,20
F_1 - Dwarf II (Amby x Delfi)	28,70±1,07*	5,60±0,43*	8,10±0,29*	5,40±1,80*
Delfi ♂	124,50±1,62	4,40±0,16	10,90±0,24	40,10±0,93

Average value ± standard deviation (n=3), * the differences are significant as compared to the parental forms (p< 0.05).

Differences between hybrid Dwarf II and parental forms for all morphological parameters were revealed. Regarding to the height of plants, length of ear and grain number in ear the significantly higher parameters for parental forms were detected. In contrast, the productive tillering of hybrid

Dwarf II was significantly higher in comparison with the parental forms.

Comparative study of capacity of the wheat root system and physiological activity of hybrid Dwarf II and parental forms are presented in Table 2.

Table 2. *Volume and absorbing surface of roots of hybrid Dwarf II and parental forms in the phase of earing.*

Samples	Volume of roots, cm³	Absorbing surface, dm²		% of the working absorptive surface on total	Surface area	
		Total	Working		Total	Working
Amby ♀	7,90 ± 0,53	1,07 ± 0,04	0,40 ± 0,02	38,0	1,36	0,51
F₁- Dwarf II (Amby x Delfi)	5,13 ± 0,20*	0,95 ± 0,01**	0,43 ± 0,01**	45,40	1,85	0,84
Delfi♂	8,60 ± 0,36	1,08 ± 0,04	0,41 ± 0,01	38,66	1,26	0,48

Average value ± standard deviation (n=3), * the differences are significant as compared to the parental forms ($p < 0.05$), ** the differences are insignificant as compared to the parental forms ($p < 0.05$).

On the base of obtained results the volume of roots of hybrid Dwarf II was significantly lower than in parental forms. Significant differences in absorbing surface were not determined between hybrid Dwarf II and parental forms. The depression, manifested by these parameters was compensated by the increase in percent of the working absorptive surface of roots (45.4%). The higher level of surface area (total and working) for hybrid Dwarf II was revealed.

The study of the intensity of photosynthesis revealed that even active chlorophyll synthesis in hybrids Dwarf II does not provide their normal viability (Table 3).

Table 3. *The intensity of photosynthesis in hybrid Dwarf II and its parent forms in earing phase.*

Samples	Intensity of photosynthesis, mg/dm²/h	The total chlorophyll content, mg/g of fresh weight
Amby ♀	30,40 ± 2,23	2,61
F₁ - Dwarf II (Amby x Delfi)	42,10 ± 2,70*	6,21*
Delfi ♂	33,10 ± 1,25	2,70

Average value ± standard deviation (n=3), * the differences are significant as compared to the parental forms ($p < 0.05$).

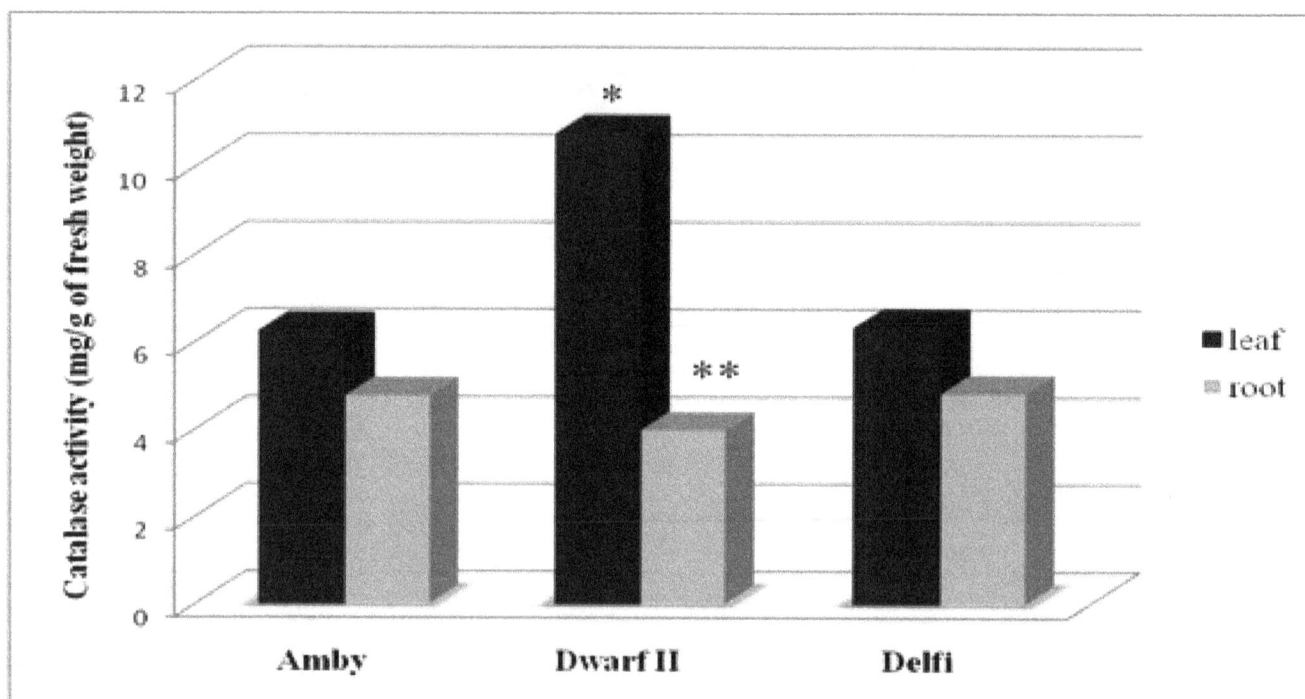

Fig. 1. *Catalase activity of leaves and roots in hybrid Dwarf II and in parental forms.*

* the differences are significant as compared to the parental forms ($p < 0.05$),
** the differences are insignificant as compared to the parental forms ($p < 0.05$).

The obtained results were shown, that in the phases of earing the content of total chlorophyll in hybrid Dwarf II was significantly increased (6.21 mg/g), compared with the parental forms (2.61 and 2.70 mg /g).

Our experiments reveal, that the hybrid Dwarf II has a short distance located polar assimilating system- roots and leaves and low level of growth inhibitors. The observed high level of chlorophyll does not provide plants vital activity. Probably, the genetic apparatus of chloroplasts does not ensure realization of the physiological capacity of photosynthetic pigments resulting low levels of chlorophyll activity.

The growth depression was manifested in both aboveground and underground organs. The catalase activity in leaves and roots was also disrupted. The metabolism of root system and, in particular, catalase activity has essential importance for hybrid viability.

As demonstrated in Figure 1 the catalase activity in leaves of Dwarf II was significantly higher, than in the parental forms. The observed differences in the roots between the hybrids and parental forms were insignificant. According to the obtained results the compensatory physiological mechanisms are more effective in the roots than in the leaves. The observed reduction of root capacity and physiological activity of the hybrid can be explained by the increase of working absorptive surface. At the same time, we have shown that closely located assimilating systems, low level of inhibitors of growth processes and high level of chlorophyll of hybrid Dwarf II do not provide normal viability.

Our data are in agreement with the results presented by Bishop and Bugbee [7], when the net photosynthetic rate and chlorophyll concentration were higher in selection dwarf wheat, compared to the taller semi-dwarfs. Earlier Morgan *et al.* [16] revealed the reduction of leaf cell sizes associated with introduction of semi dwarf stature concentrating the leaf photosynthetic machinery. In turn, the photosynthetic capacity per unit leaf area or leaf weight was increased. Thus, in hybrid Dwarf II photosynthesis occurs more intensively than in parental forms.

4. Conclusion

Investigation of the influence of complementary genes of hybrid dwarfism on the ontogenetic development of hybrid Dwarf II revealed significant changes in the root system with the depression of the aboveground organs. Changes of morphological and anatomical parameters, intensity of photosynthesis, catalase activity in leaves and roots and the reduction of of the root power were observed. Elevated level of chlorophyll and increased intensity of photosynthesis did not provide the normal development of genotypically determined hybrid organisms.

References

[1] Hermsen J. G. (1966). Hybrid necrosis and red hybrid chlorosis in wheat. Hereditas, suppl., 2: 439-452.

[2] Zeven A. C. (1966). Geographical distribution of genes causing hybrid necrosis in wheat. Euphytica, 15(3): 281-284.

[3] Zeven A. C. (1970). Geographical distribution of genes causing hybrid dwarfness in hexaploid wheat of the old world. Euphytica, 19: 33-39.

[4] Edmeadesa G.O., McMasterb G.S., Whitec J.W., Camposa H. (2004). Genomics and the physiologist: bridging the gap between genes and crop response. Field Crops Research 90: 5–18.

[5] Foulkes M. J., Snape J. W., Shearman V.J., Reynolds M.P., Gaju O., Sylvester-Bradley R. (2007). Genetic progress in yield potential in wheat: recent advances and future prospects, Journal of Agricultural Science 145: 17–29.

[6] Fischer R. A. (2011). Wheat physiology: a review of recent developments Crop and Pasture Science, 62(2): 95-114.

[7] Bishop D.L., Bugbee B.G. (1998). Photosynthetic capacity and dry mass partitioning in dwarf and semi-dwarf wheat (Triticum aestivum L.). J. Plant Physiol. Vt11. 153: 558-565.

[8] Ellis J.R. and Leech R.M. (1985). Cell size and chloroplast size in relation to chloroplast replication in light-grown wheat leaves. Planta 165:120-125.

[9] Manschadi A.M, Christopher J, deVoil P, Hammer GL (2006). The role of root architectural traits in adaptation of wheat to water-limited environments. Functional Plant Biology 33: 823–837.

[10] Palta and Watt. (2009). Vigorous crop root systems: form and function for improving the capture of water and nutrients. In. Crop physiology: applications for genetic improvement and agronomy. Eds V Sadras, D Calderini pp. 309–325.

[11] Wasson A.P., Richards R.A., Chatrath R., Misra S.C., Sai Prasad S.V., Rebetzke G.J., Kirkegaard J.A., Christopher J., Watt M. (2012). Traits and selection strategies to improve root systems and water uptake in water-limited wheat crops. Journal of Experimental Botany, 1-14.

[12] Hermsen J. G. (1967). Hybrid dwarfness in wheat. Euphytica, 16,1: 134-162.

[13] Shimshi D. (1969). A rapid field method for measuring photosynthesis with labeled carbone dioxide. J. Exp. Bot., 20: 381-401.

[14] Tretyakov N.N. (1990). Determination of total and working adsorbing surface of the root system by method of Sabinin and Kolosov. Proceedings on Plant Physiology. Moscow."Agropromizdat", 163-165, (In Russian).

[15] Luck H. (1965). Catalase. In: Bergmeyer, H.U. (Ed.), Method sin Enzymatic Analysis. Academic Press, NY, 885–894.

[16] Morgan J. A., Lecain D. R., Wells R. (1990). Semi dwarfing genes concentrate photosynthetic machinery and affect leaf gas ex-change of wheat. Crop Sci. 30: 602-608.

12

Future Fertiliser Demand and Role of Organic Fertiliser for Sustainable Rice Production in Bangladesh

Jayanta Kumar Basak[1], Rashed Al Mahmud Titumir[2], Khosrul Alam[3]

[1]Department of Environmental Science and Hazard Studies, Noakhali Science and Technology University, Noakhali, Bangladesh
[2]Department of Development Studies, University of Dhaka, Dhaka, Bangladesh
[3]Department of Economics, Noakhali Science and Technology University, Noakhali, Bangladesh

Email address:
basak.jkb@gmail.com (J. K. Basak), rtitumir@unnayan.org (R. A. M. Titumir), alam.khosrul@gmail.com (K. Alam)

Abstract: The study finds out the requirement of chemical fertilisers and suggests the role of organic fertilisers for sustainable rice production based upon projection of rice production, consumption, demand and supply of fertilisers for the years of 2020, 2030, 2040 and 2050. The total requirement for commonly used three fertilisers, Urea, Triple Supper Phosphate (TSP) and Muriate of Potash (MP) may increase significantly due to compulsions for growing increased amount of crop outputs in small fragmented parcels of land in the context of diminishing cultivable lands in Bangladesh, negatively impacting on soil fertility as well as sustainability of crop production. Since sustainable yield of crop considerably depends on balanced application of both chemical and organic fertilisers in the field level, the research suggests for increased usage of organic fertilisers.

Keywords: Rice, Fertilisers, Organic Fertilisers, Urea, TSP, MP

1. Introduction

Balanced fertilisation is a key factor for sustainable crop production. Sustainable agriculture practice is being hampered persistently due to decreased soil fertility arising out of inappropriate application of fertilisers. This is particularly important for the agriculture of Bangladesh as the country faces compulsion for growing more food to feed its sizeable population in the backdrop of diminishing cultivable land areas due to pressures stemming from increased population, habitation, industrialisation and urbanisation. Therefore, a well-thought-out strategy on balanced mix of fertilisers should receive a top priority to sustain or increase the level of production of crops.

In recent years, agricultural sector in Bangladesh has been facing a twofold major challenge of producing more food in limited lands to meet the food demand for its huge population along with keeping food prices within an accessible limit for the poor to ensure food security on the one hand, and of minimising the adverse environmental impacts to maintain a sustained level of production and profitability for the peasants along with sound conservation of resource, on the other.

Sustainable agriculture involves the processes that would enable to meet the current and long term societal needs for food, fiber and other resources optimising the benefits through conservation of natural resources and maintenance of ecosystem. Sustainable agricultural practices have to strike a balance between environmental health and economic profitability in order to promote social and economic equity [1]. Therefore, the stewardship of both natural and human resources is of prime importance. The priority of enhancing human capabilities at the individual (peasant-farmer) level and ensuring food security at the national level through efficient and equitable use of resources are compatible with the concept of sustainable agriculture. In order to strike stability between safeguarding environmental components and increased agricultural productivity, it is necessary to promote not only the use of chemical fertilisers but also usage of organic fertilisers in a balanced way.

Before the 1950's, the peasant-farmers of Bangladesh used to apply organic manures such as cow dung, bone meal etc. in aus and aman rice fields and farmyard manure (FYM), mustard oil cake and fishmeal for the mustard and vegetable crops [2]. The use of inorganic fertiliser started in the country in 1951 with the import of 2,698 tons of Ammonium Sulphate, Phosphates in 1957 and Muriate of Potash in 1960 [3]. The fertiliser was introduced at farm level in 1959 [4]. Then, in 1965, the Government launched a 'Grow More Food' campaign and provided fertilisers and low lift pump (LLP) at a highly subsidised rate with pesticide at free of cost to popularise these inputs among the peasants to meet the country's food shortage. Thus, fertiliser consumption began to increase rapidly with the introduction of HYV rice (i.e. IR5 & IR8) and use of LLPs. Now-a-days, chemical fertilisers consists of more than 75 percent of total fertilisers used for rice production [5].

A number of studies have been conducted on sustainability in agricultural sector in Bangladesh [6], [7], [8], [9], [10], [11], [12], [13], and [14]. Most of these studies, however investigated the changes in rice yield under different treatment of fertilisers by various doses in the experiment fields. It is also studied on strategies for developing the fertiliser sector in Bangladesh for sustainable agriculture [15], [16]. The farmers of Bangladesh mostly rely on chemical fertilisers for higher production, without or less application of compost. Such a fertiliser management practice leaves a massive deterioration of soil fertility, resulting loss in organic content [16].

The present research estimates requirement of chemical fertilisers and focuses on the role of organic fertiliser in sustainable rice production in Bangladesh. The estimation has relied on the different published accounts released by the Bangladesh Bureau of Statistics (BBS) and Ministry of Agriculture (MOA) such as Handbook of Agricultural Statistics and Statistical Year Book of Bangladesh [17], [18],

[19], [20], [21].

2. Contribution of Fertilisers to Crop Production in Bangladesh

Crop nutrients are the elements, or simple inorganic compounds, indispensable for the growth of crops. For rice production, 16 elements are essential - Carbon, Hydrogen, Oxygen, Nitrogen, Phosphorus, Potassium, Sulfur, Calcium, Magnesium, Zinc, Iron, Copper, Molybdenum, Boron, Manganese and Chlorine. All essential elements are required in optimum amounts and usable forms for rice plants. Nitrogen, Phosphorus, Zinc and Potassium are the most commonly applied nutrient elements in rice fields whereas Sulfur is occasionally used on the basis of soil condition and other nutrients are provided by air, water, soil and plant residues [22].

Chemical fertiliser provides the nutrient elements (Nitrogen, Phosphorus, Zinc and Potassium) for soils and plays the most vital role in crop production. It supports half of the world's grain production [23]. The contribution of fertilisers in rice production was 40 percent [24]. Results of long-term experiments conducted by Bangladesh Rice Research Institute (BRRI) also shows that fertilisers provide for 36-40 percent nutrient in rice cultivation (averaged over Boro and T. Aman crops during 1985-86 to 2006-07). In 1985-86, fertiliser has contributed 36 percent to the total soil fertility whereas in 2002-07, this percentage was 40 (Figure 1) [25]. Thus, the increasing use of fertilisers over time also indicates the degradation of soil fertility.

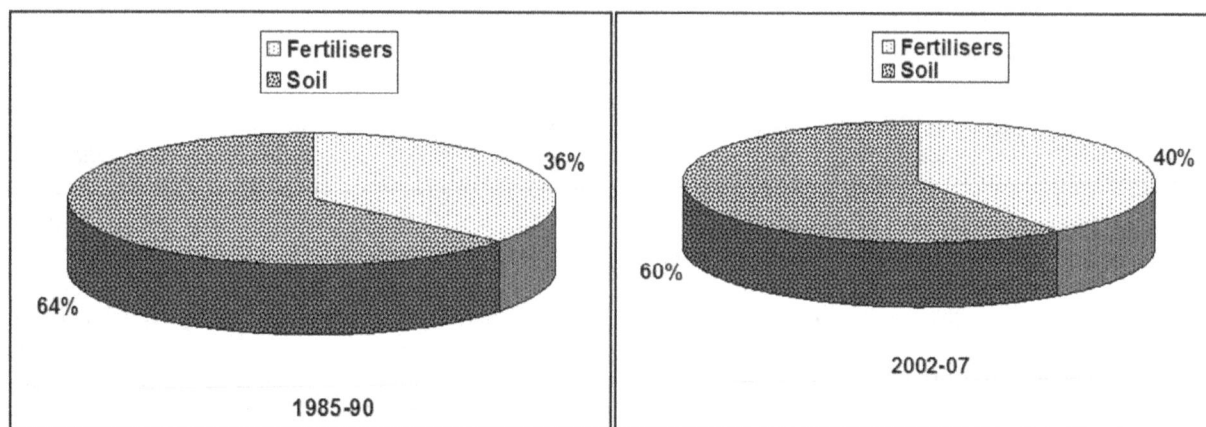

Source: Shah et al., 2008

Figure 1. Percentage of contribution in soil and fertiliser in rice yield in 1985-90 and 2002-07.

3. Demand of Fertilisers in Bangladesh

Fertiliser application mainly depends on the soil types, growing season, irrigation and the varieties that are used. The demand for fertiliser is also affected by the agro-climatic conditions. High yielding varieties (HYV) of rice are highly responsive and need adequate amount of fertiliser to achieve

the targeted production.

The use of chemical fertilisers in Bangladesh has amplified sharply after 1975. Consumption of significant amount of chemical fertiliser was noted during 1975-76. Since then, ever-increasing trend of usage of fertilisers has been observed which reached at peak value of 36.50 lakh tons during 2007-08 (Figure 2). Along with Urea, TSP and MP, the use of Gypsum, Zinc Sulphate and other micronutrients also augmented after the period of 1975-76.

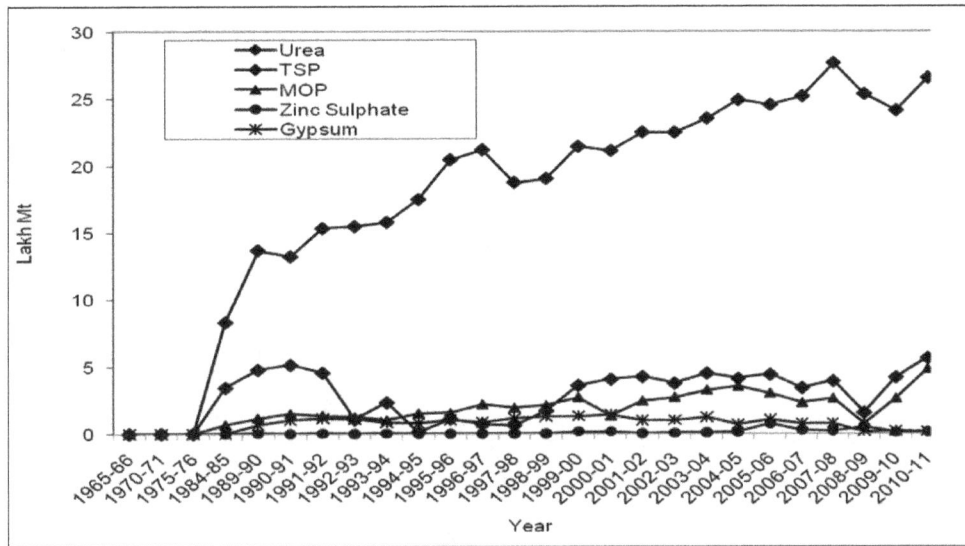

Source: Authors' calculation based on Bangladesh Fertiliser Association (BFA) data, 2009 and Bangladesh Economic Review, 2013

Figure 2. Consumption of different fertilisers in Bangladesh (year wise).

Table 1. Comparative land use under rice production of different varieties.

Seasons	Percentage of Area Coverage		
	Local varieties	HYV	Hybrid
Aus	28.67	71.33	0.00
Aman	32.11	67.89	0.00
Boro	2.29	78.56	19.16
Total	19.79	72.55	7.66

Source: Department of Agriculture Extension, 2008-09

The rate of application of urea is significantly higher compared to other fertilisers for cultivating high yielding rice varieties. Before 1975-76, most of the peasants cultivated traditional varieties and used organic manures such as cow dung, bone meal etc. The irrigated area of rice has dramatically increased and a large portion of cultivable land has been covered under minor irrigation project. In the irrigated condition, farmers cultivate HYV rice, which requires high fertiliser dose compared to the local varieties.

In 2008-09, HYVs covered more than 72 percent to the total cultivable land areas in Bangladesh (Table 1). Therefore, a large amount of fertiliser is being used to cultivate HYVs. Consequently, the demand of chemical fertilisers follows an increasing trend year after year. Like greater use of HYVs, cropping intensity has also influenced fertiliser application over the years. Cropping intensity has increased dramatically in the last decades. In 1980, the cropping intensity was 153.74 percent and intensity moved upward to reach at 176.91 in 2004-05. The Department of Agriculture Extension (DAE) claims that the current cropping intensity is 195 percent. Therefore, cropping intensity increased more than 23 percent in 15 years (1980-81 to 2004-05) (Figure 3).

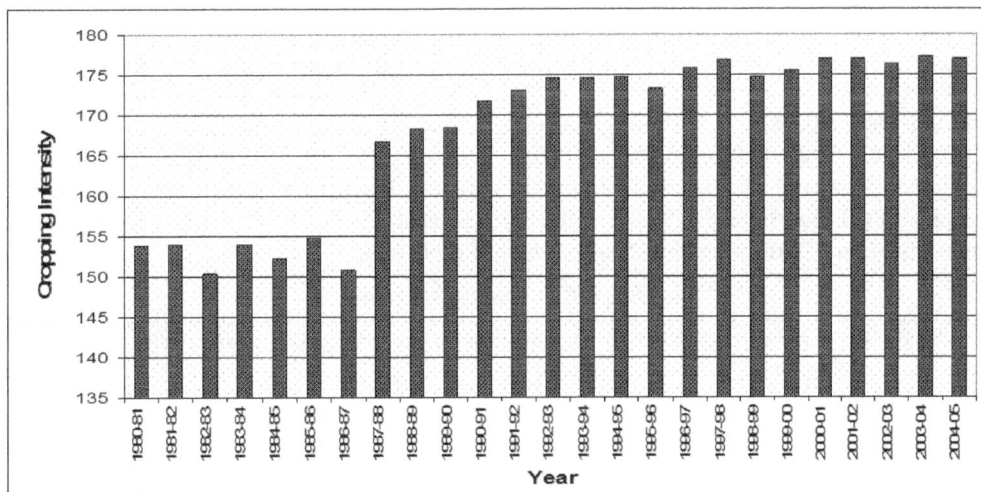

Source: Handbook of Agricultural Statistics, Ministry of Agriculture, 2013

Figure 3. Cropping intensity over time in Bangladesh, 1980-81 to 2004-05 (percentage).

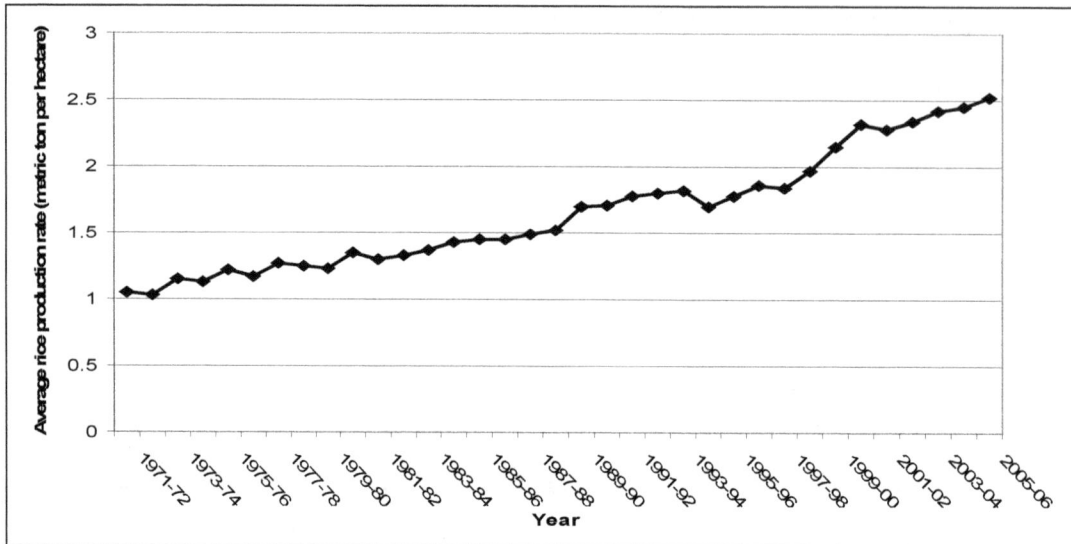

Source: Handbook of Agricultural Statistics, Ministry of Agriculture, 2013

Figure 4. *Average yield of rice from 1971-72 to 2005-06 (season wise).*

By analyzing the rice yield data for 35 years (1971-72 to 2005-06), it is found that rate of rice production is continuously increasing over the years. In 1971-72, the average rice yield was 1.05 metric ton per hectare and reached at 2.52 metric ton per hectare in 2005-06. Therefore, average rice yield increased 2.4 times in 35 years and the increased use of fertiliser has significantly influenced the yield. For instance, in Bangladesh, application of fertilisers increased several times in the same piece of land. In 1975-76, fertiliser application was 0.36 kg per hectare in agricultural land, whereas in 2007, this application was above 298 kg per hectare [26].

4. Future Requirement of Fertilisers in Rice Production

The amounts of recommended and actual dose of fertilisers for rice production in Bangladesh are given in Table 2. Urea (Nitrogen) is a major component of proteins, hormones, chlorophyll, vitamins and enzymes and is essential for rice. Rice plants require a large amount of nitrogen at the early and mid-tillering stages to maximise the number of panicles [22]. The recommended doses of other nutrients are also necessary for a potential rice yield.

Table 2. *Use of fertilisers in Bangladesh.*

Name of Crop (HYV)	Recommended dose (kg/ha)			Actual dose (kg/ha)			Use gap (percentage)		
	Urea	TSP	MP	Urea	TSP	MP	Urea	TSP	MP
T.Aus	141	101	69	135	28	17	4.26	72.28	75.36
T.Aman	166	101	69	135	30	24	18.67	70.30	65.22
Boro	269	131	121	192	47	37	28.62	64.12	69.42

Source: Handbook of Agricultural Statistics, Ministry of Agriculture, 2013

It is estimated that the total population in Bangladesh may stand at 191.65 million by 2020, 220.24 million by 2030, 253.09 million by 2040 and 290.83 million by 2050. If the annual population growth continues at a business as usual rate (1.4 percent annually), huge amount of food will be needed to meet demands of the future generations while it is assumed that a major part of the demand will be met by rice. Based upon the data of the last 40 years, the per capita rice consumption rate in Bangladesh is 153.02 kg per person per year. If the current rice consumption rate continues, the total demand for rice may reach at 68.09 million tons in 2050 (Table – 3).

Table 3. *Year-wise paddy rice demand in Bangladesh (average).*

Year	Population (million)	Paddy rice required (million tons)	Boro rice (million tons)	T.Aman rice (million tons)	Aus rice (million tons)
2007	157.75	36.93	19.20	15.14	2.59
2020	191.65	44.87	23.33	18.40	3.14
2030	220.24	51.56	26.81	21.14	3.61
2040	253.09	59.25	30.81	24.29	4.15
2050	290.83	68.09	35.41	27.92	4.77

Source: Authors' calculation based on data of World Bank and Food and Agriculture Organization (Update as of July, 2009 [27])
Population data in 2007 are taken from World Bank data, 2009
Rice consumption rate is calculated from the last 40 years data (1964 -2003)

The ten-year data (1999-2000 to 2008-09) state that *Boro* rice contributes 52 percent to the total rice production in Bangladesh, while the contribution of *T.aman* is 41 percent and *Aus* is only 7 percent. If the averaged rates of production of *Boro, T.aman* and *Aus* are considered, the total rice demand of *Boro, T.aman* and *Aus* stands at 35.41, 27.92 and 4.77 million tons respectively in 2050 (Table 3).

As per the business as usual scenario, *Boro* rice may contribute 59 percent in 2020, 60.5 percent, 61.5 percent and 63 percent in 2030, 2040 and 2050 respectively. Contribution of *T.aman* may be 34 percent, 33 percent, 32 percent, 31 percent respectively and *Aus* may be 7 percent, 6.5 percent, 6.5 percent, 6 percent respectively for the corresponding years (Table – 4).

Table 4. Year-wise paddy rice demand in Bangladesh (changing condition).

Year	Population (million)	Paddy rice required (million tons)	Boro rice (million tons)	T.Aman rice (million tons)	Aus rice (million tons)
2007	157.75	36.93	19.20	15.14	2.59
2020	191.65	44.87	26.47	15.26	3.14
2030	220.24	51.56	31.19	17.01	3.35
2040	253.09	59.25	36.44	18.96	3.85
2050	290.83	68.09	42.90	21.11	4.09

Source: Authors' calculation based on last 10 years data of Bangladesh Bureau of Statistics, 2012

The total requirement of fertilisers has been estimated on the basis of recommended dose and actual dose of fertilisers applied by the peasants. Average rice production data for the last 10 years and the changing trends of production of *Boro, T.aman* and *Aus* for the same period have also been considered. Moreover, yield of *Boro, T.aman* and *Aus* is kept constant and other factors such as irrigation management, climatic conditions, varieties etc. are also assumed unchanged to estimate the requirement of fertilisers for the specified years.

According to the fertiliser recommended dose, requirement of Urea is 39.17 lakh tons, TSP 21.77 lakh tons and MP 17.13 lakh tons (calculation based on average rice

production data) in 2050 (Table 5). On the basis of trend line analysis of rice production data, requirement of Urea stands at 38.58 lakh tons in 2050, while 20.89 lakh tons of TSP and 17.02 lakh tons of MP may be required for the same year (calculation is based on recommended dose) (Table 6).

Considering the actual dose applied in the field level, the total demand of Urea, TSP and MP in 2050 might be 30.40, 7.01 and 5.45 lakh tons, respectively (calculation based on average rice production data) (Table 7). On the basis of trend line analysis of rice production, requirement of Urea, TSP and MP might be 29.47, 6.88 and 5.35 lakh tons in 2050, correspondingly (Table 8).

Table 5. Requirement of fertiliser on the basis of recommended dose (based on average rice production data).

Year	Urea fertiliser requirement (lakh tons)			Total (lakh tons)	TSP fertiliser requirement (lakh tons)			Total (lakh tons)	MP fertiliser requirement (lakh tons)			Total (lakh tons)
	Boro	Aus	Aman		Boro	Aus	Aman		Boro	Aus	Aman	
2007	10.33	1.58	9.31	21.22	5.03	1.14	5.66	11.83	4.65	0.78	3.87	9.3
2020	12.55	1.93	11.3	25.78	6.11	1.38	6.88	14.37	5.65	0.94	4.7	11.29
2030	14.42	2.21	13	29.63	7.02	1.58	7.91	16.51	6.49	1.08	5.4	12.97
2040	16.58	2.54	14.9	34.02	8.07	1.82	9.09	18.98	7.46	1.24	6.21	14.91
2050	19.05	2.92	17.2	39.17	9.28	2.09	10.4	21.77	8.57	1.43	7.13	17.13

Source: Authors' calculation based on Ministry of Agriculture and Bangladesh Bureau of Statistics, 2012

Table 6. Requirement of fertiliser on the basis of recommended dose (on changing conditions).

Year	Urea fertiliser requirement (lakh tons)			Total (lakh tons)	TSP fertiliser requirement (lakh tons)			Total (lakh tons)	MP fertiliser requirement (lakh tons)			Total (lakh tons)
	Boro	Aus	Aman		Boro	Aus	Aman		Boro	Aus	Aman	
2007	10.33	1.58	9.31	21.22	5.03	1.14	5.66	11.83	4.65	0.78	3.87	9.30
2020	14.24	1.93	9.38	25.55	6.94	1.38	5.71	14.03	6.41	0.94	3.90	11.25
2030	16.78	2.05	10.50	29.33	8.17	1.47	6.36	16.00	7.55	1.01	4.35	12.91
2040	19.60	2.36	11.70	33.66	9.55	1.69	7.09	18.33	8.82	1.16	4.85	14.83
2050	23.08	2.50	13.00	38.58	11.20	1.79	7.90	20.89	10.40	1.23	5.39	17.02

Source: Authors' calculation based on Ministry of Agriculture and Bangladesh Bureau of Statistics, 2012

Table 7. Requirement of fertiliser on the basis of actual dose (on average rice production data).

Year	Urea fertiliser requirement (lakh tons)			Total (lakh tons)	TSP fertiliser requirement (lakh tons)			Total (lakh tons)	MP fertiliser requirement (lakh tons)			Total (lakh tons)
	Boro	Aus	Aman		Boro	Aus	Aman		Boro	Aus	Aman	
2007	7.37	1.52	7.57	16.46	1.81	0.31	1.68	3.80	1.42	0.19	1.35	2.96
2020	8.96	1.84	9.20	20.00	2.19	0.38	2.04	4.61	1.73	0.23	1.64	3.60
2030	10.30	2.12	10.60	23.02	2.52	0.44	2.35	5.31	1.98	0.27	1.88	4.13
2040	11.83	2.43	12.10	26.36	2.9	0.50	2.70	6.10	2.28	0.31	2.16	4.75
2050	13.60	2.80	14.00	30.40	3.33	0.58	3.10	7.01	2.62	0.35	2.48	5.45

Source: Authors' calculation based on Ministry of Agriculture and Bangladesh Bureau of Statistics, 2012

Table 8. Requirement of fertiliser on the basis of actual dose (on changing conditions).

Year	Urea fertiliser requirement (lakh tons)			Total (lakh tons)	TSP fertiliser requirement (lakh tons)			Total (lakh tons)	MP fertiliser requirement (lakh tons)			Total (lakh tons)
	Boro	Aus	Aman		Boro	Aus	Aman		Boro	Aus	Aman	
2007	7.37	1.52	7.57	16.46	1.81	0.31	1.68	3.8	1.42	0.19	1.35	2.96
2020	10.17	1.84	7.63	19.64	2.49	0.38	1.7	4.57	1.96	0.23	1.36	3.55
2030	11.98	1.97	8.51	22.46	2.93	0.41	1.89	5.23	2.31	0.25	1.51	4.07
2040	13.99	2.26	9.48	25.73	3.43	0.47	2.11	6.01	2.7	0.28	1.69	4.67
2050	16.47	2.4	10.6	29.47	4.03	0.5	2.35	6.88	3.17	0.3	1.88	5.35

Source: Authors' calculation based on Ministry of Agriculture and Bangladesh Bureau of Statistics, 2012

5. Importance of Organic Fertiliser

The fertility status of the soil in Bangladesh is extremely variable, having differences in thirty agro-ecological zones, each of which has specific soil and hydrological characteristics. Soils deplete and decline after crop yields, if proper fertility management is not in place. Organic fertilisers can play a vital role in restoring fertility as well as organic material composition of cultivable soils. Organic matter content in soil is found to be very low, around 1 percent in most of the soils and 2 percent in paltry amount of soils, whereas at least 3 percent is conducive for high crop productivity [28].

Bioslurry1 is one of the best organic fertilisers to rejuvenate soils since it is a rich source of both plant nutrients and organic matter. It increases physical, chemical and biological properties of soils, besides supplying essential nutrients to crop plants. It also increases organic matter content of soils and maintains health of soil. The use of bioslurry can reduce application of chemical fertilisers up to 50 percent [28]. Reduced uses of chemical fertilisers will also benefit the peasant farmers in terms of costs for cultivation in a soil environment of high fertility and productive state [29], [30].

In Bangladesh about 7 million tons of organic fertilisers are produced in every year from animal wastes, household wastes, city wastes and crop wastes [31], [32]. If this huge amount of organic fertilisers could be converted into bioslurry, a significant part of fertiliser demand could be fulfilled. If 7 million tons of organic fertiliser can be used in crop production, it could cover 5.3 percent of Urea, 19 percent of TSP and 34.13 percent of MP of the total demand of fertiliser

in FY 2008-09. Consequently, if the total organic fertiliser is used as a bioslurry, it could cover 11 percent demand of Urea, 89 percent TSP and 22.8 percent MP for the same period.

Table 9. Nutrient concentrations in commonly used organic fertilisers and bioslurry of Bangladesh.

Organic fertilisers	Nutrient Content (percentage)		
	N	P	K
Cow dung	0.51-1.5	0.40.8	0.519
Poultry manure	1.6	1.5	0.85
Compost (common)	0.40.8	0.30.6	0.71.0
Farmyard manure	0.51.5	0.40.8	0.511.9
Water hyacinth compost	3.0	2.0	3.0
Bioslurry (cow dung)	1.29	2.80	0.75
Bioslurry (Poultry litter)	2.73	3.30	0.80
Rice straw	0.52	0.25	1.20
Wheat straw	0.63	0.28	0.80
Maize stove	0.45	0.30	0.70
Sugarcane trash	0.35	0.25	0.80
Tobacco stems	0.42	0.25	1.10

Source: Islam, 2006

The contribution of organic fertilisers, therefore, may be significant if it is used as a bioslurry fertiliser. If it is possible to increase the application of organic fertilisers only by three times (21 million tons) to the present situation, it would contribute a large amount of nutrient supply to the soil for rice production (Table 11 and 12). From this basis, on average organic fertilisers may contribute more than 15 percent to the total demand of Urea, TSP and MP fertilisers. Moreover, while used as bioslurry fertilisers, it would contribute more than 30 of nutrient supply (N, P and K) for the rice production in (calculation is based on recommended dose). Considering the actual dose in field level, organic fertilisers would contribute more than 35 percent to the total demand of Urea, TSP and MP fertilisers. Whenever it is used

---

1 Bioslurry obtained from biogas plant is considered as quality organic fertiliser. This organic fertiliser is environmental friendly, has no toxic or harmful effects. Cow dung, poultry litter, compost, crop residues and green manure are commonly used in biogas plant.

as bioslurry fertilisers, it would contribute more than 85 of nutrient supply in the same period.

Table 10. Contribution of organic fertilisers to the total fertiliser demand in the FY 2008-09.

Fertilisers	Demand (Lakh tons)	Contribution of organic fertiliser (Lakh tons)	Contribution of bioslurry fertiliser (Lakh tons)	Percentage of total demand (organic fertiliser)	Percentage of total demand (bioslurry fertiliser)
Urea	28.5	1.50	3.13	5.30	11
TSP	5.0	0.95	4.45	19.00	89
MP	4.0	1.37	0.91	34.13	22.8

Source: Authors' calculation based on the nutrient concentrations in commonly used organic fertilisers and bioslurry of Bangladesh, 2012

Table 11. Contribution of organic and bioslurry fertilisers for rice production in 2050 (calculation based on recommended dose).

Fertilisers	Based upon Average Condition					Based upon Changing Condition				
	Demand (Lakh tons)	Contribution of organic fertiliser (Lakh tons)	Contribution of bioslurry fertiliser (Lakh tons)	Percentage of total demand (organic fertiliser)	Percentage of total demand (bioslurry fertiliser)	Demand (Lakh tons)	Contribution of organic fertiliser (Lakh tons)	Contribution of bioslurry fertiliser (Lakh tons)	Percentage of total demand (organic fertiliser)	Percentage of total demand (bioslurry fertiliser)
Urea	39.17	4.50	9.40	11.50	23.97	38.58	4.50	9.40	11.66	24.36
TSP	21.77	2.85	13.35	13.10	51.32	20.89	2.85	13.35	13.64	63.91
MP	17.13	4.10	2.73	24.00	17.02	17.02	4.10	2.73	24.10	16.04

Source: Authors' calculation based on the nutrient concentrations in commonly used organic fertilisers and bioslurry of Bangladesh, Ministry of Agriculture and Bangladesh Bureau of Statistics, 2012

Table 12. Contribution of organic and bioslurry fertilisers for rice production in 2050 (calculation based on actual dose).

Fertilisers	Based upon Average Condition					Based upon Changing Condition				
	Demand (Lakh tons)	Contribution of organic fertiliser (Lakh tons)	Contribution of bioslurry fertiliser (Lakh tons)	Percentage of total demand (organic fertiliser)	Percentage of total demand (bioslurry fertiliser)	Demand (Lakh tons)	Contribution of organic fertiliser (Lakh tons)	Contribution of bioslurry fertiliser (Lakh tons)	Percentage of total demand (organic fertiliser)	Percentage of total demand (bioslurry fertiliser)
Urea	30.40	4.50	9.40	14.80	30.92	29.47	4.50	9.40	15.27	31.90
TSP	7.01	2.85	13.35	40.00	190.44	6.88	2.85	13.35	4.42	194.04
MP	5.45	4.10	2.73	75.23	50.10	5.35	4.10	2.73	76.64	51.03

Source: Authors' calculation based on the nutrient concentrations in commonly used organic fertilisers and bioslurry of Bangladesh, Ministry of Agriculture and Bangladesh Bureau of Statistics, 2012

6. Conclusion and Discussion

The sustainable increase of crop production for food sufficiency requires efforts to enhance the capacity of production system. All agricultural inputs involved directly or indirectly in crop production need to be adequate and accessible at farmers' field level during the growing season. Timely supply and availability of fertilisers at reasonable prices at the doorsteps of the hard working farmers in the country is necessary for optimum supply of nutrients to the depleted soils, for successful achievement of the targeted yield.

According to the study, a linear correlation is there among the fertiliser application, the cropping intensity and high yielding varieties. In 2008-09, high yielding varieties covered more than 72 percent of the total cultivable land areas in Bangladesh. Similarly, cropping intensity has increased dramatically in the last decades and increased more than 23 percent in the last 15 years.

The study also suggests that the demand of chemical fertilisers have become significant which has a negative impact on soil fertility as well as sustainability of crop production. In 1985-86, fertiliser has contributed 36 percent to the total soil fertility whereas in 2002-07, this contribution was 40 percent. Thus, the increasing response of fertilisers over the time indicates the degradation of soil fertility. Therefore, it is urgent to conserve and add nutrients to the soil through the balance application of compost and inorganic fertilsers, which can help, maintain and increase the nutrient reserves of the soil.

Moreover, the prices of the imported fertilisers will continue to increase in the upcoming years due to high price of oil in internal market, and shrinking natural resources for fertiliser production. Therefore, mobilisation of all organic resources and recycling them into soil fertilisation programme can be considered. Besides, government can take some public awareness through media activities and advocacy to influence farmers in using balanced dose of fertiliser and can emphasise on the use of organic fertiliser for a sustainable rice production.

Acknowledgement

The authors would like to thank Dr. Jiban Krishan Biswas

and Md. Abdus Salam of Bangladesh Rice Research Institute (BRRI) for their kind supports throughout this study. We are grateful to A. Z. M. Saleh of Unnayan Onneshan for valuable inputs during this study. A version of this research has earlier been published by the Unnayan Onneshan.

References

[1]   Earth Summit, *Policy on agriculture in sustainable development*.1992.   Available   at: http://www.nda.agric.za/docs/Policy/SustainableDev.pdf

[2]   EPBS. *The Provincial Statistical Board and Bureau of Commercial Industrial Intelligence*. Statistical Abstract for East Pakistan, Dacca, 1958, p 517.

[3]   R. Ahmed, "Structure, dynamics and related policy issues of fertiliser subsidy in Bangladesh". 1987, pp 281-380.

[4]   M. A. Quasem, "Fertiliser use in Bangladesh: 1965-66 to 1975-76". Bangladesh Institute of Development Studies, Dhaka, 1978, p 37. (BIDS Research Report Series No. 25).

[5]   J. K. Basak, "Future Fertiliser Demand for Sustaining Rice Production in Bangladesh: A Quantitative Analysis". Unnayan Onneshan-The Innovators, Dhaka, Bangladesh, 2010.

[6]   K. L. Hossain, M. A. Wadud, and E. Santosa, "Effect of Tree Litter Application on Lowland Rice Yield in Bangladesh". Bul. Agron. vol.35, no.3, 2007.

[7]   S. R. Osmani, and M. A. Quasem, "Pricing and Subsidy Policies for Bangladesh Agriculture". Research Monograph: 11, Bangladesh Institute of Development Studies, Dhaka, 1990.

[8]   S. Rahman, and G.B. Thapa, "Environmental impacts of technological change in Bangladesh agriculture: farmers' perceptions and empirical evidence". *Outlook on Agriculture*, vol.28, no.4, 1999, pp. 233–238.

[9]   S. M. A. Hossain, and M. A. Kashem, "Agronomic management to combat declining soil fertility in Bangladesh". Paper presented in the 6th Biennial Conference of the Bangladesh Society of Agronomy, held on 29 July 1997 in Dhaka.

[10]   A. M. S. Ali, "Population pressure, environmental constraints and agricultural changes in Bangladesh: examples from three agro ecosystems". Agric. Ecosyst. Environ vol.55, 1995, pp. 95–109.

[11]   S.   Pagiola,   "Environmental   and   Natural   Resource Degradation in Intensive Agriculture in Bangladesh". Document of Agriculture and Natural Resource Operations Division, South Asia Region, World Bank, Washington, DC, 1995.

[12]   K. Ahmad, and S. M. Q. Hasanuzzaman, "Agricultural Growth and Environment". In: Faruquee, R.(Ed.), Bangladesh Agriculture in the 21st Century. University Press Ltd, Dhaka, 1998.

[13]   S. M. A. Hossain, M. U. Salam, and A. B. M. M. Alam, "Farm environment assessment in the context of farming systems in Bangladesh". Paper presented in the Third Asian Farming Systems Symposium on 7–10 November 1994, in Manila.

[14]   M. Asaduzzaman, "Resource degradation and sustainable development in Bangladesh: some preliminary estimates". Paper presented in seminar on planning for sustainable development of Bangladesh, held on 24-25 Sept., 1996 in Dhaka.

[15]   A.A. K. M. Quader, "Strategy for Developing the Fertilizer Sector in Bangladesh for Sustainable Agriculture". Chemical Engineering Research Bulletin 13, 2009, 39-46.

[16]   S. Z. Rashid, "Composting and Use of Compost for Organic Agriculture in Bangladesh". Proceedings of the 4th International Conference for the Development of Integrated Pest Management in Asia and Africa, 20-22 January 2011.

[17]   Ministry of Finance. *Bangladesh Economics Review*. Finance Division, Government of the Peoples' Republic of Bangladesh. Dhaka, Bangladesh, 2013.

[18]   Ministry of Finance. *Bangladesh Economics Review*. Finance Division, Government of the Peoples' Republic of Bangladesh. Dhaka, Bangladesh, 2010.

[19]   Ministry of Finance. *Bangladesh Economics Review*. Finance Division, Government of the Peoples' Republic of Bangladesh. Dhaka, Bangladesh, 2009.

[20]   Bangladesh Bureau of Statistics (BBS). *Statistical Year Book of Bangladesh*. Planning Division, Ministry of Planning, Government of the People's Republic of Bangladesh, 2012.

[21]   Bangladesh Bureau of Statistics (BBS). *Statistical Year Book of Bangladesh*. Planning Division, Ministry of Planning, Government of the People's Republic of Bangladesh, 2007.

[22]   S. K. De Datta, "Principles and practices of rice production". International Rice Research Institute. Los Banos, Philippines. 1981.

[23]   O. C. Bockman, O. Kaarstard, O.H. and I. Richards, "Agriculture and fertilizers". Agriculture group, Norsk Hydro a.s. Osl, Norway, 1990.

[24]   H L S. Tandon, "Fertiliser Guide for Extension Workers, Sales personnel, Studies, Laboratories, Dealers and Farmers". FDCO, New Delhi, India, 1992, p 158.

[25]   A. L. Shah, M.S. Rahman, and M.A. Aziz, "Outlook for fertiliser consumption and food production in Bangladesh". Bangladesh I.*Agric. and Environ*. vol.4, 2008, pp.1-8.

[26]   R. A. M. Titumir, and J. K. Basak, "A Long Run Perspective on Food Security and Sustainable Agriculture in South Asia". Dhaka University, Journal of Development Studies, vol. 1, no. 1, 2010.

[27]   Food and Agriculture Organisation (FAO), Food and Agriculture Organisation of The United Nations Statistics Division. 2009. Available at: http://faostat.fao.org/default.aspx

[28]   M.S. Islam, "Use of Bioslurry as Organic Fertiliser in Bangladesh Agriculture". Prepared for the presentation at the International Workshop on the Use of Bioslurry Domestic Biogas Programme. Bangkok, Thailand, 2006.

[29]   M. Hossain, "Fertiliser consumption, pricing and food grain production in Bangladesh". In: 1987. Fertiliser pricing policy in Bangladesh (Bruce Stone, ed.). IFPRI, Washington D.C. 1987.

[30]   S. M. H. Zaman, "Agronomic and environmental constraints in fertiliser effectiveness". In: Fertiliser pricing policy in Bangladesh (Bruce Stone, ed.). IFPRI, Washington D.C, 1987.

[31] A. Razzak, "Production Status of Organic Manure in Bangladesh". Department of Livestock Services, Dhaka, Bangladesh, 2006.

[32] S. Alam, "Production of organic manure in Bangladesh". Bangladesh Livestock Research Institute's Report, Savar, Dhaka, Bangladesh, 2006.

13

Diagnosis of the Cane Rat (*Thryonomys swinderianus*) Breeding Systems in Ivory Coast

Goué Danhoué[1, 2], Yapi Yapo Magloire[1, 3]

[1]National Polytechnic Institute Felix Houphouët-Boigny of Yamoussoukro, Ivory Coast
[2]Department of training and research in Water, Forests and Environment, Yamoussoukro, Ivory Coast
[3]Department of training and research in Agriculture and Animal Resources, Yamoussoukro, Ivory Coast

Email address:
yapimagloire@yahoo.fr (Y. Y. Magloire)

Abstract: In order to increase animal protein self-sufficiency, the government of Ivory Coast chose a policy of livestock activities diversification including the promotion of mini-livestock such as cane rat husbandry. Today, cane rat breeding has a craze among Ivorian people, but it struggles to really take off. With the aim of contributing to an optimal development of cane rat husbandry in Ivory Coast, we performed a diagnosis of the breeding systems in order to determine the factors that hinder the proper development of this activity. The diagnosis was performed using a survey questionnaire. The survey was carried out using the Participatory Rapid Appraisal Method. Sixty-six farms in 13 administrative Regions of Ivory Coast were investigated. The results showed that most of breeders (55%) were well equipped with livestock buildings in modern materials. However, the animal feeding system was inadequate, characterized by crude protein deficiency. The poor quality of the feed associated with an approximate hygiene management led to high mortality of the animals, reaching over 60% in many farms. The development of complete pelleted diets suitable to the cane rat digestive physiology, with a better control of animal health should allow cane rat breeding to take a jumpstart in Ivory Coast.

Keywords: Breeding System, Cane Rat, Diagnosis, Ivory Coast

1. Introduction

Ivory Coast imports 57% of its consumption of animal protein. To reduce its meat products deficit, this country included in its agriculture development director plan (1995-2015), the diversification of livestock activities with the promotion of mini- livestock such as cane rat breeding. Indeed, the cane rat (*Thryonomys swinderianus*) is a wild animal recently domesticated for meat production in several Sub-Saharan African countries. Bushmeat consumption from poaching is widely practiced throughout Central and West Africa [1]. Rodents including the cane rat are among the most appreciated wild animals for theirs meats [2, 3]. Cane rat consumption was estimated at 80 million animals which represents 300 000 tons of meat consumed per year in the whole West Africa [4, 5]. Cane rat breeding appears as an alternative to poaching and allows the exploitation of a rustic animal species suitable for the local environmental conditions [6, 7]. Several countries in Sub-Saharan Africa, including Ivory Coast, then began promoting cane rat breeding among their populations. This type of animal production seems to be a favorable element for sustainable development, in terms of animal resources management in Sub-Saharan Africa.

In Ivory Coast, cane rat breeding was introduced since 1995, by the Wildlife and Protected Areas School of Bouaflé. Today, cane rat breeding knows a real craze among the Ivorian people, but it struggles to really take off. Since the viability of cane rat farms is conditioned by a good control of animal husbandry techniques [8], we carried out a diagnosis of cane rat breeding systems to understand the modes of livestock management with the aim of determining the factors that hinder the proper functioning of this activity in Ivory Coast. This study should advocate recommendations for improving the breeding systems for a rational and sustainable management of captive cane rat.

2. Material and Methods

2.1. Data Collection

A survey questionnaire was used in this study. The questionnaire consisted of six sections (location of the farms, breeders and their exploitations, infrastructure and equipment, feeding system, zootechnical parameters and health status). Thus, the farms were localized according to the administrative region. Information on the socio-professional characteristics of breeders, on the characteristics of the farms and on the livestock management mode were collected. The aspects of animal housing and breeding materials have been described. The feeding system has been studied through questions concerning the nature and the mode of feed distribution to the animals. Data on animal reproduction were collected. Health aspects were addressed through studying mortality, disease and hygiene in farms.

The survey took place from November, 1999 till December, 2001. After a break, due to sociopolitical disorders intervened in Ivory Coast in 2002, the survey was continued, completed and updated until 2007. The investigation was made with two main phases (the pre-survey and the survey itself. The pre-survey was made according to an approach by successive stages which led us to the census of cane rat farms, the elaboration of survey forms and the establishment of a tour program in the farms. The survey was performed using the Participatory Rapid Appraisal Method [9]. It is a semi-open questionnaire survey with direct and semi-direct, individual or collective interviews, with breeders and various actors of the cane rat breeding sector. These interviews were completed by retrospective data and direct observations taken by photos. Thus, management monitoring sheets available in some farms have been exploited. On 105 listed farms, 66 having a staff of animals upper or equal to 5, distributed over thirteen administrative Regions of Ivory Coast were investigated.

2.2. Data Analysis

The characteristics of the production actors, the production structures and animal feeding system were studied, according to a descriptive analysis using parameters such as frequency, sums, averages and standard deviations. The parameters related to reproduction and animal health were subjected to one - factorial ANOVA with the administrative Region as the factor. When significant differences were identified, the means were compared using the Tukey's test. For the statistical analysis, the studied effects were considered significant at the significance level of 5%.

3. Results

3.1. Characteristics of the Production Actors

In Ivory Coast, cane rat breeding is practiced by people of various socio-professional groups (Fig. 1). The farmers represented more than half (56%) of breeders. Salaried people composed of state employees, officials and private workers were in second position (27%). The retirees represented 7% followed by the housewives and the removed from school (5%). The liberal professions represented only 3% of cane rat breeders.

Furthermore, cane rat breeders were predominantly men (90%). The average age of breeders was 46 years, the youngest being 16 years old and the oldest 72 years. Three types of workers were involved in livestock management: family labor, community labor and hired labor. Seventy-three percent (73%) of breeders used family labor. In this case, the farmer was responsible for livestock inspection; he was the only one who decides on the course of the activities. Community workforce (11%) was used in farms belonging to professional agricultural organizations. These groups were often informal and the execution of activities was not always well coordinated. Hired labor (16%) was observed in the farms of some economic operators. Due to the high demand and the high prices of cane rat meat in Ivory Coast, the profitability of the farm was the main motivation for many breeders (49%). For 47% of breeders, diversify livestock activities was the main motivation that led them to cane rat breeding.

3.2. Characteristics of the Production Structures

The site of the exploitation and the land status of the breeder conditioned the type of infrastructures implanted. Two categories of infrastructure were observed: definitive establishments (55%) and temporary implantations (45%). Thus, different building materials were observed, from the most modern and definitive, to the most rudimentary and temporary. Furthermore, livestock housing observed in our study could be divided into two groups: those built specifically for livestock (74%) and those constructed for other purposes and refitted to house cane rats (26%). Two types of breeding cage were observed in our study: soilless cages (3%) and floor pens (97%).

3.3. Feeding System

Table 1. Green fodder and their utilization in the farms.

Types of green fodder	Utilization in the farms (%)
Pennisetum purpureum	97.7
Panicum maximum	33.3
Stem and leaves of cassava	20.7
Stem and leaves of corn	18.2
Palm ribs	12.1
Andropogon gayanus	7.3
Coconut ribs	6.0
Imperata spp., Stems and leaves of sorghum	4.8
Pineapple, papaya, rice straw, sugarcane, Teophrosia sp	2.4
Pueraria, Brachiaria	1.2

Cane rats were fed on green fodder. Pennisetum purpureum was used as staple diet in 97.7% of the surveyed farms and Panicum maximum in 33.3% of the farms (Table 1). Forage was accompanied by a varied range of products including agricultural by-products, agro-industrial products and artisanal transformation products (Tables 2). Corn and cassava were used as energy supplements in 100% and 91.4% of the farms, respectively. The meal of dried leaves of Leucaena sp. was used as nitrogen supplement in 100% of the farms while the snail shell powder or calcined bone were used in 96.3% of the farms as mineral supplement.

Table 2. Ingredients of concentrate feed and its utilization in the farms.

Concentrate feed components	Utilization in the farms (%)
Corn kernels, Leucaena sp.	100
Table salt	98.7
Powder snail shell or calcined bones	96.3
Cassava chips	91.4
Corn bran	18.2
Powder calcined oyster shell	12.1
green Papaya	9.7
Palm seed	8.2
Refusal of attiékié, peanuts (grain, leaf)	4.8
Spent grains,	3.6
Soybean (grains, root, stem)	3.6
Dried yam peel, remaining cooking, gardening waste	2.4
Sugar cane molasses, wheat bran, taro, sweet potato, termite soil, fresh cassava, fish meal, laying food	1.2

3.4. Reproductive Parameters and Animal Health

It emerges from Table 3 that cane rat farms from the N'Zi Comoé, Marahoué, Savanes, Montagnes and Lagunes Regions had identical fertility rates (P> 0.05) ranging from 73-77% but significantly higher (P <0.05) compared to the fertility rates of the farms located in the Lacs, Fromager and Moyen Comoé Regions which were around 50%. Cane rats of the farms from the Lagunes Region had a 92% fertility rate, higher (P <0.001) than those of the farms from the Lacs, Marahoué, Savanes and Moyen Cavally Regions which varied from 80% to 84%.

In reproductive cane rats, the mortality rate in the Regions of Moyen Cavally, Lacs and Marahoué varied between 58% and 61%. It was higher (P <0.01) than those in the other Regions including the Agneby Region which recorded the lowest value of mortality rate (23%). The mortality of the young cane rats in the Region of Fromager (100%) was higher compared to the other Regions (P <0.01), where the mortality rates ranging from 28% to 41%. Mortalities were due mainly to the poor management of livestock and poor hygiene (35% mortality). The most encountered affections were alopecia, paralysis, stomach bloating and swelling of the throat (Fig. 2).

Table 3. Reproductive performance.

Regions	Fertility (%)	Fecundity (%)	Prolificacy
Vallée du Bandama	69.99 ±4.20 de	87.06 ±1.52 ef	4.18 ± 0.14 c
N'Zi Comoé	76.25 ± 23.77 d	89.96 ± 2.35 ef	4.10 ± 0.62 c
Lacs	49.83 ± 15.22 b	80.52 ± 6.32 d	6.46 ± 0.72 e
Marahoué	73.43 ± 9.98 d	80.52 ± 2.21 d	4.51 ± 0.75 c
Worodougou	00 a	00 a	00 a
Savanes	75.00 d	81.25 d	4.33 c
Montagnes	77.08 ± 1.80 d	69.27 ± 1.19 c	3.60 ± 0.11 b
Fromager	50.00 b	70.83 c	5.67 d
Agneby	60.68 ± 12.08 c	87.23 ± 10.95 ef	3.54 ± 0.14 b
Moyen Cavally	64.58 ± 2.08 e	83.75 ± 8.75 d	5.95 ± 1.45 de
Lagunes	74.26 ± 6.66 d	92.03 ± 11.43 f	4.37 ± 0.43 c
Sud Comoé	58.82 ± 29.41 c	75.29 ± 37.65 c	6.40 ± 3.20 e
Moyen Comoé	53.82 ± 13.90 b	55.47 ± 14.84 b	4.10 ± 0.05 c

a, b, c, d, e, f : For the same column, the values followed by the same letter are not significantly different at 5% level.

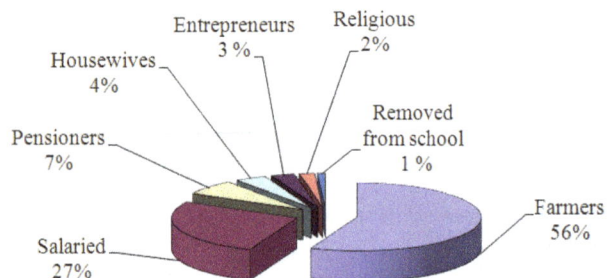

Figure 1. Socio-professional categories involved in cane-rats husbandry in Ivory Coast.

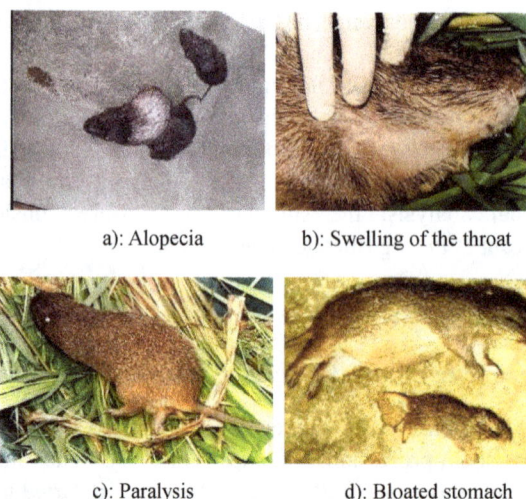

a): Alopecia   b): Swelling of the throat

c): Paralysis   d): Bloated stomach

Figure 2. (a, b, c, d) Main disorders encountered in the cane rat in Ivory Coast.

4. Discussion

The results obtained during our study showed that the socio-professional groups involved in cane rat breeding in Ivory Coast were diverse as in Benin [10]. Nevertheless, the majority of cane rat breeders (56%) were farmers, probably because the original purpose of the extension agents of cane rat breeding was to introduce this activity into the agricultural production systems. Cane rat breeding should constitute a generative sideline of income for farmers.

As for production structures, this study revealed that the majority of livestock buildings (55%) were built in modern materials. This was due to the fact that several breeders have received funding for installation through various development projects. The ANADER (the National Agency for Rural Development Support in Ivory Coast) contributed to the installation of many cane rat farms in Ivory Coast through various development projects with the National Programme for the Promotion of cane rat breeding [11].

Furthermore, the analysis of the feeding system in cane rat farms in Ivory Coast revealed two main information. The first is that cane rats were fed a forage based diet. Elephant grass (Pennisetum purpureum) was the most used forage. This grass was used in 97.7% of the farms against 33.3% for Guinea grass (Panicum maximum). Apart from these two forages, there are other forages used secondarily, especially when the first ones become scarce during the dry seasons. In Benin, Pennisetum purpureum and Panicum maximum are also the most used fodder in cane rat feeding ([10, 12]. The second information revealed by the analysis of the feeding system is that protein concentrates were absent in the diet, due to their high costs, probably. Crude protein intake was provided by the dried leaves of Leucaena sp. Indeed, the dietary supplement of the cane rat vulgarized by the supervision structures in Ivory Coast contains 8-10% dried leaves of Leucaena sp. This proportion corresponds to approximately 10 g flour dried leaves of Leucaena sp per animal and per day [11]. Leucaena sp contains only 23.5% crude protein per kg (dry matter basis). Thus, crude protein level observed in the diet of the cane rat is between 8 and 12% of the feed [3, 13], which is not able to cover animals nitrogen requirements. For comparison, the optimum rate of crude protein in the diet of the reproductive rabbits is between 17 and 18% [14]. This study showed that none of cane rat breeders in Ivory Coast uses complete pelleted diets in his farm as practiced for rabbit breeding. In fact, complete diets for cane rat do not yet exist on the market. Several studies have been carried out in order to develop such feed but nutritional requirements of the cane rat remain largely unknown. Concerning dietary fibres for instance, recent studies showed that the cane rat can digest relatively high amounts of fibres [15], due to its well developed caecum [16], but the optimum of fibres levels in the diet suitable to each physiological stage of this animal remain largely unknown.

Regarding animal health, the mortality rates recorded in the cane rat farms in our study were high. They ranged from 23% to over 60% in adult animals. In young animals, they were greater, particularly, reaching 100% in some farms. These mortality rates was very high compared to those reported by others authors in Gabon and Benin [12, 17, 18]. These authors have observed a global mortally ranging between 10 and 26 %. Only the pre-weaning mortality was high reaching 40% according to these authors. The high mortality observed in our study could be explained by the poor quality of the diets, in addition to the poor livestock management and the lack of hygiene. Furthermore, reproductive performance recorded in the animals in our study showed a high variability between the administrative Regions. Such variability can be attributed to the feeding system and the livestock management in general as zootechnical parameters related to cane rat are experiencing significant variability and are strongly linked to a rigorous breeding behavior, an improved nutrition and an accurate genetic selection [19]. However, reproductive performances recorded in the cane rat in our study are acceptable, compared to those reported by other studies. Indeed, some authors [12] recorded fertility rate between 53-66% in cane rat in Southeast Benin. Thus, despite cane rat diets characterized by crude protein deficiency, animals gave relatively good reproduction performance in our study. It has been reported that in some rodents consuming low-nitrogen diets, bacterial nitrogen fixation can be a mechanism for nitrogen supplementation [20]. For instance, it has been suggested the possibility of fixation of atmospheric nitrogen by certain caecal bacteria in voles and in the European beaver. Such fixed nitrogen is utilized nutritionally via coprophagy by these animals [20, 21, 22]. This might be the case of the cane rat, since this animal practices coprophagy and a recent study [15] revealed the great richness in Spirochaetaceae of its microbiota. In the Spirochaetaceae family, some bacterial species belonging to Treponema genus, already described in termites are able to fix atmospheric nitrogen [23, 24, 25]. These bacterial species play an important role in nitrogen nutrition of wood-eating termites, allowing these animals to meet their nitrogen requirements from quite poor food containing about 0.05% nitrogen [25]. It would be interesting if some studies are carried out to determine whether certain bacterial species in the caecum of the cane rat were involved in nitrogen nutrition of this animal, which could explain the good reproduction performance observed in our study, despite animals low nitrogen diets.

5. Conclusions

Our study showed that cane rat breeding is a model of mini- livestock fully adopted by the Ivorian population but some factors such as unavailability of quality diets constitutes a real obstacle. The development of complete pelleted diets suitable to the cane rat digestive physiology, with a better control of animal health should allow cane rat breeding to take a jumpstart in Ivory Coast.

References

[1]   Chardonnet, P., Fritz H., Zorzi, N. and Féron, E. 1995. Current importance of traditional hunting and major constrasts in wild meat consumption in sub-saharan Africa. In Bissonette J.A. and Kraussman P.R. (eds), Integrating people and wildlife for a sustainable future. The Wildlife Society, Bethesda, USA. pp. 304-307.

[2]   Hardouin, J., 1995. Minilivestock : From gathering to controlled production. *Biodiversity and Conservation* 4 : 220-232.

[3]   Van, Zyl A., Meyer, A.J. and Van der Merwe, M. 1999. The influence of fibre in the diet on growth rates and the digestibility of nutrients in the greater cane rat (Thryonomys swinderianus). *Comparative Biochemistry and Physiology*. Part A 123 : 129 – 135.

[4]   Mensah, G.A. 1993. Futteraufnahme und verdaulichkeit beim grasnager (Thryonomys swinderianus). PhD thesis. Institut 480, Université de Holenheim, Allemagne.107 p

[5]   Fantodji, A. and Mensah, G.A. 2000. Rôle et impact économique de l'élevage intensif de gibier au Bénin et en Ivory Cost. In: Actes Séminaire international sur l'élevage intensif de gibier à but alimentaire en Afrique. Libreville (Gabon) 23-24 mai 2000. Projet DGEG/VSF/ADIE/CARPE/UE. pp. 25-42.

[6]   Jori, F., Mensah, G.A. and Adjanohoun, E. 1995. Grasscutter production : an example of rational exploitation of wildlife. *Biodiverity and Conservation* 4 : 257-265.

[7]   Hardouin, J. and Thys, E. 1997. Le mini-élevage, son développement villageois et l'action de BEDIM. *Biotechnologie, Agronomie, Société et Environnement* 1(2) : 92-99.

[8]   Mensah, E. R. C. K. D., Mensah, R. M. O. B. A., Pomalegni, S. C. B, Mensah, G. A., Akpo, P. J. E. and Ibrahimy, A. 2011. Viabilité et financement des élevages d'aulacode (Thryonomys swinderianus) au Bénin. *International Journal of Biological and Chemical Sciences* 5(5) : 1842-1859.

[9]   Theis, J. and Grady, H.M. 1991. Participatory rapid appraisal for community development. A training manual based on experiences in the Middle East and North Africa. International Institute for Environment and Development. London, England, 150 p.

[10]   Mensah, G. A. and Ekué, M. R. M., 2003. L'essentiel en aulacodiculture. C.B.D.D./NC-IUCN/KIT/RéRE, République du Bénin/Royaume des Pays-Bas., 168 p.

[11]   Fantodji, A. and Soro, D., 2004. Elevage des aulacodes : expérience en Ivory Cost. Guide pratique. Agridoc. Paris : les editions du Gret., 133 p.

[12]   Adjahoutonon, K.Y.K.B., Mensah, G.A. and Akakpo, A.J. 2007. Evaluation des performances de production des élevages d'aulacodes installés dans le sud-est du Bénin. *Bulletin de la Recherche Agronomique du Bénin* 56 : 36-45.

[13]   Traoré, B. 2010. Analyse de quelques activités enzymatiques digestives et influence d'aliments complets granulés sur des performances zootechniques de l'aulacode (Thryonomys swinderianus) d'élevage. PhD thesis, Université Abobo-Adjamé, Côte-d'Ivoire, 201 p.

[14]   Lebas, F., 1989. Besoins nutritionnels des lapins. Revue bibliographique et perspectives. *Cuni-Sciences* 5, 1-28.

[15]   Yapi, Y, M. 2013. Physiologie digestive de l'aulacode (Thryonomys swinderianus) en croissance et impact des teneurs en fibres et céréales de la ration sur la santé et les performances zootechniques. PhD thesis, Université de Toulouse, France, 226 p.

[16]   Yapi, Y.M., Gidenne, T., Farizon, Y., Segura, M., Zongo, D. and Enjalbert, F., 2012. Post-weaning changes in the digestive physiology and caecal fermentative activity in the greater cane rat (*Thryonomys swinderianus*). *African Zoology*, 47(2): 311–320.

[17]   Jori, F. and Chardonnet, P. 2001. Cane rat farming in Gabon. Status and perspectives. 5th International Wildlife Ranching Symposium, Pretoria, South Africa, march 2001.

[18]   Jori .F, Cooper, J.E. and Casa,l J. 2001. Postmortem findings in captive cane rats (Thryonomys swinderianus) in Gabon. *Veterinary Record*, 148: 624-628.

[19]   Jori, F. 2001. La cria de roedores tropicales (*Thryonomys swinderianus* y *Atherurus africanus*) como fuente de alimento en Gabon, Africa central. PhD thesis. Facultat de Veterinaria Universitat autonoma de Barcelona.150 p.

[20]   Varshavskii, A.A., Puzachenko, A.Y., Naumova, E.I., Kostina, N.V., 2003. The enzymatic activity of the gastrointestinal tract microflora of the greater mole rat (Spalax microphtalmus, Spalacidae, Rodentia). *Dokl Biol Sci* 392:439–441

[21]   Meshcherskii, I.G., Naumova, E.I., Kostina, N.V., Varshavslii, A.A., Umarov, M.M., Ylur'eva, O.S., 2004. Effect of deficiency of dietary nitrogen on cellulose digestibility and nitrogen-fixing flora activity in the sibling vole (*Microtus rossiaemeridionalis*). *Biol Bull* 31:457–460.

[22]   Vecherskii, M.V., Naumova, E.I., Kostina, N.V., Umarov, M.M., 2009. Assimilation of biological nitrogen by European beaver. *Biol Bull* 36:92–95.

[23]   Lilburn, T.G., T.M. Schmidt, and Breznak, J.A. 1999. Phylogenetic diversity of termite gut spirochaetes. *Environmental Microbiology*, 1: 331–345.

[24]   Lilburn, T.C., Kim, K.S., Ostrom, N.E., Byzek, K.R., Leadbetter, J.R., Breznak, J.A., 2001. Nitrogen fixation by symbiotic and free-living spirochetes. *Science*, 292:2495–2498

[25]   Breznak, J.A. 2002. Phylogenetic Diversity and Physiology of Termite Gut Spirochetes. *Integrative and Comparative Biology*, 42: 313–318.

14

Rural Households Willingness to Pay for Small Ruminant Meat in South-Western Nigeria

Otunaiya Abiodun Olanrewaju[1, *], Ologbon Olugbenga A. Chris[1], Adigun Grace Toyin[2]

[1]Department of Agricultural Economics and Farm Management, College of Agricultural Sciences, Olabisi Onabanjo University, Yewa Campus, Ayetoro, Ogun State, Nigeria

[2]Department of Agricultural Economics and Extension, College of Agricultural Sciences, Landmark University, Omu-Aran, Kwara State, Nigeria

Email address:

aootunaiya@gmail.com (A. O. Otunaiya)

Abstract: Subsistent level of indigenous small ruminant meat production in Nigeria limits its supply and consequently, accounted for its high prices. The study examined the willingness to pay for indigenous small ruminant meat in Ijebu division of Ogun state, Nigeria. A multistage sampling technique was used to select 120 rural households used for the study. The results of the descriptive and Logit regression analyses revealed that rural households head in the study area are mostly females, attained appreciable level of literacy with a means of livelihood and mostly in their middle age. These rural households consumed small ruminant meat regularly, well educated about the nutritional value of the ruminant meat, and mostly, willing to pay the market price of small ruminant meat whenever they are ready to consume the meat. Factors influencing their willingness to pay for small ruminant meat are age of the household head, occupation, distance from market, and price of small ruminant meat. The study recommends the establishment of slaughter houses and small ruminant meat market to consumption areas and in various villages to broaden the market as well as providing improved market access to producers.

Keywords: Small Ruminant, Meat, Consumption, Logit, Nigeria

1. Introduction

Food production in the form of meat, milk and other livestock products constitute a major group of livestock outputs. Indigenous sheep and goats are among the most important species of livestock (Shalander and Pant, 2002). About 14% of the total value of livestock output was contributed by small ruminants (Degefu, 2003). Small ruminant production in sub Saharan Africa is an important aspect of the livestock economy; where more than 80 % of rural families especially women and children keep sheep and goats (Jaitneret al., 2001; Kumar et al., 2003; Boyazoglu et al., 2005). In Nigeria, rural households keep 37.4%and 8.3% of goats and sheep respectively with an average number per owner being 6.5 sheep and5.2 goats (International Livestock Centre for Africa, ILCA, 1991).

The goat being an animal that survives most ecological zones is considered to be one of the most important protein producing animals and they provide 30-36% of the total meat consumption of the Nigerian populace annually (Gomna & Rana, 2007; York & Gossard, 2004)

In Nigeria small ruminants contribute an estimated 35% to the total meat supply; they are more important in the north than in the south, and more important in rural than in urban areas. Population estimates suggest there are roughly 34.5 million goats and 22.1 million sheep in the sub-humid zone of Nigeria. The major breed of sheep is the Yankasa; the West African Dwarf is the major goat breed (Nwosu, Madu, & Richards, 2007). Due to poor management, inbreeding and inadequate nutrition, these goats are usually predisposed to a range of health problems (Kathiravan, Thirunavukkarasu, &Michealraj, 2007).

The level of small ruminant meat consumption is expected to increase geometrically and double the current consumption level by the year 2020 (De Vries, 2008; Juma, et.al 2010). Structural change in the economy as a result of increases in urbanization, population and income growth will promote the demand for small ruminant meat; hence, create markets for animal products and encourage commercialization of livestock production (Idowu, et.al., 2012). The extent of this commercialization depends on the consumption of the products by consumers (Otunaiya & Shittu, 2014). Meat consumption behavior is the deciding factor for the development of the livestock sector in general and small ruminants in particular (Thammi and Suryanarayana, 2005). Consumer tastes and preferences are reflected in the market. These are revealed through purchase decisions and price premiums that consumers pay for both visible (Langyintuo et al., 2004) and invisible characteristics of meat.

In most part of Nigeria, small ruminant are raised at subsistence scale, hence, resulting in inadequate quantity of sheep and goat meat in the market, especially during festival. This shortage in supply accounted for the high price of small ruminant meat when compared with the prices of its substitutes. This study, therefore, examined the small ruminant consumption and the willingness to pay among respondents in Ijebu division of Ogun State, Nigeria. The study specifically describes the socio economic characteristics of respondents in the study area and examines the determinants of respondents' willingness to pay for small ruminant meat.

2. Methodology

The study area was carried out in Ijebu division of Ogun State, Southwest geo-political zone of Nigeria. Ijebu division is one of the four divisions of the State; it comprises of five Local Government Areas (LGA). The Ijebu tribes inhabit the South-Central part of Yoruba land - a territory that is bounded in the North by Oyo State, in the East by Ondo State, and the West by Egba land.

Primary data was collected from the respondents with the aid of well structured questionnaire. Multi-stage sampling technique was employed for the purpose of this study. In the first stage, three (3) Local Government Areas were randomly selected from the division; namely: Ijebu North East LGA, Ijebu –Ode LGA and Ijebu North LGA. In the second stage, three (3) Villages/Towns from each of the selected LGA were randomly selected; namely: Ijebu North East LGA (Ilese, Atan, and Isonyin), Ijebu–Ode LGA (lgbeba, Obalende, and Olisa) and Ijebu North LGA (Ijebu Igbo, Oru and Ago-lwoye). In the final stage, households were sampled proportional to the population of the village/town. The procedure gave a total sample size of one hundred and twenty (120) respondents used in the study.

2.1. Analytical Procedure

Descriptive statistics and economic model were used for the purpose of the study. Descriptive statistics such as percentages, means and frequency were used to discuss the socio-economic characteristics of the respondents. The empirical analysis utilized a contingent valuation dichotomous choice methodology. A single-bounded logit model was used to explore factors affecting willingness to pay for indigenous ruminant meat. The mean price that households were willing to pay for indigenous ruminant meat was also estimated. A consumer's decision to continue buying indigenous ruminant meat (willing to pay) or not to buy given some increase in price is a binary or dichotomous mode, in which probabilistic modeling frame work can be applied. In this case, both logit and probit models can be used. Theory states that the difference in the models is related to the assumption of the error term distribution and that both yields roughly the same results for the variables of interest. Independent distributed error term in logit regression assumed to be normally distributed with zero mean and constant variance. (Hosmeret.al. 2013).Consequently a logit model was used to examine the factors affecting willing to pay for indigenous ruminant. The logit model was implicitly specified as follows:

2.2. Logit Model

A logistic regression model allows us to establish a relationship between a binary outcome variable and a group of predictor variables. It models the logit-transformed probability as a linear relationship with the predictor variables. More formally, let y be the binary outcome variable indicating unwillingness to pay/willingness to pay for indigenous small ruminant with 0/1 and p be the probability of y to be 1, $p = prob(y = 1)$. Let $X_1, X_2, . X_k$ be a set of predictor variables. Then the logistic regression of y on $X_1, X_2, . X_k$ estimates parameter values for $\beta_0, \beta_1 ... \beta_k$ via maximum likelihood method of the following equation.

$$logit(p) = log(p/(1-p)) = \beta_0 + \beta_1 X_1 + \beta_k X_k + \varepsilon_i \quad (1)$$

In terms of probabilities, the equation above is translated into

$$p = exp\ (\beta_0 + \beta_1 X_1 + \beta_k X_k)/1 + exp(\beta_0 + \beta_1 X_1 + \beta_k X_k) + \varepsilon_i \quad (2)$$

where:
  y_i = Willingness to pay (1= if willing, 0 = otherwise)
  X_1 = Level of education (in years)
  X_2 = Age (in years)
  X_3 = Household income (N/per month)
  X_4 = Dependency ratio (Ratio of non-working members to household size)
  X_5 = Marital Status (Married =1; Others = 0)
  X_6 =Sex (Male =1; Female = 0)
  X_7 = Consumer Preference (Small ruminant =1; others = 0)
  X_8 = Religion
  X_9 = Occupation (Livestock farming =1; others = 0)

  X_{10} = Distance from market (Km)
  X_{11} = Availability of ruminant meat (Available within community =1; Not available = 0)
  X_{12} = Small ruminant price (N/ kg)
  X_{13} = Household size (Number)
  X_{14} = Small ruminant meat quality perception (Better than other meat =1; otherwise = 0)
  β_i = Unknown parameters to be estimated
  ε_i = Independent distributed error term

3. Results and Discussion

3.1. Socio-Economic Characteristics of Rural Farm Households

The study examines the socio-economic characteristics of the sample households both in terms of the characteristics of the household heads, household composition, living conditions and access to basic social amenities.

3.2. Household Characteristics

Table 1. Distribution of sampled households by Socio-economic Characteristics.

Age-Grouped	Frequency	Percentage
Above 30 years	29	24.2
31-40 years	36	30.0
41-50 years	50	41.7
51-60 years	5	4.2
Sex		
Male	47	39.2
Female	73	60.8
Marital status		
Married	90	75.0
Single	30	25.0
Educational status		
Secondary school	44	36.7
O.N.D/H.N.D	33	27.5
University	43	35.8
Religion		
Christians	77	64.2
Muslim	43	35.8
Main occupation		
Banker	14	11.7
Business	15	12.5
I T student	4	3.3
Civil servants	26	21.7
Teacher	15	12.5
Farmer	8	6.7
Marketer	1	.8
Clerk	3	2.5
Receptionist	2	1.7
Trader	24	20.0
Butcher	1	.8
Cleaner	3	2.5
Tailor	4	3.3
Experienced		
Below 5 years	112	93.3
6-10 years	6	5.0
11-15 years	2	1.7
Family size		
1-3	32	26.7
4-6	70	58.3
7-9	18	15.0
Total	120	100.0

Table 1 presents the distribution of the sampled households by size as well as personal characteristics of the household heads; including age, sex, education level, marital status, religion, and main occupation. As shown in the table, a typical household in the sample is made up of 4-6 members. It is also worthy of note that most (93.3 per cent) of the household heads do not have farming as their main occupation.

In terms of age, evidence on Table 1 shows that majority (41.7 per cent) of the sampled households have the age of their household heads falling between 41 and 50 years. As much as 4.2 per cent were aged, while youths (30 years or younger) featured less prominently (24.7 per cent) among the respondents. This confirms the commonly reported aging rural farm population in Nigeria (Bah, et. al. 2003), and suggests that availability of off-farm livelihood options might be necessary to retain youths within the rural farm sector.

Table 1 further shows that 60.8 per cent of the heads of the households were females and majority (75.0 per cent) were married. This implies that most the household heads in the study area are either divorced or separated mothers. Meanwhile, all the household heads (100 per cent) possess some formal education, which is predominantly at the primary and secondary level; an appreciable level of literacy exists among the respondents. This is expected to enhance their ability to take full advantage of extension services, thus affecting their income generation and reduce their poverty.

3.3. Households Living Conditions and Access to Basic Amenities

Access to basic social amenities and living conditions are important indicators of household welfare and poverty status. Thus, this study examined these key indicators among the sampled rural farm households. The results are summarized in Table 2.

As shown in Table 2, majority of the sampled households (55%) used both Ibadan Electric Distribution Company (IBEDC) power supply and generating set; 45% depends on IBEDC alone as the source of power. This also have implications on the level of value addition that is possible in the rural farm sector, given that most agro-processing activities requires stable electricity supply. The results in Table 2 further show that majority of the survey communities have no access to tap water but depend mainly on borehole. Majority of the respondents 56.7% used kerosene as their cooking fuel. The result also shows that most of the sampled households, in the study area, live in houses with water closet system of toilet. Majority 62.5% make use of private hospital for their health treatment and made use of designated refuse dump site to dispose their refuse.

Table 2. Distribution of households by access to social amenities and living conditions.

Electric source	Frequency	Percent
IBEDC	54	45.0
Both generator and IBEDC	66	55.0
Water sources		
Well	6	5.0
Pipe borne water	18	15.0
Borehole	64	53.3
Pipe borne water and borehole	24	20.0
Well and borehole	1	0.8
Well and pipe borne water	7	5.8

Electric source	Frequency	Percent
Cooking fuel		
Kerosene	68	56.7
Charcoal	3	2.5
Gas	5	4.2
Kerosene and gas	19	15.8
Kerosene, gas and electricity	16	13.3
Kerosene and charcoal	8	6.7
Charcoal, gas, electricity and kerosene	1	0.8
Toilet		
Bush toilet	2	1.7
Pit latrine	29	24.2
Water closet	89	74.2
Treatment access		
Primary health clinic	25	20.8
Private hospital	75	62.5
Traditional healing	5	4.2
Self medication/chemist shop	3	2.5
Self treatment with herbs	4	3.3
Primary health clinic and private hospital	5	4.2
Private hospital and self treatment with herbs	3	2.5
Waste disposal means		
Bush disposal	29	24.2
Refuse dump	44	36.7
Government waste disposal system	34	28.3
Bush disposal and refuse dump	4	3.3
Bush disposal and government waste disposal	1	0.8
Refuse dump and government waste disposal	8	6.7
Total	120	100.0

Table 3. *Distribution of Households by their Attitude to Small Ruminants Consumption.*

Attitude to Small Ruminant Consumption	Frequency	Percent
Do you consumed small ruminant meat		
Yes	111	92.5
No	9	7.5
Why Do You Consumed Small Ruminant Meat		
no response	9	7.5
Taste	40	33.3
Flavor	17	14.2
Nutritional value	54	45.0
Are you aware of the nutritional value of small ruminant meat		
Yes	112	93.3
No	8	6.7
Are you willing to pay the market price of small ruminant meat when ready to consumed		
Yes	107	89.2
No	13	10.8
In case your income increases will you consumed more of small ruminant meat		
Yes	105	87.5
No	15	12.5
Do you have any close substitute you do use instead of small ruminant meat		
Yes	110	91.7
No	10	8.3
Total	120	100.0

The above table (Table 3) shows that majority of the respondents 92.5% regularly consumed small ruminant meats. About 93.3% of the respondents were aware of the nutritional value of the meat but only about 45% of those who regularly consumed the meat did so because of its nutritional value. About 89.2% of the respondents are willing to pay the market price of small ruminant meat whenever they are ready to consume the meat. Majority of the respondents (91.7%) had no substitute for small ruminant meat; but only about 87.5% of the respondents considered the meat as normal goods; while it was inferior goods to about 12.5% of the sample.

Table 4. *Proportion of expenditure/income that goes into small ruminant's consumption.*

	Mean Expenditure on small ruminants N	Proportion of expenditure/income that goes into small ruminant's consumption (%)
Mean expenditure on small ruminant meat	9810.72	14.98
Mean income of the households	65471.67	

Table 4 shows the ratio of respondents' expenditure on small ruminant to the income. The result shows a ratio of 0.1498. This implies that consumers of small ruminant in the study area spend as much as about 15% of their total income to purchase small ruminant meat.

3.4. Determinants of Willingness to Pay for Small Ruminant Meat

This section presents results of Logit regression analysis of the determinants of the willingness to pay for small ruminant meat among households in the study area. The results are presented in Table 5.

The results of the maximum likelihood coefficients from the logit estimation (table 5) indicate that four (4) out of the 14 variables included in the model have a statistically significant influence on the probability of willingness to pay for small ruminant meat. These variables are age of the household head, occupation, distance from market, and price of small ruminant meat.

The coefficient of age (0.121) is positive and significant at $p < 0.05$. This shows that advancement in age of a household head tends to increase the likelihood of the household's willingness to pay for small ruminant meat. The marginal effect of one year advancement in the age of a household head on willing to pay for ruminant meat is 0.20. The coefficient representing the occupation of the respondent (-0.18) is negative but significant at $p < 0.05$. This shows that the likelihood of a household's willingness to pay for small ruminant meat reduces with increase involvement of the household in livestock farming.

Table 5. Logit regression result of willingness to pay for small ruminant meat.

Variable	Parameters	t value	Marginal effect
Constant	3.37	0.70	0.55
Educational level	-0.15	-0.94	-0.25
Age (year)	0.12**	2.13	0.20
Household income	-0.86	-0.35	-0.14
Dependency ratio	-1.37	-1.58	-0.22
Marital status	1.10	1.20	0.18
Sex	0.27	0.45	0.44
Consumer preference	-0.79	-0.01	-0.13
Religion	-0.63	-1.04	-0.10
Occupation	-0.18**	-2.15	-0.30
Distance from market	-0.91***	-3.94	-0.15
Availability of ruminant meats	-0.16	-0.02	-0.27
Price of ruminant meat	0.42**	1.97	0.70
Household size	-0.18	-1.09	-0.30
Quality of ruminant meat	-0.94	-0.95	-0.15
Sigma	0.15		
Log likelihood function	59.85		

Note: ***, ** and * denotes p-value at 1%, 5% and 10% level respectively

The coefficient of the distance from respondent's home to the market (-0.91) is also negative and significant (p<0.05) in explaining the probability of willingness to pay for small ruminant meat. This implies that the longer the distance a household will cover before buying small ruminant meat, the lower the likelihood of household (-0.15) to be willing to pay for small ruminant meat. The coefficient of the current price of small ruminant meat (0.42) is positive and significant (p<0.05) in explaining the probability of willingness to pay for small ruminant meat. This implies that the higher the current prices at which household are buying small ruminant meat, the higher the likelihood of household to be willingness to pay for small ruminant meat. This result is contrary to the a-priori expectation of change in quantity demanded to a change in price of normal goods. Also, contrary to results from similar study (Jumaet.al, 2010) who reported a negative coefficient for price of small ruminant in Kenya; this study shows that factors other than the current price of small ruminant meat influence the demand of the meat in the study area. Hence, small ruminant meat can be said to be luxury goods.

4. Conclusion and Recommendations

The study concludes that most rural households in the study area are headed by females. The household heads have an appreciable level of literacy with a means of livelihood and mostly in their middle age. These rural households consumed small ruminant meat regularly, well educated about the nutritional value of the ruminant meat and mostly willing to pay the market price of small ruminant meat whenever they are ready to consume the meat. Factors influencing their willingness to pay for small ruminant meat are namely: age of the household head, occupation, distance from market, and price of small ruminant meat. Based on the fore going, the following recommendations were made:

The study results indicate that household consuming small ruminant are mostly headed by women who will not necessarily

turn to a substitute when the price of small ruminant meat increases. This would mean that consumption of ruminant meat is an issue of preference and not price change. This would mean that there is a niche market for small ruminant meat. More so, the distance to market play a significant role in the willingness to pay for small ruminant meat. Therefore there is a need to properly identify and establish this market niche and assist both producers and consumers to access it. This would help improve the marketing of small ruminant meat product, providing opportunities for rural communities to generate greater incomes; thus improving the livelihoods of these small livestock keepers. A potential policy option is to establish a number of slaughter houses and small ruminant meat market to consumption area and in various villages to broaden the market as well as providing improved market access to producers.

References

[1] Bah, M., Cissé, S., Diyamett, B., Diallo, G., Lerise, F., Okali, D, & Tacoli, C. (2003). Changing rural–urban linkages in Mali, Nigeria and Tanzania. Environment and Urbanization, 15(1), 13-24.

[2] Boyazoglu, J.; Hatziminaoglou, I. & Morand-Fehr P. (2005). The role of the goat in society: Past, present and perspectives for the future. Small Rumin. Res., 60:13-23.

[3] De Vries, J. (2008). Goats for the poor: Some keys to successful promotion of goat production among the poor. Small Ruminant Research, 77(2), 221-224.

[4] Degefu, G.T. (2003). The Nile: Historical Legal and Developmental Perspectives. New York, USA. Pp. 429.

[5] Gomna, A., & Rana, K. (2007). Inter-household and intra-household patterns of fish and meat consumption in fishing communities in two states in Nigeria. British Journal of Nutrition, 97(01), 145-152.

[6] HosmerJr, D. W., Lemeshow, S., & Sturdivant, R. X. (2013) .Applied logistic regression (Vol. 398).John Wiley & Sons.

[7] Idowu, A. O., Ambali, O. I., & Otunaiya, A. O. (2012). Microfinance and small scale pig business in Osun State, Nigeria. Asian Journal of Business and Management Sciences, 1(9), 1-8.

[8] International Livestock Centre for Africa ILCA(1991).A hand Book of African Livestock Statistics. Working Document No. 15. August 1991. Addis Ababa, Ethiopia.

[9] Jaitner, J.; Sowe, J.; Secka-Njie, E. & Dempfle, L (2001).Ownership pattern and management practices of small ruminants in the Gambia - implications for a breeding programme. Small Rum. Res., 40:101-8, 2001.

[10] Juma, G. P., Ngigi, M., Baltenweck, I., & Drucker, A. G. (2010).Consumer demand for sheep and goat meat in Kenya. Small Ruminant Research, 90(1), 135-138.

[11] Kathiravan, D. G., Thirunavukkarasu, M., & Michealraj, P. (2007). Willingness to pay for annual health care services in small ruminants: The case of south India. Journal of Applied Sciences, 7(16), 2361-2365.

[12] Kumar, S.; Vihan, V. S. & Deoghare, P. R. (2003).Economic implication of diseases in goats in India with references to implementation of a health plan calendar. Small Rum.Res., 47:159-64, 2003.

[13] Langyintuo, A.S., Ntoukam, G., Murdock, L., Lowenberg-DeBoer, J., Miller, D.J. (2004).Consumer preferences for cowpea in Cameroon and Ghana. Agricultural Economics 30: 203-21

[14] Nwosu, C. O., Madu, P. P., & Richards, W. S. (2007). Prevalence and seasonal changes in the population of gastrointestinal nematodes of small ruminants in the semi-arid zone of north-eastern Nigeria. Veterinary parasitology, 144(1), 118-124.

[15] Otunaiya, A. O., & Shittu, A. M. (2014). Complete household demand system of vegetables in Ogun State, Nigeria. Agricultural Economics (Zemědělská Ekonomika), 60(11), 509-516.

[16] Shalander, K and Pant, K.P (2002).Goats in India: Status and Technological Possibilities for Improvement In: Birthal, P. and ParthasarathyRao, P. (eds) (2002) Technology options for sustainable livestock production in India: proceedings of the Workshop on Documentation, Adoption, and Impact of Livestock Technologies in India, 18-19 Jan 2001, ICRISAT Patancheru, India. New Delhi 110 012, India and Patancheru 503 324, Andhra Pradesh, India: National Centre for Agricultural Economics and Policy Research and International Crops Research Institute for the semi-Arid Tropics. 220 pp.

[17] ThammiRaju, D. and Suryanarayana, M.V.A.N. (2005). Meat consumption in Prakasam district of Andhra Pradesh Livestock Research for Rural Development 17 (11) 2004

[18] York, R., & Gossard, M. H. (2004). Cross-national meat and fish consumption: exploring the effects of modernization and ecological context. Ecological economics, 48(3), 293-302.

15

Plantlet Regeneration of Somatic embryos from Leaf Explants of *Mentha arvensis*(L.) A medicinally important Plant

Sammaiah D.[1], Odelu G.[2], Venkateshwarlu M.[3], Srilatha T.[4], Anitha Devi U.[5], Ugandhar T.[6]

[1]Department of Botany, Govt. Degree College, Huzarabad, (T.S)
[2]Department of Botany, Govt. Degree College, Jammikunta, (T.S)
[3]Department of Botany, Kakatiya University, Warangal, (T.S)
[4]Department of Botany, Govt. Degree & PG College for Women, Warangal, (T.S)
[5]Department of Botany, Govt. Degree & PG College for Women, Karimnagar, (T.S)
[6]Department of Botany, SRR Govt. Degree & P.G College, Karimnagar, (T.S)

Email address:

tugandharbiotech@gmailcom.in (Ugandhar T.)

Abstract: The present study was conducted with the aim of evaluating some of the factors that influence induction and plantlet regeneration of somatic embryos in *Mentha arvensis* (L) var piperascens Holmes (menthol or Japanese mint)since there are no available reports on Somatic embryogenesis and regeneration of plantlet in this medicinally important plant species. Leaves from plants growing under temporary shed were cultured on Murashige and Skoog medium fortified with (0.5-5.0mg/L) Napthlene acetic acid (NAA)+(0.5 mg/L)Thidizuron(TDZ). High frequency of somatic embryo formation was found at (2.5mg/L) NAA+ (0.5mg/L) TDZ in leaf explants respectively, Secondary somatic embryogenesis was also observed when primary somatic embryos were sub cultured same somatic embryo induction medium well developed cotyledonary stage embryos were germinated on MS medium supplemented with (0.5mg/L) (NAA) + (0.5-5.0mg/L)TDZ maximum 80% of somatic embryos germination and plant let formation was found at (2.5 mg/L) NAA+ (0.5 mg/L) (TDZ). The post translation survival rate of plants was 80% plants and flowers formation were morphological similar to the mother plants.

Keywords: Leaf Explants NAA (Napthlene Acetic Acid), TDZ (Thidizuron), *Mentha arvensis* (L), Somatic Embryos and Regeneration

1. Introduction

Mentha ravensis (Linn). var piperascens Holmes (menthol or Japanese mint) is an industrial crop that is widely cultivated for its essential oil from which menthol is crystallized. The essential oil, menthol and terpenes of the dementholated oil of *M. arvensis* are variously used in the food, perfumery and pharmaceutical industries. Improvement in the pest and disease tolerance and other adaptive characters determining the yield and quality of essential oil will make mint cultivation more economical. Construction of the desired *M. arvensis* genotypes will require transfer of specific foreign genes into the crop. Efficient procedures are required to regenerate plants from the transformed cells and for rapid micro-propagation of plant(s) of selected genotype(s). *In vitro* high efficiency procedures for cell and callus cultures and shoot regeneration from axillary buds and leaf explants have been reported in some species of the genus *Mentha*, especially the commercially important species *M. piperita* and *M. spicata* (Lin and Staba 1961; Cellarova 1992).

Total production of *Mentha* matches more than 32,000 MT per year and India is considered one of the largest producer and exporter with 27% commodity market share. Corn Mint is valued as multimillion cash crops for its multipurpose uses in the field of pharmaceuticals cosmetics as well as flavouring of food and beverages, oral preparations, such as toothpastes, dental creams and mouth washes. Green leaves of plants are used formaking chutney and for flavouring culinary preparation vinegar, flavour liqueurs, breads, salads, soups, cheese jellies, spice mixtures for many processed foods as well as in herbal tea (Moreno *et al.,* 2002; Kofidiset al., 2006; Yadegarinia *et al.,* 2006).

Mint plant is used as insect repellent, anaesthetic, galactofuge, refrigerant, stimulant, stomachic and excellent gastric organic stimulant (Budavari *et al.,* 1989; Gulluce*et al.,* 2007). Many authors worked on *in vitro* production of *Mentha* species using different explants viz., nodes, internodes, axillary buds and leaf discs. (Karasawa and Shimizu 1980, Rech and Pires, 1986, Van Eck and Kitto, 1990, Kukreja *et al.,* 1992, Sato *et al.,* 1993, Caissard *et al.,* 1996, Reed, 1999). Preliminary work on in vitro regeneration of *M. arvensis* was carried out by several investigators but the results exhibited low efficiency of shoot regeneration from nodal explants.

Somatic embryo genesis offers great potential in plant multiplication and crop improvement for efficient cloning and genetic transformation (Ammiratio 1987, Roberts *et.al.,* 1995). It is an alternative and efficient method for plant propagation over regeneration via organogenesis. Somatic embryos are believed to originate from single cell while organogenesis is through collective organization of cell. Therefore the plants derived from somatic embryos tend to be genetically alike; in addition somatic embryos are bipolar structures with root and shoot apices that can easily be developed into complete plantlets.

Embryos formed in cultures have been variously regenerated as accessory embryos, adventive embryos, embryoids and supernumerary embryos. Kohlenbach (1978) has proposed the following classification of embryos.

1. Zygotic embryos- those formed by fertilized egg or the zygote.
2. Non- Zygotic embryos- those formed by cells other than the zygote.
   i. Somatic embryos- those formed by the sporophytic cells (except zygote) either in vitro or in vivo. Such somatic embryos arising directly from other embryos or organs (stem embryos in carrot and butter cup) are termed adventive embryos.
   ii. Parthenogenetic embryos – those formed by unfertilized egg.
   iii. Andorogenic embryos-those formed by the male gametophyte (microspore pollen grains).

In Somatic embryogenesis the embryo regenerate from somatic cells, tissues or organs either de novo or directly from tissues (Adventive origin) which in the opposite of zygotic or sexual embryogenesis. Various terms for non-zygotic embryos have been reported in literature such as adventive embryo (somatic embryos arising directly from other organs or embryos) parthenogenetic embryos (those formed by the unfertilized egg) and androgenetic embryos (formed by the male gametophyte). However in general context somatic embryos are those which are formed from the somatic tissue in culture, i.e. *In vitro* conditions. In sexual embryogenes is the act of fertilization triggers the egg cell to develop into an embryo, however it is not the monopoly of the egg to form an embryo. Any cell of the gametophytic (embryo-sac) or sporophytic tissue around the embryo-sac

may give rise to an embryo, cells of the nucellus or linear integument may develop into embryos in the members of Rutaceae family. There are examples of embryos arising from endospermal cells also. However, the rate of multiplication is very low. In the present communication, we report reproducible method for *in vitro* plant regeneration via somatic embryogenesis in *M. arvensis* var piperascens Holmes (menthol or Japanese mint) from leaf explants using various cytokinins especially TDZ.

2. Methodology

2.1. Plant Material

Seeds of *M. arvensis* var piperascens Holmes (menthol or Japanese mint) were collected. From CIMAP (Central Institute of Medicinal Aromatic Plant Uppal. Hyderabad). Seeds which were initially soaked overnight and then washed with running tap water for 30 min to remove adherent particles, thoroughly washed seeds were then immersed in 5% (v/v) Teepol for 10 min and then rinsed 3 times with sterile double distilled water. This was followed by the surface sterilization with 05 % (m/v) HgCl$_2$under the sterile conditions for 5 min. these were rinsed 5 times in sterile double distilled water to remove all traces of HgCl$_2$ the sterilized seeds were then placed on to the basal Murashige and Skoog (1962) medium for germination.

2.2. Culture Media and Culture Conditions

Leaf (6 weeks old) explants from axenic seedlings were placed on MS medium supplemented with 30 gm/L sucrose along with different combinations of NAA (0.5 – 5.0 mg/L) + 0.5 mg/L TDZ respectively. The pH of medium was adjusted to 5.8 prior to autoclaving 121° C for 15 -20 min. All the cultures were incubated under 16/8 h light / dark photo period at 25±2 ° C. a light intensity of 40 μ mol m^{-2}s^{-1} was provided by cool- white fluorescent tubes. The cultures were transferred to fresh medium after an interval of 4 weeks.

For germination and plantlet formation somatic embryos were transferred to MS medium supplemented with (0.5 mg/L) NAA + (0.5–5.0 mg/L) TDZand incubated under the same culture conditions.

3. Results

Results on somatic embryogenesis in *M. arvensis* are presented in (Table-1) leaf cultured on various concentrations of NAA (0.5-5.0 mg/L) in combination with (0.5mg/L) TDZ become swollen and generally dedifferentiated and developed friable callus after 8-10 days of culture. Maximum number of somatic embryos / explant and higher percentage of response for somatic embryos formation have been found at (2.5 mg/L) NAA + (0.5 mg/L) TDZ in leaf explants of *M.arvensis* (Fig –I) with the increase of NAA concentration up to 5.0 mg/L with (0.5 mg/L) TDZ within 25-30 days of culture, globular, cotyledonary, and heart shaped embryos have formed directly on the surface of callus. But when the concentration of NAA was increased above (3.0 mg/L), percentage of response and somatic embryo induction were found decreased at higher concentration of NAA (5.0 mg/L) in combination with (0.5 mg/L) TDZ. When the explants of primary somatic embryos were cut into fragments and cultured on the same induction medium secondary somatic embryos were induced within two weeks.

Table 1. *Effect of various concentrations of NAA and (0.5 mg/L) TDZ onSomatic embryo genesis from leaf explants of M.arvensis.*

Hormone conc (mg/L)	% of cultures responding	% of Standard Error formation	Average no of Standard Error /explants (S.E.)*
NAA+TDZ			
0.5+0.5	73	31	6.7 ± 0.3
1.0+0.5	80	50	6.9 ± 0.5
1.5+0.5	90	60	10.7 ± 0.3
2.0+0.5	92	76	12.0 ± 0.54
2.5+0.5	94	80	15.0 ± 0.5
3.0+0.5	85	70	13.0 ± 0.6
3.5+0.5	75	65	11.0 ± 0.4
4.0+0.5	70	60	9.0 ± 0.5
4.5+0.5	65	58	7.0 ± 0.4
5.0+0.5	60	50	5.0 ± 0.5

*S.E. Standard Error

Table 2. *Effect of NAA + TDZ on the conversion of somatic embryoids into plantlets in M.arvensis.*

Hormone conc (mg/L)	% of cultures responding	Average no of shoots/embryods (S.E.)*
NAA+TDZ		
0.5+0.5	52	2.6 ± 0.63
0.5+1.0	56	3.0 ± 0.44
0.5+1.5	58	3.2 ± 0.45
0.5+2.0	60	3.8 ± 0.43
0.5+2.5	65	4.0 ± 0.32
0.5+3.0	70	6.0 ± 0.34
0.5+3.5	52	3.6 ± 0.34
0.5+4.0	54	2.5 ± 0.45
0.5+4.5	50	2.0 ± 0.43
0.5+5.0	40	1.5 ± 0.23

*S.E. Standard Error

Fig. 1. *Somatic embryogenesis and plant regeneration in Leaf explants cultures of M.arvensis L. a) Embryogenic callus induction b) Maturation of Somatic embryos c) germination of somatic embryos, d) plant with normal shoot and root system developed from somatic embryos E) regenerated plantlets growing in Pot*

Thus proliferation of somatic embryos occurred in two ways:

1. Multiplications of somatic embryos from the explants through primary somatic embryogenesis and

2. Proliferation of secondary somatic embryos from already formed

3.1. Germination of Somatic Embryos

For germination, globular, heart and torpedo shaped embryos (a mixture) were transferred to MS medium supplemented with different concentration of auxin such as NAA (0.5mg/L) in combination with TDZ(0.5-5.0mg/L).Highest (70%) frequency of embryo germination was noticed on a medium containing (0.5mg/L) NAA in combination with (3.0mg/L) TDZ. TDZand NAA were better for germination of somatic embryos in M.arvensis. When the concentration of BAP was increased above (3.0 mg/L), percentage of response and somatic embryo germination frequencies were decreased it was found that at higher concentration of TDZ (5.0 mg/L) in combination with (0.5 mg/L) NAA. Maximum number of somatic embryos germination and higher percentage of response for somatic embryos germination have been found at (3.0 mg/L) TDZ + (0.5 mg/L) NAA in leaf explants derived somatic embryogenesis of M.arvensis (Table -2) (Fig-1-A,B and C).

3.2. In Vitro Rooting

Fully elongated healthy shoots were transferred on MS medium fortified with different concentration of IAA (0.1mg/L) or IBA (1.5 mg/L).Profuse rhizogenesis was observed on MS medium produced roots (Fig-D).

3.3. Acclimatization

Rooted plantlets were removed from the culture medium and the roots were washed under running tap water to remove agar. Then the plantlets were transferred to polypots containing pre- soaked vermiculite and maintained inside a growth chamber set at 28°C and 70 – 80 % relative humidity. After three weeks they were transplanted to poly bags containing mixture of soil + sand + manure in 1: 1: 1 ratio and kept under shade house for a period of three weeks. The potted plantlets were irrigated with Hogland's solution every 3 days for a period of 3 weeks (Fig-E).

4. Discussion

Somatic embryogenesis is an important step in any successful plant transformation scheme. Stable transformation required that a single cell gives rise to a plant. The ideal transformation scheme is that via somatic embryogenesis, because from callus each transformed cell has the potential to produce a plant. Somatic embryogenesis and subsequent plant regeneration has been reported in most of the major crop species (Evans and Sharp, 1981). Soybean and cotton proved to be the most difficult to regenerate.

In the present investigation, the results on somatic embryogenesis have shown that auxin such as NAA (0.5-5.0mg/L0 along with cytokininTDZ are essential for inducing the somatic embryogenesis from leaf explants of M.arvensis.A major factor for somatic embryogenesis is the nature of growth regulators used in the induction medium. The type of auxin or auxin in combination with cytokinin

used in the medium can greatly influence somatic embryo frequency. The requirement of cytokinin in addition to auxin was observed in Sapindus trifoliatus (Desai et.al., 1986), Termineliaarjuna (Kumari et.al., 1998) and Psoralea corylifolia (Sahrawat and Chand, 2001) as it was observed in the present studies. Somatic embryo genesis was induced on medium containing NAA alone in Solanum melogena (Matsuoka and Hinata, 1979; Gladdle et.al., 1983; Sharma and Rajam, 1995).

Matsuokaand Hinata(1979) and Gleddle et.al., (1983) also observed a stimulatory effect of NAA on embryogenesis in eggplant hypocotyls explants as well as in leaf explants with carrot hypocotyls explant where somatic embryos were formed in response to a wide range of auxins including (IAA: NAA, 2,4-D, 2,4,5-T).

The embryogenesis action of NAA in egg plants were also different in another aspect where the role of auxin in embryogenesis species other than eggplant, induction and maturation of somatic embryos to the cotyledon stage was achieved on the same medium. The removal of NAA was required only for further development of embryos to plant lets and cotyledon explants grown on NAA exhibited globular shaped heart shaped and torpedo shaped stages. The embryogenic potential was markedly dependent on genotype isolated embryogenic callus from M.arvensis.

The sugars which supported callus proliferation on cotyledon explant i.e. sucrose, fructose, and glucose also supported embryogenesis, Somatic embryogenesis in NAA treated leaf explants was inhibited by cytokinins. Matsuoka and Hinate (1979) also observed cytokinin-induced embryogenesis in eggplant hypocotyls cultures. Ethylene also typically inhibited somatic embryogenesis (Ammirato 1983). The molecular aspects of embryogenesis, embryo specific proteins in carrot have been studied in rice (Chen and Luthe, 1987).Embryo specific proteins in somatic embryogenesis has been studied in alfaalfa (Stuat et.al., 1985). These developmental regulated genes are now being isolated from many plants (Choi et.al., 1987). BAP has been used for shoot induction of melon (Kathal et.al., 1986; Suesmatsu et.al., 1986; Dirks and Buggenum, 1989). The effect of BAP on somatic embryogenesis was tested by Oridate and Oosawa (1986) and the most efficient embryo formation was obtained at a concentration of (0.1 mg/L). The adventitious shoot formation and somatic embryogenesis in melon can be controlled by the ratio of auxins and cytokinin in the medium. In Coffea arabica somatic embryos also developed only when callus grown on 2,4 –D containing medium was transferred to 2,4-D free medium (Sondal and Sharp, 1977) and in pumpkin, NAA and IBA favored embryogenesis (Jelaska, 1974). In nucellus cultures of Vitis embryo formation occurred in the presence of NAA and BAP (Molling and Srinivasan, 1976).

The high frequency of multiple shoots was recorded from direct organogenesis of nodal explant. MS medium supplemented with benzyl adenine (4.4µM), kinetin (4.6µM), and 3% sucrose promoted the maximum number of shoots as well as beneficial shoot length in Paederia *foetida* (L.) (Thirupathi *et.al.,* 2013)

Somatic embryogenesis is also preferred because it allows production of plant without somaclonal variation and in efficient cloning and genetic transformation (Sharp et.al., 1980). Synthetic seeds can also be developed by encapsulating somatic embryos in sodium alginate complex with calcium chloride as it was developed in Solanum melogena (Lakshmana Rao and Singh, 1991). Whereas Binzel et.al., (1996) reported that the entire process of induction and maturation of the embryos was completed on the same MS medium containing auxins and cytokinins (NAA+ TDZ) in M arvensis it was observed the requirement of both the hormones in the present investigations. Similarly somatic embryos maturation on MS medium containing the combination of auxins (NAA) and cytokinins (BAP) was observed in Cajanus cajan (Mallikarjuna et.al., 1996), Prunu saxivum (Garin et.al., 1997) and Hardiewickia binate (Chand and Singh 2001). Solanum surattensce (Ugandhar.2002). NAA and cytokinin (TDZ) was observed in Mentha piperita (Saha et.al.,2010)

Thus somatic embryogenesis always appeared to be dependent on the types of auxin/cytokinin/auxin+cytokinin and their concentration in the medium. The type of phytohormone and its concentration also varies from genotype to genotype. High concentration of auxin in combination with less concentration of cytokinin induced the somatic embryogenesis and maturation of somatic embryos in *M.arvensis*. But in the present study, a highest germination rate 70% was achieved. Plant lets with well developed shoot and root systems were washed carefully with tap water and transferred to plastic cups containing a mixture of vermiculite and soil for hardening. The acclimatized plantlets were successfully transplanted to pots under field conditions. Regenerated plants showed no observable morphological alteration.

During the present investigations, it was found that the high concentration of auxin in combination with less concentration if cytokinin induced the somatic embryogenesis and maturation of somatic embryos in *M.arvensis*. In conclusion, for induction of in vitro somatic embryo genesis the type of primary explant, choice of genotypes and hormonal concentration plays on important role (Patel *et.al.,* (1994).

References

[1] Ammirato, P.V. (1983). The regulation of Somatic embryos development in plant cell cultures, suspension cultures technique and hormone requirements. Bio. Technol., 1: 68-74.

[2] Ammirato, P.V. (1987).Organisation events during Somatic embryogenesis. In: Plant Tissue & Cell Culture, (Eds.). Green, C.E., Somers, D.A., Hackett, W.P. & Biesboer, D.D. Alan R. Liss, Inc., New York, pp. 57-81.

[3] Binzel, M. L., Sanhla, A. N., Joshi, S. and Sankhla, D., (1996).:In vitro regeneration in chilli pepper (Capsicum annuum L.) from half-seed explant. Plant Growth Regulation, 20: 287-293.

[4] Budavari, S., O'Neil, M. J., Smith, A. and Heckelmen, P. E. (1989).The Merk Index.An Encyclopedia of Chemicals, Drugand Biologicals.(11thEdition). Merck and Co., Rahway.

[5] Caissard,J.C., Faure, O., Jullien, F., Colson, M and Perrin, A.(1996).Adirect regeneration in vitro and transient GUS expression in Mentha piperita. Plant Cell Rep.16: 67-70.

[6] Cellarova, E.(1992).Micro-propagation of Mentha L.; In: Biotechnology in agriculture and forestry (ed.) Y P S Bajaj (Berlin, Heidelberg: Springer-Verlag) vol. 19, pp. 262-276

[7] Chand, S. and Singh, A.K. (2001).Direct somatic embryogenesis from zygotic embryos of a timber-yielding leguminous tree. Hardwickia binata Roxb.Curr. Sci., 80: 882-888.

[8] Chen, L.-J. and Luthe, D.S. (1987) Analysis of proteins from embryogenic and non-embryogenic rice (Oryza sativa L.) calk Plant Sci. 48, 181-188.

[9] Choi, J., Liu, L.S., Borkid, C. and Sung, Z.R.(1987). Cloning of developmentally regulated genes" Proc.Nat Acad. Sci 84: 1906- 1910.

[10] Choi,Y.E., Kim,J.W and Soh, W.Y.(1987) Somatic embryogenesis and plant regeneration from suspension cultures of Acanthopanax koreanum. Plant Cell Rep., 17: 84-8.

[11] Desai,HV. Bhatt, P.N and Metha,A.R (1986) Plant Regeneration of Sapindus trifoliatus L. (Soap nut) through Somatic embryogenesis Plant Cell Rep.,3: 190-191.

[12] Dirks, R. and Buggenum, M.V. (1989). In vitro plant regeneration from leaf explants of Cucumis melo L, J. Plant. Physiol. 132: 373-377.

[13] Evans, D.A., Sharp, W.R. and Flick, Ch.E. (1981). Hand Book of Plant Tissue Culture Methods and Appl. in Agr.Culture. New York, London, pp: 45-113.

[14] Garin, E., Grenier, E. and Granier G.D. (1997) Somatic embryogenesis in wild cherry (Prunusauium).Plant Cell Tissue and Org. Cult.,48: 83-91.

[15] Gleddle, S., Keller, W. and Setterfield, G. (1983). Somatic embryogenesis and plant regeneration from leaf explants and cell suspensions of Solanum melongena (egg plants). Can. J. Bot., 61:656-665.

[16] Gulluce, M., Sahin, F., Sokmen, M., Ozer, H., Daferera, D.,Sokmen, A., Polissiou, M., Adiguzel, A. and Ozkan, H. (2007).Antimicrobial and antioxidant properties of the essential oils and methanol extract from Mentha longifolia L.. Food Chemistry,103: 1449-1456.

[17] Jelaska, S. (1974).Embryogenesis and organogenesis in Pumpkin explants.Physiol . Plant. 31: 257-261.

[18] Karasawa, D. and Shimizu, S. (1980).Triterpene acids in callustissues from Mentha arvensis var. piperascens Agric. Biol. Chem. 44: 1203-1205.

[19] Kathal, R., Bhatnagar, S.P. and Bhojwani, S.S (1986). Regeneration of plants from leaf explants of Cucumis meloc.v. pusasharbati. Plant Cell Reports.7: 449-451.

[20] Kofidis, G., Bosabalidis, A. and Kokkini, S. (2006).Seasonal variations of essential oils in a linalool-rich chemotype of Mentha spicata grown wild in Greece.Journal of EssentialOil Research, 16; 469-472.

[21] Kohlenbach, H.W. (1978).Camparative Somatic embryogenesis In: Frontiers of Plant tissue culture (Ed. Thorpe T.A.) Universal Calagry Press. pp. 59-66.

[22] Kukreja,A. K., Dhawan, 0.P., Ahuja, P. S., Sharma, S. and Mathur,A. K.(1992). Genetic improvement of mints: On the qualitative traits of essential oil of in vitro derived clones of Japanese mint (Mentha arvensis var. piperascens Holmes).J. Essent. Oil Res., 4: 623-629.

[23] Kumari,K.G., Ganesan,M and Jayabalan, N. (1996). Somatic organogenesis and plant regeneration in Ricinus communis. Biol. Plant.52:17-25.

[24] Lakhmana Rao, P.V. and Singh B. (1991).Plantlet regeneration from encapsulated somatic embryos of hybrid Solanum melongena L Plant Rep.10: 7-11.

[25] Lin M and Staba E J (1961) Peppermint and spearmint tissue cultures, callus formation and submerged culture; Lloydia (24) 139-145

[26] Mallikarjun, N., Reena, M.J.T., Sastri, D.C. and Moss, J.P. (1996).Somatic embryogenesis in pigeon pea (Cajanus cajan). Indian J. Exptl. Biol., 34: 282-294.

[27] Matsuoka, H. and Hinata, K. (1979).NAA–induced organogenesis and embryogenesis in hypocotyls callus of Solanum melogena. L.J. Exptl. Bot., 30: 363-370.

[28] Mollins, M.G. and Sreenivasan, C. (1976). Somatic embryos and plantlets from on ancient clone of the Grape vine (Cv Cabernet Sauvignon) by apomixes In Vitro J.Exp. Bot., 27: 1022-1030.

[29] Moreno, L., Bello, R., Prime-Yufera, E., Esplugues, J. (2002).Pharmacological properties of the methanol extract from Mentha suaveolens Ehrh. Phytotherapy Research. 16: 10-13.

[30] Murashige, T. and Skoog, F.(1962).A revised medium for rapidgrowth and bioassay with tobacco tissue cultures.Physiol. Plant., 15: 473-497.

[31] Oridate, T. and Oosawa, K.(1986).Somatic embryo genesis and plant regeneration from suspevsion callus culture in melon (Cucumis melo L.).Ja J. Breed., 36: 424 – 428.

[32] Pandya, H. A. and Saxena, O. P. (2000). "Role of PGRs on tissue culture raised plantlets of Gladiolus, Chrysanthemum and Lily" (abstr) Role of plant growth regulators and plant biotechnology to improve growth and productivity of plants, Botany Department, Gujarat University, Ahmedabad, India.

[33] Patel, M.B., Bhardwaj, R. and Joshi, A. (1994).Organogenesis in Vigna radiates L. Wilczek" Indian J. Exp. Biol. 29: 619-622.

[34] Rech,E.L. andPires, M. J. P.(1986) Tissue culture propagation of Mentha sps.by the use of axillary buds Plant Cell Rep. 5:7-18.

[35] Reed, B. M. (1999).In vitro storage conditions for mint germplasm. Hort. Sci. 34: 350-352.

[36] Roberts, A.V., Yokoya, K., Walker, S. and Motley, J. (1995). Somatic embryogenesis in woody plants (Eds) Jain, S., Gupta, P. Newton, R.) Kluwer Academic Publishers, The Netherlands, pp. 277-289.

[37] Sato, S., Newell, C.Kolacz, K. and Tredo, L. (1993). Stable transformation via particle bomardment in two different soybean regeneration systems.Plant cell Rep.12: 408-413.

[38] Saha. S Ghosh P. D and Sengupta. C (2010) "In Vitro Multiple Shoot Regeneration of Mentha piperita," Journal of Tropical Medicinal Plants, Vol. 11, pp. 89- 92,

[39] Sahrawat AK and Chand S (2001) Continuous somatic embryogenesis and regeneration from hypostyle segments of Psoralea corylifolia Linn: An endangered and medicinally important Fabaceae plant. Curr. Sci. 81:1328-1331.

[40] Sharma, P. and Rajam M.V. (1995). Genotype, explant & position effects on organogenesis & Somatic embryogenesis in eggplant (Solanum melongena L.) J. Exptl. Bot., 46: 135-141.

[41] Sharp, W.R., Sondehl, M.R., Caldas, L.S. and Maraffa, L.S. (1980). The physiology of In vitro a sexual embryogenesis Hort. Rev. 2: 268-310.

[42] Sondahl, M.R. and Sharp W.R. (1977). High frequency of induction somatic embryos in cultured leaf explants of coffea arabica L. Z. Pflanzen Physiol., 81: 395-408.

[43] Stuart, D.A., Nelsen, J., Strickland, S.G. and Nichol, J.W. (1985). Factors affecting developmental processes in alfa alfa cell culture. In: Tissue culture in Forestry & Agriculture. R.R. Henke, K.W. Hughes, M.P. Constantin and A. Hollaender (Eds). Pp. 58-73.

[44] Suesmatso, N., Ootsuka, H. and Toda, M. (1986). Bull. Schizuoka.Agr. Exp.Stone. 31: 31-38.

[45] Thirupathi1.M Srinivas.D and Jaganmohan Reddy (2013) High frequency of multiple shoots induction in Paederia foetida (L.)- A rare medicinal plant 2013 Science publishing Group 1(5):60-65

[46] Ugandhar T. (2002). Tissue culture studies in Solanam surattense Ph.D.Thesis, Kakatiya University, Warangal, (T.S).

[47] Van Eck,J. and Kitto,S.L. (1992).Regeneration of peppermint and orange mint from leaf disks. Plant Cell Tiss.Org. Cult., 30: 41-46.

[48] Yadegarinia, D., L. Gachkar, M. B. Rezaei, M. Taghizadeh, S. A.Astaneh, I. and Rasooli. (2006). Biochemical activities of Iranian Mentha piperita L. and Mentha communis L., essential oils. Phytochemistry, 67, 1249-1255.

16

The Contribution of Ethiopian Orthodox Tewahido Church in Forest Management and Its Best Practices to be Scaled up in North Shewa Zone of Amhara Region, Ethiopia

Abiyou Tilahun[1,*], Hailu Terefe[1], Teshome Soromessa[2]

[1]Department of Biology, College of Natural Science, Debre Berhan University, Debre Berhan, Ethiopia
[2]Department of Environmental Science, Science faculty, Addis Ababa University, Addis Ababa, Ethiopia

Email address:
abiytila22@gmail.com (A. Tilahun)

Abstract: This research was conducted in selected moasteries of Ethiopian Orthodox Church (EOTC) in North Shewa zone. The main objective of the study is to identify main constraints which hinder the society to learn from religious instituions and apply the best practices and habit of experiences on forest conservation. Accordingly six monasteries were selected at different altitudes and sites purposively. Once the study forest areas were identified, 10 X 10 m quadrats were laid systematically in the forests for vegetation sampling. For the socioeconomic survey, 112 individuals were selected. Moreover, focus group discussion and key informant interviews were employed. Church forests enveloped in this study have an area ranging from 1.6 ha to 100 ha. The total number of species and families in each of the six churches ranged from 17 to 60 and 15 to 39 respectively. Different regeneration status was revealed from the height and diameter class distribution for some of the woody species. The height and diameter class distributions for all individuals in each studied church showed that the forests are at different secondary stages of development. The classification of the species group by ordination techniques showed the differentiation in species group types has a strong relationship with altitude. These church forests didn't come to exist just by chance. Results indicated that it is by the commitment of the church based on strong theological thoughts and biblical basis. It was found that the local community respects and protects church forests, and considers the church as a central institution and platform. However, the community is not strongly committed to adopt forest management culture of the church due to: 1) the church leaders didn't teach more to their followers to plant trees and to transfer the knowledge; 2) limited knowledge of the community about the benefits of forests to their livelyhood. 3), In general, the result of this study revealed that, forests conserved by EOTC and its tradition provide an opportunity to establish insitu and exsitu conservation sites for forest resources. They also have greater prospects in implementing forestry conservation, development, research and education programs with some avoidable threats and constraints for which recommendations were presented. Hence, it willl be worthy to include the church and mosque communities when delivering trainings and sharing responsibilities in aforestation programes.

Keywords: Monasteries, Indigenous Knowledge, Natural Forest, Conservation, Sustainability

1. Introduction

1.1. Background

Forests have multidimensional functions and uses to mankind and other living organisms. They play indispensable roles in the life support systems on our planet. However, due to rapid population growth and natural factors the existing natural forests depleted in Ethiopia and brought significant decline in their biodiversity to the extent that some species are on the verge of local extinction. The problem of deforestation and loss of biodiversity is more pronounced in the Central highlands of Ethiopia, particularly in the North Shoa Zone, where forests are downscaled to patches and strips on the tops of hills and heads of streams. As a result, very little of the natural forest of the Central Ethiopian highlands remains today. The deterioration of natural resources not only destroys the environment, but also undermines the very foundation on which economic growth and long-term prosperity depends (Tadesse Woldemariam, 1998). In such devastated areas, conserving and maintaining species diversity has been a very challenging task. The only areas where one can observe trees in Central Ethiopia are in the surroundings of churches. These patches of natural forest have survived as a result of the traditional conservation effort of the Ethiopian Orthodox Tewahido Churches (Yeraswork Admassie, 1995). These forests are still sanctuaries of many plant and animal species that have almost disappeared in most parts of northern Ethiopia (Alemayehu Wassie, 2002).

In north Shewa, The Ethiopia Orthodox Tewahido Church has about 406,706 followers, 1800 Churches and 27,980 Priests. The church has a long history of conservation of forest resources arround the churches. Church compounds are serving as conservation sites and hot spots of biodiversity, mainly indigenous trees and shrubs of Ethiopia, which, in turn, give prestige to the religious sites (Alemayehu Wassie et al., 2005).

These church forests, however, are threatened because disturbances such as cutting, grazing, droughts, and fires occur at increasing intensity and frequency. Environmental sectors promote restoration (Yeraswork Admassie, 1995), but the current results suggest that area closure alone cannot restore the original species composition and diversity of the natural ecosystem, such as encountered in some larger church forests. Church forests still harbor a large number of indigenous forest species and at the same time conserve the soil and microclimate. The forest and their species could be used as new sources of regeneration (including via their soil seed banks).

1.2. Religious Institution and Their Culture of Forest Conservation

Many tree species surrounding religious areas have relationship with the term sacred groves in most literature. The tradition of the sacred grove is well known in Ethiopian tradition in the experience of traditional religion as the Oromo sacred Adbar or in the clump of trees that customarily envelope the Debr (Bahru Zewde, 1997). Sacred groves are smaller or larger ecosystems, set aside for religious purposes. The origin of sacred groves can be attributed to the slash and burn system of agriculture, where several forests patches was left standing around farmlands. These groves came to be institutionalized as centers for culture and religious life. Taboos and social sanctions protect the sacred groves from deterioration due to human interference. These habitat patches may be the only primary forests remaining locally. Several endemic and endangered species have been recorded from sacred groves. Sacred groves, which form in situ conservation sites and act as a refuge for species, are becoming ecologically important in the light of the current rates of deforestation and species loss. They buffer against the depletion of genetically adapted local variants and overall biodiversity in a region. They can serve as important recruitment areas to surrounding ecosystems. Sacred groves are of great economic significance too (Alemayehu Wassie, 2002).

The doctrine of the religions behind those sacred groves may vary but ultimately the experience of conserving trees for religion purposes is apparent worldwide. Thus, trees not only meet the economic and ecological needs of the people, but also form an integral part of their culture and spiritual tradition (Yeraswork Admassie, 1995).

The Ethiopia Orthodox Tewahido Church has long history of planting and conserving tree species. Church compounds are the monasteries of trees and other biodiversity resources where one can animate trees escaped from being destroyed forever under the shelter of the church value and esteem. Many indigenous trees and shrubs destroyed completely over the last century, are still found standing in the compounds of rural churches (Taye Bekele, 1998). The area of forest cover preserved by the Ethiopian Orthodox churches in some parts of the country has been declined and found in patches. These patches of forests are used as sources of seeds for raising seedlings in nurseries (EFAP, 1996).

Church compounds are serving as in situ conservation and hot spot sites for biodiversity resources, mainly indigenous trees and shrubs of Ethiopia, which in turn give prestige for the religious sites. As a result, these forests are sanctuaries for different organisms ranging from microbes to large animals, which will have almost disappeared elsewhere. Historically, most of the church forests was destroyed and burned with the churches and other precious heritages by the anti-Christian expedition led by Ahmed Ibn Ibrahim also called 'Gragn' meaning 'left handed' in the beginning of the 16th century. After 'Gragn' has been killed in 1543, most of the churches and monasteries were reconstructed together with their forests (Aymro Wondmagegnehu and Motovu, 1970). In the process of nationalization of private properties during the socialist regime the EOTC was left without its land holdings, including the forests, which have been preserved for centuries. The fate of those forests was ruthless exploitation and destruction, which in turn brought a severe reduction in biological diversity and ecological imbalance as well.

2. Materials and Methods

This study was conducted in North Shoa Zone, Amhara Region. The topography comprises uneven and ragged mountainous highlands, extensive plains and also deep gorges and cliffs. This features provided the zone with spectacular scenic beauty and stunning views across the Yifat and great rift valley up to the Awash river in the Afar (Zelalem Tefera, 2001).The rainfall is characterized by a bimodal distribution with the major rainy season being from June-August and the 'Belg' from March-May. The annual average rainfall varies between 400 – 700 mm and the annual average temperature ranges between -8 - 35.7°C.

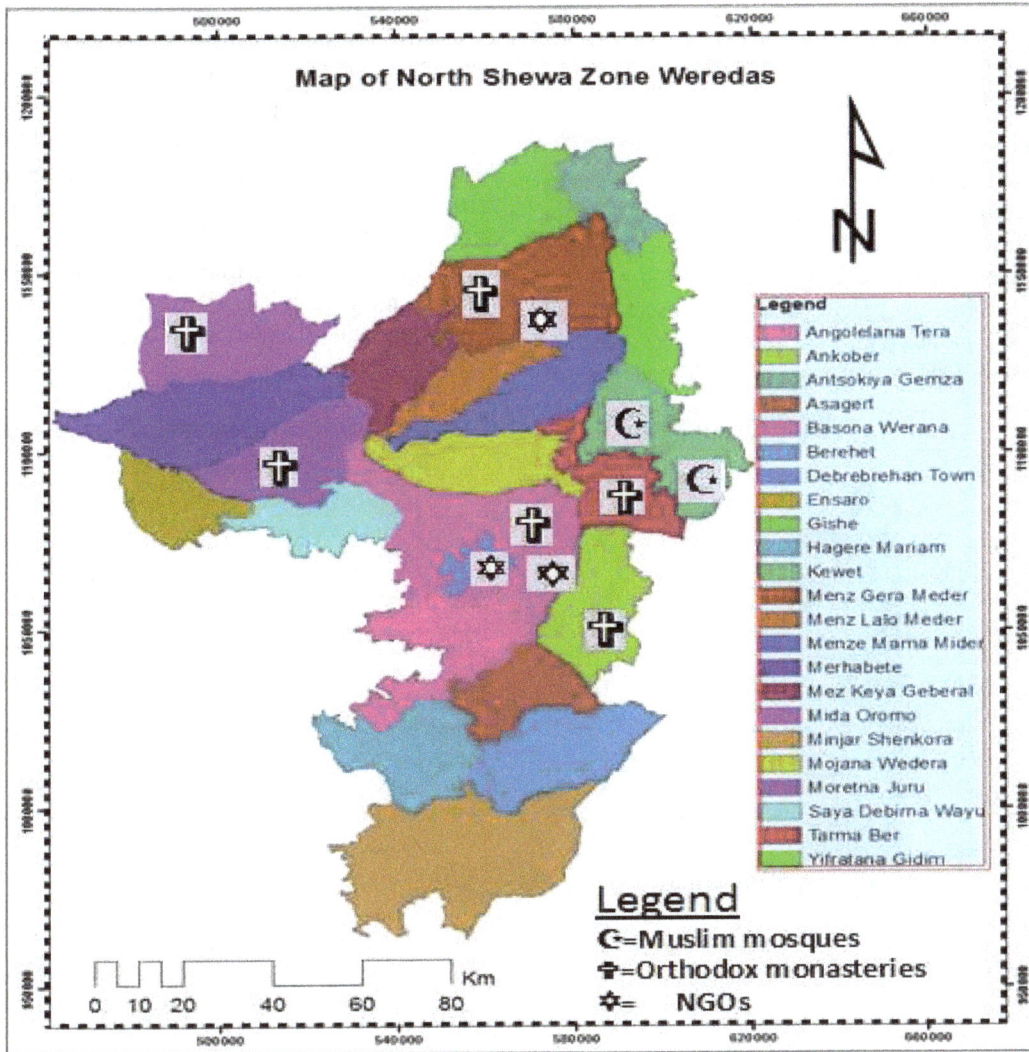

Figure 1. *Map of the study areas.*

Selection of study sites was based on representativeness of ecological and administrative regions, Emphasis on old sites and sites of religious significance, Site accessibility and feasibility for the study and Sites with high conservation value and mixture of sites with and without threats. Six monasteries of Ethiopian Orthodox Tewahido church were selected purposively from six district of North Shoa Zone for the socio-economic survey as a study site (Getahun Kassa and Eskinder Yigezu, 2015):

I.   Washa Debre Medihanit Abune Melke tsiediek monastery in Mida Woremo District
II. Debre Bisrat Abune Zena Markos Monastry in Moret and Jiru District
III. Debre Menkirat Mitak Amanuel Monastery in Ankober and Basona Worena District
IV. Debre Hail Kidus Rufael church in Menz Gera District
V.   Debre Menkirat Gashu Amba Merkoriwes in Tarmaber District
VI. Debre Kerbe Dagmawi Goligota in Basona werana District

3. Methods

The participatory rural appraisal (PRA) method was employed to generate the socio-economic information and government intervention following Martin (1995) and Cunningham (2001). Primary data was collected by semi-structured, key-informant interviews and group discussion (Muluken Mekuyie, 2014). Descriptive statistics were used to analyze the data included percentages and frequencies. Based on the participatory rural appraisal result analysis on the traditional conservation, natural regeneration and current status of forest resources was drawn. All the stakeholders such as District office of agricultural experts, administrative bodies, EOTC representatives, Church scholars, community elders, NGOs were discussed on the practices of the Church on biodiversity conservation. Semi structured questionnaires was pretested before use on a wider scale, and some improvement was made to the questionnaire. The results were presented in the forms of tables and graphs (Siboniso M. Mavuso, Absalom M. Manyatsi, and Bruce R. T. Vilane, 2015).

4. Result

4.1. Socio-economic Survey of Ethiopian Orthodox Tewahido Church

4.1.1. Participatory Rural Appraisal

Initial discussion was held with selected District agricultural bureau representatives and religious leaders to explain the purpose of the study and to get permission in order to conduct the study in the area (Elias Endale, 2003).

Based on the information generated from the discussions at various levels, key participants were selected. The selection criteria was their long residence, better knowledge of environment and natural resources and ability to articulate the functioning of religious institute on the forest. The extraction of information on forest status was made by using Participatory Rural Appraisal (Abeje Eshete, et al., 2005).

A total of 112 respondents, of which 94(83.9%) male and the remaining 18(16.07%) female were involved in the study. 71(63.39%) of the respondents were from EOTC Monks and clergies, 13(11.6%) from Muslim Imams and Sheks, 6(5.35%) from NGOs, 12(10.7%) from community leaders who live in the area more than 30 years, 10(8.9%) from kebele and District agricultural expert and administrative bodies (Fig. 2).

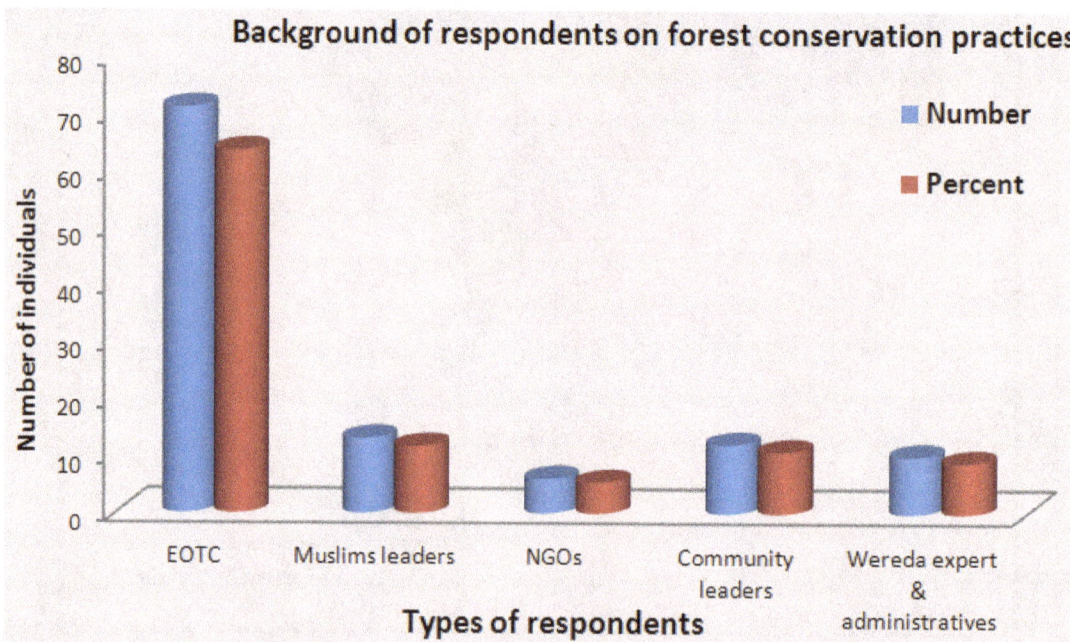

Figure 2. Discussant stakeholders in forest conservation , 2005 E.C.

Group discussions with multiple stakeholders and Participatory Rural Appraisal which allow us to explore a range of concerns and raise awareness on the resources were the main methods to collect the information. During the discussions an attempt was made to encourage the stakeholders in such a way that their cooperation is of great benefit to themselves and the country. The selection of appropriate time and location were encouraging for the participants of the group discussion.

In these discussion and participatory rural appraisal, information gathering were focused on indigenous knowledge of each group in relation to methods of sustainable use and conservation and giving priority for tree species, showing very low regeneration potential. Pair-wise ranking technique were used to identify and make comparison among plant species identified for different uses such as construction, farm implements, beehive making, bee fodder, fuel wood, woodcarving, medicinal use and others. In addition, the most threatened species were also identified based on the number of dead standings, number of stumps, seedlings and saplings.

Participants thought that forest protection and forest genetic resource conservation seems to be the duty and responsibility of the monastery and the government alone. But the efforts of EOTC monastery to protect their natural forest against irresponsible deforestation are indispensible.

Ninety eight percent (110) of the respondent believes that religious institutions have indigenous knowledge and practice of forest resources conservation but, the local communities cannot exploited the experience to apply on their surrounding due to different constraints. Eighty one percent(91) respondent believes that this conservation practice is gradually decline and is being eroded in the near future because of government administrative complexity, 95% of the respondent agreed that, lack of education to the young generation by the religious leaders about forest conservation and population growth and farm land expansion(99.5%) (Table 1).

Table 1. Percentage of the respondent on the forest conservation practice in North Shewa zone.

Discussion points/items	Agree	Disagree	None
There is religious institution that has long history of planting, protecting and preserving of trees species	99%	1%	0%
This practice is still operational in the institution at your locality	23%	77%	0%
The cause for the decline of forest conservation practices			
• Government involvement in the traditional practice and territory	88%	12%	0%
• Decline the acceptance of religious leaders on resource conservation by the new generation	62%	38%	0%
Local community exploit and apply the best practice of religious institution	7%	93%	0%
Reasons, why local people cannot learn and apply such practice in government owned and private forest resources			
1. Poverty			
• Lack of grazing land for their domestic animals	98%	2%	0%
• Illegal agricultural land expansion	95%	5%	0%
2. Lack of knowledge dissemination and awareness on conservation to the local people from the institution	76%	24%	0%
3. Poorness interms of knowledge transfer to the generation about the experiences of EOTC on forest conservation	68%	20%	12%
There is an effort to incorporate the best practice by the government and agricultural experts with the modern conservation and management strategies	17%	83%	0%
It is possible to integrate the religious conservation and management practice with modern conservation system to rehabilitate degraded areas and sustain the remaining few natural forests in the zone	94%	6%	0%

Table1. shows that 99% of the participant conclude that religious institution has long history of planting, protecting and preserving of trees species. But such interesting nature conservation habit is not sustainably managed due to factors like anthropogenic pressure, farm expansion, charcoal production, grazing by domestic animals, low awarness on the forest and alien invasive species (Figure 3). The local people, agricultural expertise, governmental representatives have little effort to learn and apply best practice of surrounding monastery on forests and other natural resources conservation.

4.1.2. Uses of Forest Plant Species Identified by the Local Communities During PAR

The forests grown in monasteries are considered as sacred because they are growing in God's compound and besides that they are symbolic to the presence of angels guarding each church, surrounding it. Besides the spiritual aspects, trees in monasteries provide several material benefits. They are sources of fuel wood that is required for services of the church; they provide shade from the scorching sun for the clergy and the laity during mass and religious festivals and the stems of the standing trees give support to individuals during prayers. Besides, the trees add aesthetic value to the church. As it is listed in Table 2, in monasteries, forests play other important roles in addition to the aforementioned services. Hermits and monks use the forests as praying sanctuary and also feed on the leaves, fruits and other parts of the wild plants. Conservation of forest around monasteries helps to give grace for the monasteries, serve as traditional learning and teaching under the shade of trees, resting places/sanctuaries for saints, provide sweet aroma to the church and trees around churches symbolise the fact that God created Adam and and Eve and placed him in Eden. The participants point out the Material benefits of trees to local people and monastery community identified during

PRA. They have grouped in to four basic uses: I. Provisioning Services (Food, Fiber and Fuel, Genetic Resources, Biochemicals, Construction materials for churches), II. Regulating Services (Invasion resistance, Herbivory, Pollination, Seed dispersal, Climate regulation, Pest regulation, Disease regulation, Natural hazard protection, Erosion regulation, Water purification), III. Cultural Services (Spiritual and religious values, Wood for making sacramental objects such as drums, crosses and support sticks for church services, Knowledge system, Education / inspiration, Recreation and aesthetic value, National heritage) and IV. Supporting Services (Primary production, Provision of habitat, Nutrient cycling, Soil formation and retention, Production of atmospheric oxygen, Water cycling and Climate balance).

4.1.3. Threats and Major Causes of Forest Depletion

Based on PRA discussion, collected data and physical observation threatened tree and shrub species in the considered study site were identified. Almost all of the participants identified that, *Juniperus procera* as most threatened species. *Hagenea abyssinica* was identified as second most threatened species followed by *Podocarpus falcatus* and *Olea europaea* sub sp.*cuspidata*. Both species are highly preferred for woodcarving, local construction and fuel wood. *Ekebergia capensis* and *Pittosporum viridiflorum* which is used for timber, woodcarving and fuel wood, was identified as third most threatened tree species by both groups. The different choice of second most threatened tree species by men and women shows that both groups have got different perspective in examining and interpreting problems in a different way based on their own experience and roles in society. Therefore, it is important to accommodate the ideas and experiences of all stakeholder groups to come to an effective and practical forest conservation mechanism.

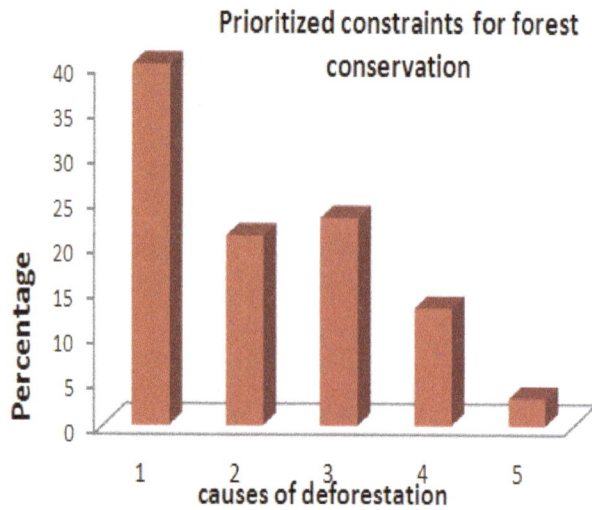

Figure 3. *Major causes of deforestation and losses of biodiversity.*

Key: (1= Anthropogenic pressure, 2=Farm expansion and Charcoal production, 3=Grazing by domestic animals, 4= Low awarness on the forest and 5= Alien invasive species.
Source: Participants of the PRA discussion in forest, 2005 E.C.

(i). Anthropogenic Pressure

High human pressure with negative impact on the natural resources and biodiversity richness weighs on the forest today: intensive human impact from livestock grazing and fuel wood and timber use. As figure 3. describes, Anthropogenic pressure accounts the largest share for forst destruction followed by grazing by domestic animals and then farm expansion and charcoal production. Human pressure is highly intesified in Mitak Amanuel and Abune Zena Markos natural forest. Farm expansion is also the main problems of Mida Abune Melketsiediek monastery forest.

(ii). Invasive Species

The destructive impacts of invasive alien species was identified in Kewet,Yifratana Gidim districts towards Afar and Dessie. Among the invesive species *lantana camara* was replacing all of the shruby plant species with a very rapid rate.This is intensified to the right of Tarmaber to Menz Guassa plains and terrains from Shewarobit to Ataye, in these areas it quickly takeover valuable grazing lands and most of the mountainous areas which was covered by many indigenous herbs and shrub species.

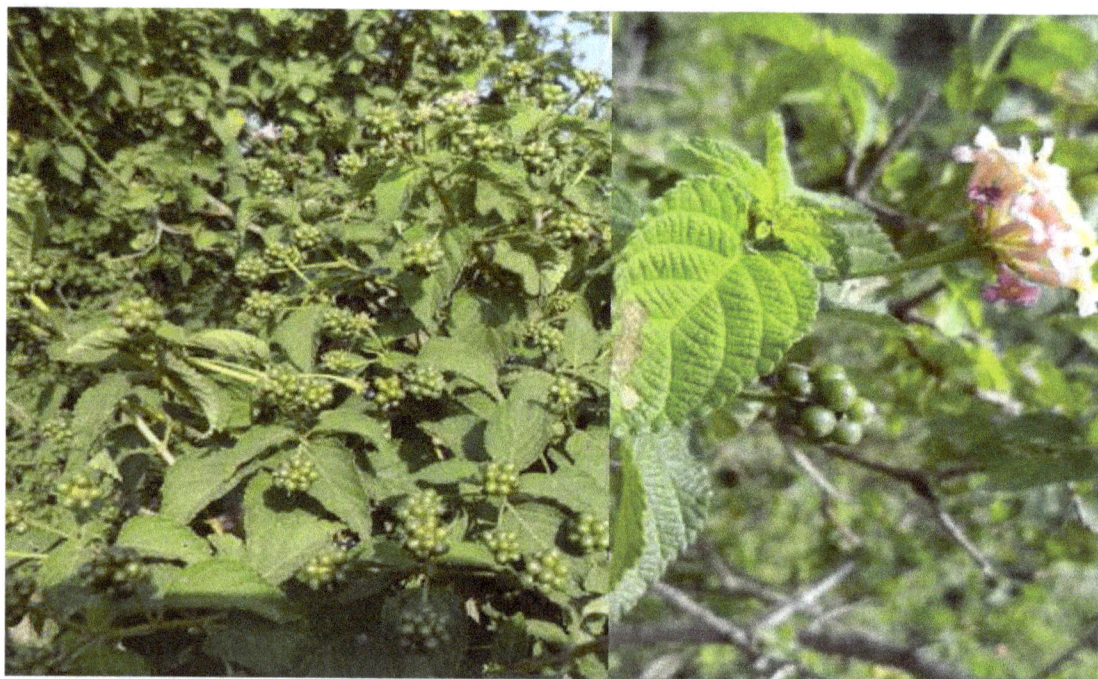

Plate 1. *Photo by Ablyou Tilahun and Hailu Tereje, 2005E.C.*

(iii). Over Grazing by Livestock

Our results confirmed that livestock grazing is the major factor limiting seedling establishment and seedling survival and growth in church forests(Table 3). Studies in Ethiopian highlands showed that heavy grazing pressure significantly increased surface runoff and soil loss and reduced infiltration capacity of the soil which in turn undermines suitability of sites for germination (Mwendera and Mohamed Saleem, 1997). This suggests that the effect of grazing on seedling survival and growth depends on the species. In general controlling livestock grazing is of paramount important for

both the internal regeneration of church forests and for restoration of the degraded surroundings.

Table 2. *Localy threatened tree species identified by stakholders in the PRA.*

Rank	Identified species	Percentage of priority
1st	*Juniperusprocera*	65
2nd	*Hageneaabyssinica,*	21
3rd	*Oleaeuropaeasubsp.Cuspidata*	11
4th	*Podocarpusfalcatus*	2
5th	*Ekebergiacapensis*	1

Source: Participants of the PRA discussion, 2005 E.C

According to the participants, population growth, grazing by domestic animals farm expansion and natural forest fire (lightening) are the major causes of the depletion of Religion natural forest, in this order of importance (Table.2). All the problems facing natural forest are strongly interrelated to each other.

4.2. Analysis of the Study Areas

4.2.1. Debre Medihanit Abune Melketsedik Monastery in Mida Woremo District

Washa Debre Medihanit Abune Melketsedik monastery is found in Mida weremo District, and was built in the 14th Century in the name of Abune Melke tsiediek. It is located at $10^0 14' 19$" N and $039^0 05' 07$"E at altitudinal range of the forest varies between 2299 to 2580 m a.s.l. The mean annual minimum and maximum temperature of the area is 16^0 and $30^0 C$ respectively. The mean annual rainfall of this area is 1427 mm. From this information it could be concluded that these monastery forests lie within the moist "Weina Dega" Climatic Zone Classification of Ethiopia. The actual forest covers 45 ha.The most important both religional and cultural features are the Monastery and homesteads of the Mida weremo people particular interest even in the national context, this is due to its thousands of old intact mummy. Various churches in the area have old and extremely valuable artefacts, which are of interest to tourists.

Woody species composition:The monastery has a very large and old Weira (*Olea euroapea sub sp cuspidata*) trees boardered by farmlands in the South,North and West are large cliffs in the East and plain in the South East.The upper canopy of these forests was dominated by *Albizia schimperiana, Millettia ferruginea, Croton macrostachyus, Celtis africana* and *podocarpus falcatus*. Smaller trees of the forest included *Ackocanthera schimperi, Dombeya torrida, Olea europaea sub sp cuspidata, Allophylus abyssinicus, Ekebergia capensis* and *Pittosporum viridiflorum*. The shrub layer was dominated by *Budleja polystachya, Carissa spinarum, Euclea schimperi, Maytenus arbutifolia* and *Calpurnia aurea*.

Threats:The monastery forest has faced much pressure from the local people. Areas other than owned by monastery are totaly degraded and gradually eroding the edges of the monasteries particularly that of western and Northern part of the monastery.

4.2.2. Debre Bisrat Abune Zena Markos Monastry in Moret and Jiru District

The Monastery is found in Moretena Jiru district, and was built 12th century in the name of Abune Zena Markos. It is located at $09^0 52' 05$"N and $039^0 04' 29$"E at an altitude of 2866 m.a.s.l. The actual forest area covers 23 ha. The monastery forest resources are declining in both quantity and quality. The natural forest is composed of mainly over-mature trees with poor prospect for regeneration. The destruction of the forest aggravated soil erosion to the southern part the District.

Woody species composition: The characteristic species of Abune Zena markos Monastery natural forest are *Juniperus procera* and *Olea europaea ssp cuspidata*. The density of these species having DBH above 10 cm is 130, 28 and 20 stems/ha respectively. *Maytenus arbutifolia* is the main under-storey species and also the regenerating species in the natural forest. There are 16 indigenous tree and shrub species each representing different families. This forest also contains relatively large plantations of *Cupressus lusitanica* and *Eucalyptus globulus*. The complete floristic list of indigenous tree and shrub species of Abune Zena markos monastery forest is given in Appendix 1. There are old aged huge tree species more than 2000 years old planted by St. Abune Teklehaimanot which can be considered as the key stone species for other organisms, especially *Juniperus procera* and *Olea euroapea sub sp cuspidata*.

Threats: The major threats on this forest are mainly from human pressure for wood products and for expansion of agricultural land. There is severe illegal cutting of trees for fuel wood and construction purposes. The local people are settling in the forest and converting the forest in to agricultural land. Grazing by domestic animals especially, cattle, goats and sheep in the forest are common practice of the local community.

4.2.3. Debre Menkirat Mitak Amanuel Monastery in Between Ankober and Basona Worana

This monastery is found at the boarder of both Ankober and Basona werana District and built in the first half of the 16th C in the names of St. Amanuel, St. Teklehaimanot and St. Gebriel . It is located at $09^0 34' 27$" N and $039^0 41' 07$" E with an altitude of 2837 m.a.s.l. The actual forest area is 78 ha surrounding a sloping hill. The forest has become a small refuge for a large variety of fauna and flora and numerous native species. The forest is an important source of both seeds and seed dispersers vital to traditional shifting cultivation practices, and of herbs for local medicinal, social, and religious purposes.

The forest has been protected and managed by religious leaders and villagers for centuries. When it was first demarcated, unwritten regulations were put in place by the Monastery monks and priest and other village leaders regarding forest land use in and around the forest. Over time, some of these rules have been amended to ensure their continued relevance and effectiveness. Today, they protect the original forest by regulating the behavior of the people of area and to some extent of the residents of neighboring communities are violating this regulation. All forms of deforestation and grazing in the around the forestare prohibited. Entrance into the forest area is permitted only during annual Holy days.

Shortage of fuelwood, farm implements, and grazing lands exist in and around the forest despite the care evident in the protection of the area. Women walk up to 5 kilometers to collect fuel wood; one full day of searching is commonly required to gather enough fuel wood for three days. Residents around the monastery know they could make use of the forest resources in the forest, but their faith and their forest management regulation prohibites them from doing so.

Woody species composition: The upper canopy of this

forest was dominated by characteristic species such as *Juniperus procera* and *Olea europaea sub sp cuspidata*. Smaller to medium sized tree species of this forest, which posses the middle canopy, included *Croton macrostachyus, Rhus glutinosa, Nuxia congesta, Pittosporum viridiflorum* and,

Ekebergia capensis. The shrub layer is dominated by *Ross abyssinica, Calpurnia aurea, Carissa edulis, Dodonaea angustifolia, Euclea schimperi, Maytenus addat, Maytenus arbutifolia, Myrsine africana, Osyris quadripartita,* and *Premna schimperi.*

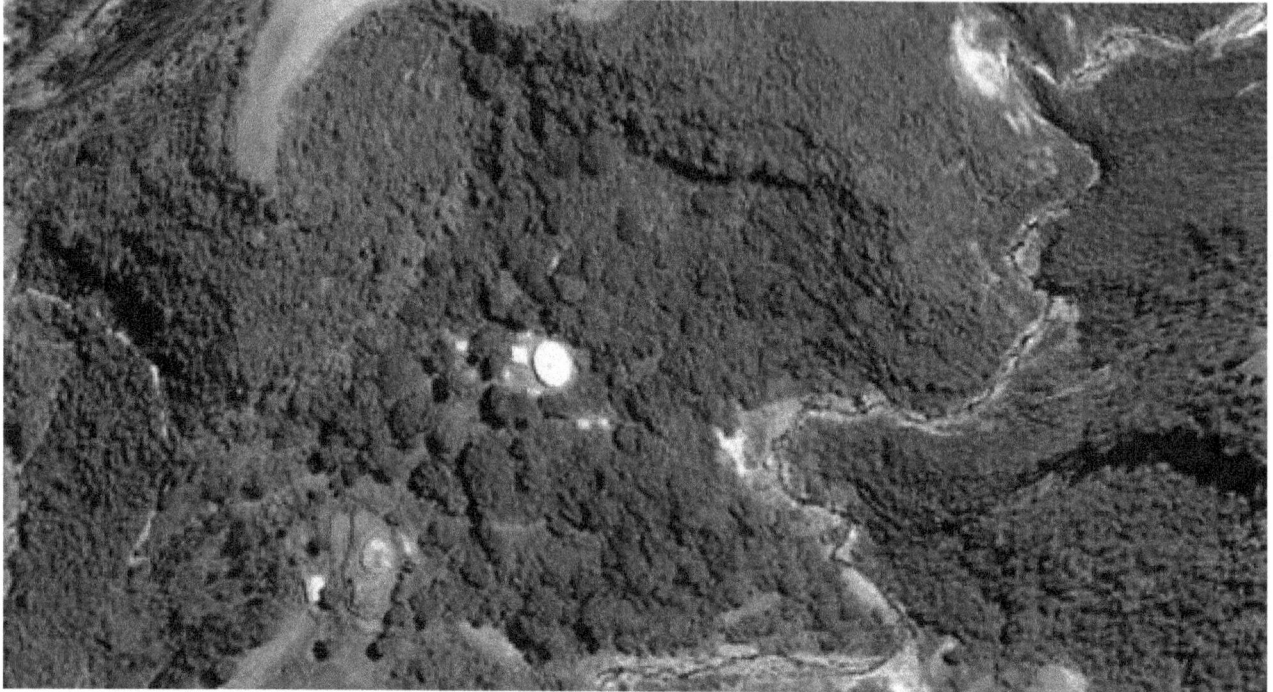

Plate 2. Well protected part of Mitak Amanuel forest Source from Google earth of imagery date 1/29/2014.

Threats: The major threat to this forest is mainly from human pressure for wood products. There is severe illegal cutting of trees for plumbering, fuelwood and construction. Most of the surrounding local people depend on the forest for their living particularly by selling timber for house construction. The Mitak Amanuel Monastery forest is found in an area where there is a serious shortage of wood for energy source and other uses. According to the monastery people forest products are transported all the way to Ankober, Debele and Debre Birhan. Deforestation rate is very high and the kebele administration and the concerned bodies have not taken appropriate measures. There is a fear that the monastery forest can be lost within a short period of time unless drastic measures are taken.

4.2.4. Debre Hail Kidus Rufael Church in Menz Gera District

This church is found in Menz Gera District, and was established in 1924 by Atse Be-Ede Mariam as both spiritual and palace. It is located at $10^0 15'32''$ N and $39^0 32'24''$ E at an altitude of 2921 m.a.s.l. The actual forest area covers 40 ha. This site is highly encroached by the local and nearby dwellers. According to the church scholars and dwellers, there are two sections of the forest. The inner section, next to the church demarcated by a stone fence, is entirely forbidden for animals while in the outer section of the forest, animals are free to rest under the shade. It is forbidden to cut down trees.

4.2.5. Debre Menkirat Gashu Amba Merkoriwes in Tarmaber District

This church is found in Tarmaber District, and was established during the era of Amha Iyesus as spiritual site. It is located at $09^0 51'52''$ N and $039^0 50'25''$ E at an altitude of 1947 m.a.s.l. The actual forest area covers 3 ha which is surrounded by farmlands in all direction.

Woody species composition: The upper storey of the forest is non-uniform and non-compact layer of trees like *Celtis africana, Juniperus procera* and *Euphorbia abyssinica*. The most dominant trees in the forest are *Olea europaea sub sp cuspidata, Ziziphus mucronata, Terminalia brownii, Syzygium guineense, Rhus glutinosa, Carissa spinarum, Croton macrostachyus, Calpurnea aurea, Euphorbia abyssinica, Opuntia ficus indica, Lippia adoensi, Premna schimperi, Adhatodia schimperiana, Eucalyptus camaldulensis, Euclea schimperi, Pterollobium stellatum, Dodonaea angustifolia* and *Euphorbia tirucalli*. There are significant numbers of species which are very aged and serve as flagship species for more *species of animals. Among which Ficus vasta* and *Juniperus procera* are the most dominant. In addition to planed tree plantation by the church administration, there are volunteers who plant trees around the church holdings. Volunteers plant trees that they choose for the purpose of the memory of themselves even after they had gone. For example a passed father called Aba Tariku planted Warka *(Ficus vasta)* 250 years ago at the front side of the gate where holiday celebrations and mass gatherings for various purposes going on today. Everybody call the tree father Tariku's tree. The tree has huge and deep roots, circular canopy with wide branches. It is the grace of the vicinity when seen from distant(Plate .3).

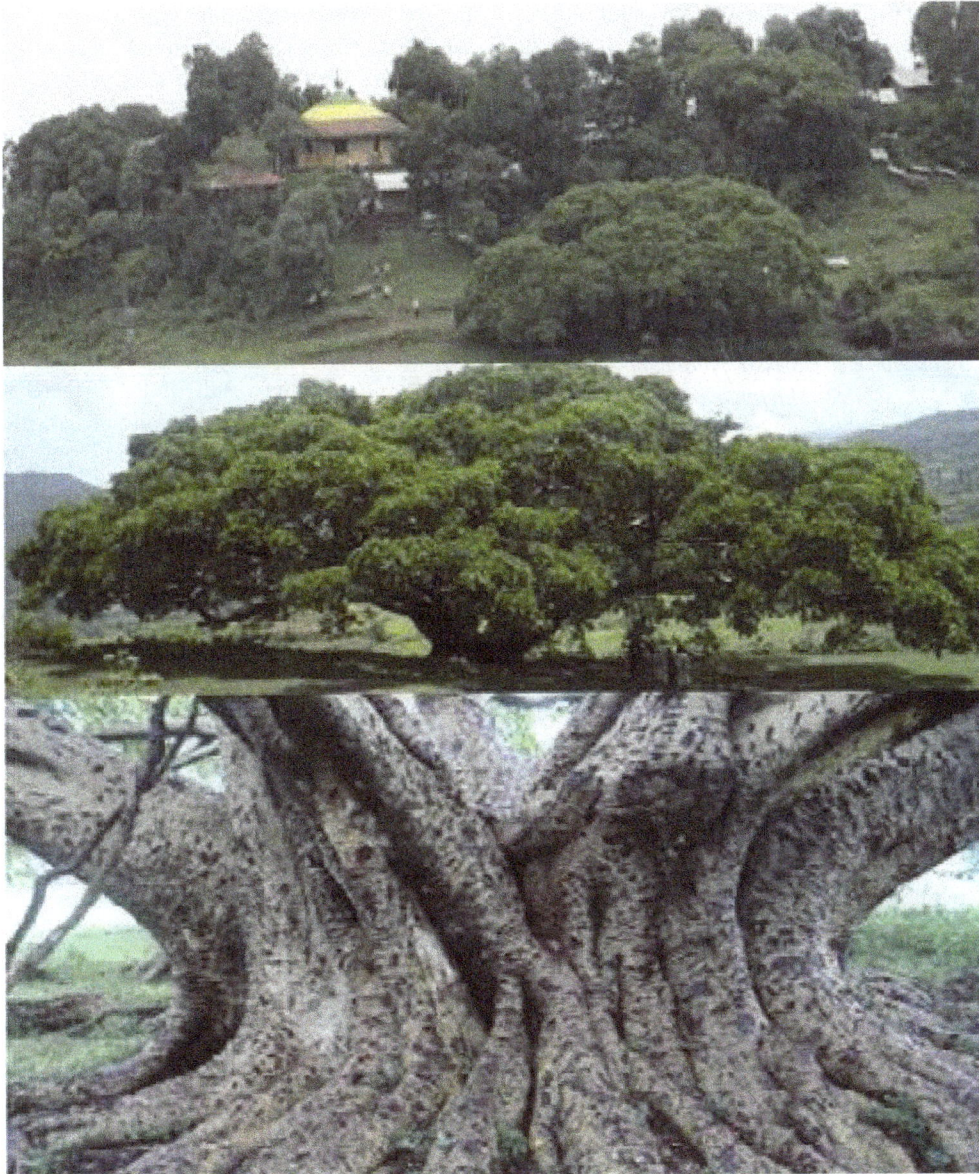

Plate 3. Ficus vasta planted by Aba Tariku 250 years ago, Photo by Abiyou T. and Hailu T., 2005E.C.

The area has relatively low humidity and limited rainfall where there is a prolonged dry season. Forests have diminished due to human interference and have been replaced by bushlands in most areas. Soils have become shallow as a result of soil erosion that has been taking place for centuries.

The church community has comments on the current soil and water conservation and rehabilitation programs of the government. According to the result of PRA, the effort made by the government to reclaim degraded areas and non-productive lands is very good. However, it would be more effective if government give us the training and try to address the people through priests since every farmer is the son/daughter in God of them. Furthermore, the seedlings that are brought to a certain area should be selected seriously for their adaptation to that specific locality.

There is no deforestation practice around the church; so that many aged trees among with the forest diversity are kept for long periods of time. The reason for this, according to the church community, is that people face problems when they cut and take plants. Some people also do not want to cut trees from the church vicinity because of religion honors. Even though those ancient churches with their forest diversity live in the community for centuries, people couldn't learn from them and plantation of indigenous trees around the homesteads of the farmers is not practiced. There are two reasons for this situation in Gashu Amba. First, Church fathers didn't give due attention and advice to their son/daughter in God to plant trees and conserve nature; second, farmers plant Eucalyptus trees for log, construction, firewood and other economical use. They focus mainly on the short term benefits of Eucalyptus than the long term sustainable environment by planting indigenous trees.

4.2.6. Debre Kerbe Dagmawi Goligota- in Basona Werana District

The forest is found in Basona werana with seven killo meter distance in the North from Debre Birhan town. Its geographical location is at 9^0 49' 23"N and 039^0 27' 54" E with an altitude of 2,786 m.a.s.l. and has an area of 18 ha forest cover.

Woody species composition: The upper canopy of this forest was dominated by *Olea europaea sub sp cusidata*, *Croton macrostachyus*, *Allophyllus abyssinicus*, *Juniperus procera* and *Eucalyptus globulus*. Smaller trees in this forest included *Nuxia congesta*, *Rhus glutinosa* and *Hypericum revolutum*.The shrub layer was dominated by *Rossa abyssinica*, *Calpurnia aurea*, *Carissa spinarum*, *Dovyalis abyssinica*, *Bersama abyssinica*, *Maytenus arbutifolia*, *Osyris quadripartita*, *Maesa lanceolata* and *Myrsine africana*. There were 56 indigenous tree and shrub species representing 31 families. The complete floristic list of indigenous tree and shrub species of Golgota forest is given in Appendix 6.

Threats:Though local people that reside around the forest exert some pressure on the forest for fuelwood and construction wood, it is not as severe as other monastery forests.

5. Discussion

5.1. Expperience of EOTC Forest Conservation Practiceses and the Habit of Government to Apply Traditiona Religious Knowledge

The main consraints obserbed which hinder to use local ecological knowledge on both the government and locl community were:

5.1.1. Poverty

This can be reflected by lack of grazing land for their domestic animals due to agricultural land expansion as a factor of population pressure. The farming community focus on the short term benefits like production of food crops, feed for their livestock and fast growing trees like Eucalyptus tree for construction and fire wood. Hence, the holdings of the farming community including the communal lands are severly degraded and devoid of diverse tree species.

5.1.2. Lack of Attention to Knowledge Dissemination on Forest Conservation Practices

The central objectives of religious education is about dogmatic and cannonical doctrine of the church which is about what they are allowed to do and not to do. They know that plant species are bases for the life of human beings in spiritual, medicinal, social, economical, environmenal and ecological perspectives. But there is no more experience and habit of teaching to the followers about conservation. As it is listed in Table 1 76% of the PRA participant agreed that there is lack of education about knowledge and best habit transfer to the young generation and natural resources expertise. The monasteric community revealed "nature conservation knowledge transfer" as a constraint practice in the church.

They address their concern that many religious institutional education systems do not value to knowledge transfer. They request cooperation and training from government and outsiders toward their sacred plants conservation and utilization. EOTC monks and clergies recognized the importance of experience sharing among other elders because they believe that this strengthens the conservation habits, instead of emphasizing the differences among religious institutions. They acknowledge the need to disseminate the experiences and knowledge of their ancestors among our peoples.

5.1.3. Lack of Co Operatiernal Bodies to Monastic Communities

According to the result of PRA, there should be signed formal agreements between religious organizations and the governmental representatives in terms of respection of organizational structure of monastic community. It is important to maintain the religious institutional arrangements and decision-making processes and avoid the inclination to interfere in some decision-making processes. With respect to participation, the religious community request to be consulted on time and efficiently by the governments and the private sector regarding any action or program that may directly or indirectly affect their vegetation and other resources.

5.1.4. Gradual Decline of Cultural Integrity of Some Young Generations

Monastric communities that have strong historical continuity, cultural and spiritual heritage should be supported. These communities are known to preserve, develop, and transmit their innate identity and ancestral territories to future generations as the basis of their continued existence. The church community in accordance with their own cultural patterns spritual belifs and legal systems need recognition and support in order to pass their heritage to the next generation; because contributions to indigenous knowledge and spiritual beliefs can reflect improved conservation efforts. Most of the religious leaders recognize that all members of their community are united because they have a common origin and a spirituality that is founded on love and respect for nature. Their knowledge is practical and collective and is directly linked to people's daily lives. They also acknowledge the respect that they have for their elders as they are the carriers of knowledge and history and they remind them of conserving and preserving their own culture and history. But, by nowadays the young generation is moving to the imported fashions than the domestic and original cultures and spiritual heritages due to unwise utilization of current and global products.

5.1.5. Lack of Clear Demarkation of Territories of Sacred Sites and Experience Sharing

According to the result of PRA, the first thing that 95% of the monastery community mentioned is the importance of demarcation to their survival and that they understand how important it is to live in harmony with nature. They have preserved certain areas within their territories that are sacred, such as rivers, caves, lakes, springs, islands, etc. They acknowledge their common history of the loss of territories. For this reason, they think that the land cannot be sold and they request common titles to their lands through presidential decrees or other similar means. They call to all humanity so that together we can preserve nature and avoid water and air pollution, deforestation of our forests, and the indiscriminate exploitation of resources.

5.2. Major Threats to Church Forests in the Selected Mnasteries of North Shewa Zone

This study has got experiences of the religious institution as the major contributor for forest conservation and origin of endangered species like *Ackocantra schimperi*, "etsezewie", "etsehiwot" and etc, in north Shewa. But there are root causes for the loss and destruction of such species in particular and forest resources in general.

Clearing for Agriculture: Conversion of natural forests to agriculture is the greatest threat due to population growth, low productivity of agriculture, unsustainable agricultural systems due to soil erosion.

Unsustainable levels of harvest: Over-harvest of forest products are occurring on a widespread basis leading to deforestation. The main causes are demographic pressures and growing demand, especially urban demand, for firewood, charcoal and other products.

Over-grazing:Overgrazing results in decreased soil cover, increased erosion, elimination of the natural regeneration of plant species and loss of biodiversity. Root causes include demographic growth and lack of incentives for sustainable use

Alien invasive species: are one of the major factors threatening biodiversity resources in north Shewa especially Kewet District. The prominent alien species that cause damage across the country include *Parthenium hysterophorus*, *Prosopis juliflora, Eichornia crassipes* and *Lantana camara. Lantana camara* is becoming a major problem by replacing natural vegetation in to shrubby areas. This problem is intensified from Shewarobit to Kombolicha.

Extreme care is required in the selection of species to be introduced to minimize any impact on native species.

5.3. Relative Success of the Ethiopian Orthodox Church in Protecting Natural Forest

5.3.1. Strong Beliefs of Local People and Religious Leaders

The small pocket of forest in the monastery is protected primarily because it is the sanctuary of the Monks and nuns. Indeed, it is difficult to isolate the remnant patches of forest-related activity from religious beliefs and practices surrounding the monastery / churches. In the people's minds, the defamation of the forest would dishonor the monk's community and bring misfortune to the offending individual and the community. By protecting the forest, the people of Mitak derive many benefits in addition to the spiritual rewards of serving their faith. The people's strong beliefs in the religion, ensures that only few villagers violate the rules safeguarding the sacred grove; indeed villagers are key to protecting it from non-believers. Community vigilance rather than formal policing is sufficient to enforce the regulations.

5.3.2. Locally Accepted Protection Guidelines for the Church Forest

The successful establishment and preservation of the forests trace their roots to informal regulations and practices founded on the spritual worship. For centuries, local people have adhered to specific guidelines that restrict land use in and around the church and have performed activities to secure the forest from human interferences. By respecting the church's privilege in entering the forest, keeping their compounds, farms, and livestock out of the church forest and protecting the holy site from fires and other threats, residents have a stake in local adherence to these rules.

5.3.3. Regional Collaboration and Support

Church forests are becoming increasingly important in the area, adding strength to church representative's beliefs and protection efforts. As religious beliefs, small communities are weakened and their pressures on forest resources increase, natural forests in many communities are encroached upon and forests are degraded. This is especially the case in areas Mitak Amanuel, Abune Zena Markos and Gashu Amba Qidus Merkorios due to the proximity of the forest to urban centers where religious systems and commonly-held natural resources are exceptionally vulnerable to modernization and large-scale development. Such forest areas are threatened by encroachment, trampling or destruction and most local resource base has been so degraded that the forests are under an urgent need of reforestation

5.4. Forest Conservation Strategies of Monasteries of EOTC

According to the result of the interview and PRA, EOTC apply two conservation strategies, namely religious holiness and/or legal protection.

5.4.1. Religious Holiness

The main mode of protection is achieved through creating religious commitment and respect among the followers. As the church is believed to be the house of God, everything in the compound is sacred and respected. Every follower is expected to respect and protect the house of God together with the forest enveloping the church. Cutting a tree in the church compound is considered as denying the presence of God unless; it is for the special purpose of the church. It is believed that cutting in and smuggling of trees from the church compound would bring a misery and the one who did it is considered as a person who has violated the Kingdom of God and would be isolated from the church communities. A person that cuts atree or even a dead branch for personal use would be presented to the church community / church scholars and asked to repent and be committed not to repeat the mistake again. If the person fails to admit his/her mistake voluntarily He/she would be alienated from the church community which is said to be 'Gizit'. Hence, since Orthodox Christians fear such conditions and protects the trees in church forests.

5.4.2. Legal Protection Methods

Monastery forest resources are found around people of different attitudes and perspectives of which force to use guards and civil law to conserve and protect their forest resources. At present the demand forest product is increasing the encroachers and outlaws destruct the forest. Therefore, EOTC assigned guards to prevent such conditions before it happens and to bring to the civil courts for the appropriate measures.

6. Conclusion

Religion and cultural practices have contributed in the conservation of resources through the ascription of spiritual powers to soil, water, forest and etc.

Monastery and church forests are very important repositories for both faunal and floral resources of Ethiopia, especially for endangered and endemic species as sources of germplasm particularly of tree seeds for the conservation of these species. They are indicators of species to be selected for conservation programs in their respective localities. In regions where the original vegetation cover has gone, the islands of monastery and/or church forests are indicators of the Agroecological Zones and of past natural vegetation.

Monastery and/or church forests are excellent centres of learning and research. They are ideal sites for studies on vegetation history, ecology, taxonomy and other fields of biology and forestry. *In situ* forest conservation sites can be established in several of the monastery forests. Monastery and church forests are sources of knowledge on biodiversity including its uses. Moreover, they could serve as models of sustainable forest management with a minimum amount of human intervention.

Monastery forests are under severe threat and are declining in their regenerating potentials. Conservtion of monastery forests by the churches becoming beyond its capacity to save them from the pressure of the local people for agricultural land expansion, grazing and destructive timber production. The issue of ownership aggravated the rate of deforestation as the surrounding people. The Ethiopian Orthodox Tewahido Church should be entitled of its forests. This may require the recognition of monastery and church forests as one of the forest ownership categories by the government. Anthropogenic pressure is byfar the most devastating challenges in the conservation of monastery forest. The traditional concept of forest conservation is in need of shift to a new concept of forest genetic resource conservation and the customary practices are also in need of modification with advanced technologies as well.

To prevent the loss of invaluable biological and natural resources, a rescue operation should be launched by all concerned bodies and the intervention of the Ethiopian government is crucial to save the remaining monastery and church forests. *In situ* conservation sites should be delineated and managed in appropriate monastery / church forests and seeds of threatened species should be collected for *ex situ* conservation measures.

Research institutes, IBC, Debre Birhan University, Forest research center, Amhara forest enterprise, and Agriculture and Natural resource conservation experts need to be integrated to use the accumulated forest conservation practice, knowledge and germplasms. Further studies are recommended in other aspects of monastery and church forests.

Appendix

Appendix 1. List of plant species collected from selected monasteries of North Shewa.

No	Scientific Name	Family	Ha	Local name(Am)
1	*Acacia abyssinica* Hochst.exBenth.	Fabaceae	T	Bazragirar
5	*Acokanthera schimperi*	Apocynaceae	T	Merenz
2	*Albizia schimperiana*	Fabaceae	T	Sessa
3	*Allophylus abyssinicus*	Anacardiaceae	T	Embis
6	*Apodytes dimidiata* E.Mey.ex.Arn	Icacinaceae	T	Donga
7	*Arundinaria alpina* K. Schum	Poaceae	T/S	Kerkeha
8	*Asparagus africanus* Lam.	Asparagaceae	S	Sriti
9	*Balanites aegyptiaca*	Balanitaceae	T	Bedeno
10	*Bersama abyssinica* Fresen.	Melianthaceae	S	Azamir
11	*Buddleja polystachya* Fresen.	Loganiaceae	T	Anfar
12	*Calpurnia aurea*(Ait.) Benth.	Fabaceae	S	Digita
14	*Capparis tomentosa*	Capparidaceae	T	Gumero
13	*Carissa spinarum* L.	Apocynaceae	S	Agam
15	*Celtis Africana*	Ulmaceae	T	Kawoot
17	*Clematis simensis* Perr. And Guill.	Ranunculaceae	C	Azohareg
18	*Clerodendrum myricoides* Vatke.	Lamiaceae	S	Misirch
19	*Clutia lanceolata*Forssk.	Euphorbiaceae	S	Fiyelefej
20	*Cordia Africana*	Boraginaceae	T	Wanza
21	*Crotalaria laburnifolia* L.	Fabaceae	S	x

No	Scientific Name	Family	Ha	Local name(Am)
22	*Crotalaria mildbraedii Bak.f*	Fabaceae	S	x
23	*Croton macrostachyus Del.*	Euphorbiaceae	T	Etsezewie
24	*Cupressuslusitanica Mill.*	Cupressaceae	T	Yeferenjitsid
25	*Cyathula uncinulata(Schrad)*	Amaranthaceae	S	x
26	*Discopodium penninervium Hochst.*	Solanaceae	S	Ameraro
93	*Dodonaea angustifolia*	Sapindaceae	T/Sh	Kitkita
27	*Dombeya torrida*(J.F. Gmel) P. Bamps	Sterculiaceae	T/S	Wolikfa
28	*Dovyalis abyssinica Warb.*	Flacourtiaceae	S	Koshim
29	*Dracaena steudneri*	Boraginaceae	T	Etsepatos
30	*Ekebergia capensis Sparrm.*	Meliaceae	T	Lol
31	*Erica arborea L.*	Ericaceae	S	Asta
90	*Eucalyptus camaldulensis*	Myrtaceae	T	Key bahirzaf
32	*Eucalyptus globules Labill.*	Myrtaceae	T	Nech Bahir zaf
91	*Eucalyptus globulos*	Myrtaceae	T	Nechi bahirzaf
33	*Euclea schimperi*	Ebenaceae	S	Dedeho
34	*Euphorbia abyssinica*	Euphorbiaceae	T	Kulkual
94	*Euphorbia tirucalli*	Euphorbiaceae	S	Kulkual
35	*Ficus sur*Forssk.	Moraceae	T	Shola
36	*Ficus vasta*	Moraceae	T	Warika
37	*Galiniera saxifrage*	Rubiaceae	T	Yetotakula
38	*Grewia ferruginea*	Tiliaceae	T	Lenkoata
39	*Hagenia abyssinica J.F.Gmel.*	Rosaceae	T	Kosso
40	*Halleria lucida L.*	Scrophulariaceae	T/S	Masinkero
41	*Hypericum revolutumVahl.*	Hypericaceae	S	Amija
42	*Inula confertiflora A. Rich.*	Asteraceae	H	Weinagift
43	*Jasminum abyssinicum*Hochst.ex DC.	Oleaceae	Li/C	Tenbelel
44	*Juniperus procera Hochst ex Endl.*	Cupressaceae	T	Tsid
45	*Justitia schimperiana*(Hochst ex Nees) T.Anders.	Acanthaceae	S	Simyiza
46	*Leucas stachydiformis. Briq.*	Lamiaceae	S	Limich
47	*Lippia adoensis Hochst .ex Walp*	Lamiaceae	S	Kesie
48	*Maesa lanceolata Forssk.*	Myrsinaceae	T/S	Abeye
50	*Maytenus arbutifolia(A.Rich.) Wilczek*	Celastraceae	T/S	Atat
51	*Maytenus obscura(A.Rich.) Cuf.*	Celastraceae	T	Kumbel
54	*Myrica salicifolia Hochstex.A.Rich.*	Myricaceae	T	Kelewa
55	*Myrsine africana L.*	Myrsinaceae	S	Kechemo
56	*Nuxia congesta R. Br. Ex Fresen.*	Loganiaceae	T/S	Anfar
57	*Ocimum lamiifolium*Hochst. exBenth.	Lamiaceae	S	Damakesie
58	*Olea capensis*	Oleaceae	T	Damot weira
59	*Olea europaea subsp.* Cuspidata Wall.ex.G.Don.	Oleaceae	T	Weira
61	*Olinia rochetiana A. Juss.*	Oliniaceae	T	Tifie
62	*Opuntiaficus-indica (L.) Miller*	Cactaceae	S	Beles
63	*Osyris quadripartitaDecn.*	Santalaceae	S	Keret
64	*Periploca linearifolia*A. Rich & Quart.-Dill.	Asclepiadaceae	L	x
65	*Phoenix reclinata*	Arecaceae	T	Selen/zenbaba
66	*Phytolacca dodecandra*L "Herit	Phytolaccaceae	Li/C	Endod
67	*Pinus patula*D.Don.	Pinaceae	T	Pachula
68	*Pittosporum viridiflorum*Sims	Pittosporaceae	T	Weil
69	*Podocarpus falcatus (Thunb.) Mirb.*	Podocarpaceae	T	Zigiba
16	*Premna schimperi*	verbenaceae	T	Chocho
70	*Prunus africana (Hook.f) Kalkm*	Rosaceae	T	Tikurenchet
71	*Rhamnus prinoides*LHer.	Rhamnaceae	S	Giesho
72	*Rhamnus staddo*A.Rich.	Rhamnaceae	S	Tsedo

No	Scientific Name	Family	Ha	Local name(Am)
74	*Rhus glutinosa*Gilbert	Anacardiaceae	T	Tilem/talo
73	*Rhus vulgaris*Meikle	Anacardiaceae	T	Kimo
75	*Ricinus comminus L.*	Euphorbiaceae	S	Gulo
76	*Rosa abyssinica*Lindley	Rosaceae	S	Kega
77	*Rubus steudneri*Schweinf.	Rosaceae	C	Enjori
78	*Rubus volkensii*Engl.	Rosaceae	C	Enjori
79	*Rumex nervosus*Vahl	Polygonaceae	S	Embuacho
80	*Salix subserrata*	Salicaceae	T	Wonzadmik/akaya
82	*Sideroxylon oxyacanthum*Baill.	Sapotaceae	S	x
81	*Solanecio gigas*	Asteraceae	H	shikokogomen
83	*Sparmannia ricinocarpa*(Eckl. & Zeyh) O.Ktze	Tiliaceae	H	Welikifa
84	*Stephania abyssinica (*Dill. Rich.) Walp.	Menispermaceae	Li/C	Ayithareg
96	*Syzygium guineense*	Myrtaceae	T	Dokima
95	*Terminalia brownii*	Combretaceae	T/Sh	Abalo/weyba
85	*Thymus schimperi* Ronniger	Lamiaceae	H	Tosign
86	*Urera hypselodendron*(A.Rich.) Weed.	Urticaceae	Li/C	Lankuso
60	*Vernonia amygdalina*Del.	Asteraceae	T	Girawa
88	*Vitis vinifera*	Vitaceae	Li/C	Wein
89	*Zehneria scabra*	Cucurbitaceae	H	Etse sabek
92	*Ziziphus mucronata*	Rhamnaceae	T	Geba/kurkura

Key: (Ha=habit, T=tree, S=shrub, T/S=tree/shrub, C=climber, SCs=scandent shrub, L=Liana, E=epiphyte, Am=Amharic, and F=fern).

References

[1] Abeje Eshete, Demel Teketay and Hulten, H. H. (2005). The Socio-Economic Importance and Status of Populations of *Boswellia papyrifera* (Del.) Hochst in Northern Ethiopia: The Case of North Gondar Zone. *Forests Trees and Livelihoods* 15: 55-74.

[2] Alemayehu Wasie (2002). Opportunities, Constraints and Prospects of Ethiopian Orthodox Tewahido Churches in Conserving Forest Resources: The Case of Churches in South Gonder,Northern Ethiopia. MSc thesis, Swedish University of Agricultural Sciences, Skinnskatterberg, Sweden.

[3] Alemayehu Wasie, Demel Teketay and N. Powell (2005). Church forests in North Gondor Administrative Zone, Northern Ethiopia. Forests, Trees and Livelihood 15(4): 349-374.

[4] Aymro Wondmagegnehu and Motovu, J., (1970). *The Ethiopian Orthodox Church*. The Ethiopian Orthodox Mission, Addis Ababa, Ethiopia.

[5] Azene Bekele, (1993). Useful trees and shrubs for Ethiopia: Identification, Propagation and Management for Agricultural and Pastoral Communities.

[6] Bahru Zewde, (1997). *The Forests of WalloIn Historical perspective*: With a Focus on the Yagof State forest. Paper prepared for the Conference on "Environmental and Development" Debre Zeit, 12-14 June 1997.

[7] Cunningham, A.B. (2001). *Applied Ethnobotany: People, wild plant use and conservation*. Earthscan, London.

[8] Dagnachew Gebeyehu, (2001). Sampling strategy for assessment of woody plant species diversity: a study in Tara Gedam natural forest, Ethiopia. MSc. Thesis submitted to Swedish University of Agricultural Science.

[9] Demel Teketay, (1997). Seedling population and regeneration of woody species in dry Afromontane forests of Ethiopia. *Forest Ecology and Management* 98 (1997), 149-165.

[10] Dereje Denu (2007). Floristic composition and ecological study of Bibita forest (Gurda Farda), Oromia National Regional State, west Ethiopia. Unpublished M.Sc Thesis, Addis Ababa University, Addis Ababa.

[11] Elias Endale (2003). Socio-economic data of Agriculture and Natural Resource, Agricultural Development, Gamo Gofa Zone of SNNP.

[12] Getahun Kassa, Eskinder Yigezu (2015). Women Economic Empowerment through Non Timber Forest Products in Gimbo District, South West Ethiopia. American Journal of Agriculture and Forestry. Vol. 3, No. 3, 2015, pp. 99-104. doi: 10.11648/j.ajaf.20150303.16

[13] Gadgil, M., and Vartak, V.D., (1976). The sacred groves of Western Ghats in India. *Economic Botany* 30:152-160.

[14] Kent, M. and Coker, p., (1992). *Vegetation Description and Analysis*: A Practical Approach. Belwill haven press,London.

[15] Martin, G. J. (1995). *Ethnobotany: A Methods Manual.* Chapman and Hall, London.

[16] Muller-Dombis, D. and Ellenberg, H. (1974). Aims and Methods of Vegetation Ecology, John Wiley and Sons. Inc. New York. pp 547.

[17] Muluken Mekuyie Fenta (20140. Human-Wildlife Conflicts: Case Study in Wondo Genet District, Southern Ethiopia. Agriculture, Forestry and Fisheries. Vol. 3, No. 5, 2014, pp. 352-362. doi: 10.11648/j.aff.20140305.14

[18] Mwendera, E.J. and M.A. Mohamed Saleem (1997). Hydrologic response to cattle grazing in the Ethiopian highlands. Agriculture, Ecosystems and Environment 64: 33-41.

[19] Siboniso M. Mavuso, Absalom M. Manyatsi, Bruce R. T. Vilane(2015). Climate Change Impacts, Adaptation and Coping Strategies at Malindza, a Rural Semi-Arid Area in Swaziland. American Journal of Agriculture and Forestry. Vol. 3, No. 3, 2015, pp. 86-92.   doi: 10.11648/j.ajaf.20150303.14

[20] Tadesse Woldemariam (1998). *Diversity of Woody Plants and Avifauna in a Dry Afromontane Forest, On the Central Plateau of Ethiopia*. SLU: Ethiopian MSc in Forestry Program thesis work.

[21] Taye Bekele, (1998). *Biodiversity Conservation: Experience of the Ethiopian Orthodox Tewahido Church*, MSc in Forestry Graduates Magazine, 1994-1998.

[22] Yeraswork Admassie(1995). *Twenty Years to Nowhere: Property Rights, Land Management* and *Conservation in Ethiopia*. PhD. Dissertation, Uppsala: Uppsala University.

[23] Zelealem Tefera (2001). *Common Property Resource Management of an Afro-Alpine Habitat: Supporting a Population of a Critically Endangered Ethiopian Wolf* (*Canis simensis*), Ph.D. Dissertation, Durrel Institute of Conservation and Ecology, University of Kent, Canterbury.

[24] Zewge Teklehaimanot (2001). Biodiversity conservation in ancient church and monastery yards in Ethiopia.

17

Impact of Mycorrhiza Fungi from Grassland Rhizosphere and Liquid Organic Fertilizer to the Growth and Yield of Sweet Corn on Ultisols in South Konawe, Indonesia

Halim[1], Makmur Jaya Arma[1], Fransiscus S. Rembon[2], Resman[3]

[1]Specifications Weed Science, Department of Agrotechnology, Faculty of Agriculture, Halu Oleo University, Southeast Sulawesi, Indonesia
[2]Specifications Soil Nutrition, Department of Agrotechnology, Faculty of Agriculture, Halu Oleo University, Southeast Sulawesi, Indonesia
[3]Specifications Soil Science, Department of Agrotechnology, Faculty of Agriculture, Halu Oleo University, Southeast Sulawesi, Indonesia

Email address:

haliwu_lim73@yahoo.co.id (Halim), makmurarma@gmail.com (M. J. Arma), fransrembon@yahoo.com (F. S. Rembon),
resman_pedologi@yahoo.com (Resman)

Abstract: This research was conducted in the Abenggi of village, District of Landono, Regency of South Konawe, Province of Southeast Sulawesi and Laboratory of the Faculty of Forestry and Environmental Science Halu Oleo University Kendari, Indonesia took place from November, 2014 untill April, 2015. This study aims to determine the effect of mycorrhiza fungi from granssland and liquid organic fertilizer to the growth and yield of sweet corn on Ultisols Abenggi. This research is compiled using a randomized block design (RBD) with factorial pattern. The first factor are mycorrhiza fungi (M) with three levels: without mycorrhiza fungi (M_0), mycorrhiza as 20 g each planting hole (M_1), mycorrhiza as 40 g each planting hole (M_2) and second factor are liquid organic fertilizer (P) which comprises three levels of treatment: without liquid organic fertilizer (P_0) liquid organic fertilizer as 50 ml L^{-1} water each plot (P_1), liquid organic fertilizer as 100 ml L^{-1} water each plot (P_2). The combination of these two factors obtained 9 combinations of each treatment was repeated three times in order to obtain the amount of 27 experimental units. The observed variables in this study were plant height, stem diameter, cob weight with husk, corn cob without husk, cob length, cob diameter, number and percentage of mycorrhiza infection on the roots of the sweet corn. The results showed that the interaction between mycorrhiza fungi and liquid organic fertilizer has a significant effect on the plant height age 42 days after planting and cob weight without husk with the best results obtained in the treatment of mycorrhiza as 40 g hole^{-1} with liquid organic fertilizer as 50 ml L^{-1} water. Application the mycorrhiza fungi as 40 g hole^{-1} gives the best results on the growth and yield of sweet corn.

Keywords: Sweet Corn, Mycorrhiza Fungi, Liquid Organic Fertilizer, Ultisols

1. Introduction

Corn is a crop that plays an important role as the main food of Indonesian society after the rice. In addition to corn as food demand continues to increase along with the development of the livestock sector which uses corn as feed material [17]. Importance of corn plants to be developed extensively since corn is one commodity that has a strategic role in fulfilling the nutritional needs, improve food security as well as the needs of industrial raw materials [14]. One of the main obstacles encountered in efforts to increase the production of corn, especially in Southeast Sulawesi is a type of soil which was dominated by Ultisols. Ultisols is a land that has a fairly advanced level of development, characterized by the cross section of deep soil, clay fraction rise along with the depth of this soil has the potential to Al toxicity and poor organic matter content. It also poor soil nutrient content, especially P and exchangeable cations such as Ca, Mg, Na, K, Al content is high, low cation exchange capacity and vulnerable to erosion [25].

Ultisols have a fairly extensive distribution covering almost 29.7% of the total the mainland of Indonesia that makes the land has an important role in the development of dryland farming in Indonesia [24]. Texture Ultisols usually varies greatly depending on parent material of soil. Ultisols are derived from granite that is rich in mineral quartz generally has a rough texture like sandy clay [26]. Ultisols reaction generally sour to very sour (pH 5 to 3.10), except Ultisols of limestone that have a neutral to slightly acid reaction (pH 6.80 to 6.50). Cation exchange capacity in Ultisols of granite, sediments and tuffs are low respectively cmol ranged from 2.90 to 7.50 kg-1, 6.11 to 13.68 cmol kg-1 and 6.10 to 6.80 cmol kg-1, while that of the material volkan andesitic and limestone is high (> 17 cmol kg-1). The results of research [16], shows that some of the material volkan Ultisols, calcareous tuffs and limestone have a high cation exchange capacity. According to the results of research [8], that condition of Ultisols Abenggi villages are: pH 5.77, C-Organic 1.92%, 0.17% Nitrogen, 12.75 ppm Phosphorus and 0.22 me100g-1 Potassium. Such circumstances may result in less optimal crop production in general and in especially sweet corn. To cope with things like this, it is necessary to touch the utilization of biotechnology such as mycorrhiza fungi and liquid organic fertilizer.

Mycorrhiza fungi are fungi that are mutualistic symbiosis with plant roots [6]. While indigenous mycorrhiza fungi is a type of mycorrhiza are found in association with plant roots naturally without human intervention in the process of initial infection between mycorrhiza with host plants, for example the mycorrhiza fungi from grassland rhizosphere [10]. The results of research [9] showed that in the area of secondary forest plant root were found the kinds of mycorrhiza fungi such as Glomus sp, Gigaspora sp sp and Acalauspora sp with different number of spores. Similarly, the results of research [11], that there are 15 species of plants are found in association with mycorrhiza fungi with infection rates of between 60% - 90%. According [22], that's infection of mycorrhiza fungi can enhance plant growth and the ability to utilize available nutrients in the soil. The colonization of mycorrhiza fungi on plant roots can expand the field of root uptake in the presence of external hyphae grow and develop through the roots [3]. The results of research [10], that the application of mycorrhiza fungi at a dose of 1 ton ha-1 is able to increase the production of corn to 10 tonnes ha-1 on Ultisols. Similarly, the results of research [8], application of mycorrhiza fungi can increase the production of corn ranged from 7.90 to 8.60 tonnes ha-1 compared to no provision of mycorrhiza fungi of 5.50 tonnes ha-1.

Organic fertilizer is the collective name for all kinds of organic materials of plant and animal origin that can be reformed into nutrients available to plants. The use of organic fertilizer could be a solution to reduce the excessive application of inorganic fertilizers due to the organic materials that can improve the physical, chemical and biological soil. Function organic fertilizer to the chemical properties of improving the cation exchange capacity, increase the availability of nutrients and increase the mineral weathering process. As to the biological properties that made the source of food for soil microorganisms such as fungi, bacteria and other microorganisms [7]. The liquid organic fertilizer containing potassium, which are crucial elements in any process plant metabolism, namely in synthesis and plays a role in maintaining good turgor pressure so as to allow the

smooth metabolic processes and ensure the sustainability of cell elongation [18]. The advantages of liquid organic fertilizer is to quickly overcome nutrient deficiency, it does not matter in nutrient leaching and able to provide rapid nutrient [28].

2. Methodology

2.1. Place and Time

This research was conducted in the Abenggi of village, District of Landono, Regency of South Konawe, Province of Southeast Sulawesi and Laboratory of the Faculty of Forestry and Environmental Science Halu Oleo University Kendari, Indonesia took place from November, 2014 untill April, 2015.

2.2. Materials and Tools

The materials used in this study were *raffia* ropes, water, soil, mycorrhiza fungi, sweet corn seed, polybag (size 20cm x 30cm, 40 cm x 50cm)), aquades, 30% sucrose, Acero Formalin Alcohol (FAA), 10% KOH solution, a solution of hydrogen 10% alkaline peroxide (H_2O_2), a solution of HCl 1%, dyes carbol fuchin 0.05%, laktogliserol, filter paper and paper labels. The tools used were tillage tools, vernier caliper, machetes, meter, digital cameras, filter to see mycorrhizal spore size (mesh size of 500lm, 250lm, 90lm, 60lm, and 50lm), analytical balance, autoclave, binocular microscope, glass measuring, petridish, pipettes, scissors, and stationery.

2.3. Research Design

This research is compiled using a randomized design block (RDB) with two factors. The first factor is doses of mycorrhiza fungi (M) with three levels: without mycorrhiza fungi for each planting hole (M_0), 20 g mycorrhiza fungi for each planting hole (M_1), 40 g mycorrhiza fungi for each planting hole (M_2) and the second factor is a liquid organic fertilizer (P) which comprises three levels of treatment: without liquid organic fertilizer (P_0), 250 ml of liquid organic fertilizer for each plot (50 ml L^{-1} water) (P_1), 500 ml of liquid organic fertilizer for each plot (100 ml L^{-1} water) (P_2). From these two factors obtained 9 combinations of each treatment was repeated 3 times in order to obtain the amount of 27 experimental units.

2.4. Procedure

Implementation procedures of the research are as follows:
(a) Preparation of land; land preparation includes cleaning, soil tillage, as well as the making of the plot.
(b) Application of mycorrhiza fungi; application of mycorrhiza fungi conducted simultaneously planting sweet corn plants given to each planting hole. The positions of mycorrhiza fungi are under seed of sweet corn [10]. Liquid organic fertilizer plant is given at the age of 7 days after planting. Sprinkled with the liquid organic fertilizer in the planting hole

evenly on experimental plots.

(c) Planting; planting be done by drill, each planting holes gets 2 seed of sweet corn, with a spacing of 20 cm x 75 cm.

(d) Maintenance; plant maintenance includes weeding, watering and replanting. Weeding was done by removing or cleaning the grass or weed that grows on a plot, in order to avoid competition and nutrient uptake by weeds. Watering was done to maintain the condition of field capacity. Watering was done twice a day, morning and evening, if it does not rain. Watering is done starting from the beginning of planting until the plants germinate or grow normally. Stitching is done at the time the plant was 7 days after planting.

2.5. Observation of Variables

The variables were observed in this study include:

(a) The plant height; plant height is measured from the base of the stem to the tip of the highest leaf by using a meter at the age of 14, 28 and 42 days after planting (DAP).

(b) The stem of diameter; stem diameter were measured using calipers at the age of 14, 28 and 42 DAP. The measurements were made approximately 15 cm from the ground.

(c) The weights of the cob with husk: weighed before corn cobs separated from its husk.

(d) The weights of the cob without husk; weighed after corn cobs separated from its husk.

(e) The length of the cob; measured from the base to the tip of a corn cob that contains the seeds.

(f) The diameter of the cob; measured by using a vernier caliper. The portions were measured, namely the base, middle and lower end portions cob containing seeds.

(g) The percentage of mycorrhiza fungi infection; preceded with staining roots. The steps in staining roots are as follows: (1) washing the roots with water, (2) saving the FAA for fixation prior to painting, (3) soaking in 10% KOH and heat with an autoclave for 15-20 minutes at 121^0C, (4) washing with distilled water 3 times, (5) soaking in hydrogen peroxide outsmart 10% (H_2O_2), (6) washing with distilled water 3 times, (7) soaking with HCl 1%, (8) wasting HCl without washed with distilled water, (9) soaking in carbon fuchin with concentration of 0.05% w/v in laktogliserol and heat at 900C for several hours or in an autoclave at 1210C for 15 minutes, (10) removing the paint and soak the roots in laktogliserol, and (11) observing the roots sample using a microscope (Brundrett, 2004) in [11]. The Observations were carried out using a dissecting microscope at a magnification of 40 times. Furthermore, mycorrhiza fungi infection was calculated by using the formula proposed by Brian and Schults (1980) in [10]: $IP = \frac{r1}{r1+r2} \times 100\%$.

Note:

IP = the percentage of mycorrhiza fungi infection

r1 = the number of root infected examples

r2 = the number of root not infected examples

2.6. Data Analysis

Data of each variable were observed were analyzed by variance of analysis. If the F count is greater than the F table, then continued with Duncan Range Multiple Test (DRMT) at 95% confidence level.

3. Result and Discussion

3.1. Plant Height and Stem of Diameter

Application independently of mycorrhiza fungi significantly affect to the average height of sweet corn plants at 14 DAP and significantly affect to stem of diameter at 14 and 42 DAP (Table 1).

Table 1. *Effect of application of mycorrhiza fungi to the average height of sweet corn plants (cm), stem of diameter (cm) at the age of 14 and 42 DAP.*

Treatment	Parameter		
	average height of sweet corn plants (cm)	stem of diameter (cm)	
	14 DAP	14 DAP	42 DAP
without mycorrhiza fungi (M_0)	22.62 b	0.23 b	2.59 a
20 g mycorrhiza fungi (M_1)	22.78 b	0.24 b	2.48 ab
40 g mycorrhiza fungi (M_2)	27.23 a	0.27 a	2.30 b
$UJBD_{0.05}$			
2=	2.330	0.021	0.222	0.033
3=	2.443	0.022	0.233	0.035

Note: The numbers are followed by the same letters in the same column, no significant based DRMT 95%

Table 1, the highest average of sweet corn height on 14 DAP obtained in the treatment of M_2 as 27.23 cm were significantly different with treatment of M_1 and M_0, but M_1 does not differ significantly with treatment of M_0. The highest average of sweet corn stem diameter on 42 DAP obtained in the treatment M_0 is 2.59 cm were significantly different with treatment of M_2, but did not differ significantly with treatment of M_1. [22], states that the granting of mycorrhiza fungi cause the growth of the plant stem is greater than the plants that are not infected by mycorrhiza fungi. The results of resesarch showed that mycorrhiza fungi and organic liquid fertilizer is very real effect on plant height and stem diameter of sweet corn plants at the age of 14 DAP and 42 DAP, while at the age of 28 DAP provides no real influence. It is thought to during the vegetative plant roots have not been infected by mycorrhiza fungi perfectly so that the role of mycorrhiza fungi helping the roots to expand nutrient absorption area has not gone well. According [27], mycorrhiza fungi develop in the cortex, where the infection is influenced by anatomical roots and the age of the plant.

Although mycorrhiza fungi are given in plants but available nutrients phosphorus is still below the optimal requirements for the growth of corn plants, this is because the phosphorus content at a lower marginal land. On the other hand, needs a lot of plants to phosphorus relative to the process of photosynthesis, respiration and energy metabolism. [26], stating that the nutrient content in soil is generally low due to marginal alkaline leaching takes place intensively.

Application of liquid organic fertilizer independently significantly affect the average plant height at 14 DAP, whereas at 28 and 42 DAP did not affect significantly (Table 2).

Table 2. Effect of application of organic liquid fertilizer to the average height of corn plants (cm) at the age of 14 DAP.

Treatment	average height of sweet corn plants (cm)	DRMT 0.05
Without liquid organic fertilizer (P$_0$)	22.68 b	2=2.330
50 ml L^{-1} water (P$_1$)	24.11 ab	3=2.443
100 ml L^{-1} water (P$_2$)	25.84 a	

Note: The numbers are followed by the same letters in the same column, no significant based DRMT 95%

Table 2 shows that the highest average of sweet corn height obtained at the treatment of P$_2$ is 25.84 cm were different unreal with treatment of P$_1$, but significantly different with treatment of P$_0$. The treatment of P$_1$ did not differ significantly with treatment of P$_0$ and P$_2$. The treatment of P$_1$ had no significant to treatment of P$_0$, but significantly different from treatment of P$_2$. The results of research showed that the application of liquid organic fertilizer significantly affected to stem diameter, corncob with husk, cob length, cob diameter and the number of seed rows cob^{-1}. [19] declare that the liquid organic fertilizer containing elements of potassium which play an important role in every metabolic process plants, namely in synthesis and plays a role in maintaining good turgor pressure so as to allow the smooth metabolic processes and ensure the sustainability of cell elongation. In the vegetative phase liquid organic fertilizer only to give effect to the corn plant height age of 14 DAP. It is suspected liquid organic fertilizer dose given is still very low, so it has not significantly affected the growth of sweet corn plants.

Table 3. Effect of applications of mycorrhiza fungi and liquid organic fertilizer to the average height of corn plants (cm) at the age of 42 DAP.

mycorrhiza fungi	Liquid organic fertilizer			DRMT$_{0.05}$
	Without liquid organic fertilizer	50 ml L^{-1}	100 ml L^{-1}	
	(P$_0$)	(P$_1$)	(P$_2$)	
Without mycorrhiza fungi (M$_0$)	202.13 a	192.97 b	193.07 b	2=12.747
	p	p	p	3=13.368
20 g mycorrhiza fungi (M$_1$)	193.07 a	212.57 a	198.43 ab	
	q	p	q	
40 g mycorrhiza fungi (M$_2$)	202.80 a	212.57 a	208.73 a	
	p	P	p	

Note: The numbers are followed by letters are not the same in the column (ab) and rows (p-q) the same significantly different based DRMT 95%

Application of mycorrhiza fungi and liquid organic fertilizer significantly affect the average plant height at 42 DAP, while at the age of 14 and 28 DAP does not affect significantly (Table 3).

Table 3 shows that the average height of the highest plant found in the combination treatment of the mycorrhiza fungi as 20 g planting hole^{-1} and liquid organic fertilizer as 50 ml L^{-1} and 40 g planting hole^{-1} respectively as 212.57 cm. At the level of interaction without liquid organic fertilizer and without mycorrhiza fungi on plant height and stem diameter at the age of 42 DAP provides no real influence. It is presumably related to the conditions of weather especially rainfall is quite high at the study site. This is supported by results of research [1], that information on climate characteristics was an important factor for determination of areas for maize cultivation.

3.2. The Weight of Cob Without Husk

Application of mycorrhiza fungi and liquid organic fertilizer significantly affect to weight of cob without husk (Table 4).

Table 4. Effect of application of mycorrhiza fungi and liquid organic fertilizer to the average weight of cob without husk (kg).

mycorrhiza fungi	Liquid organic fertilizer			DRMT$_{0.05}$
	Without liquid organic fertilizer	50 ml L^{-1}	100 ml L^{-1}	
	(P$_0$)	(P$_1$)	(P$_2$)	
Without mycorrhiza fungi (M$_0$)	0.20 b	0.20 c	0.23 b	2=0.036
	p	P	p	3=0.037
20 g mycorrhiza fungi (M$_1$)	0.28 a	0.28 b	0.28 a	
	p	P	p	
40 g mycorrhiza fungi (M$_2$)	0.25 a	0.32 a	0.32 a	
	q	P	p	

Note: The numbers are followed by letters are not the same in the column (ab) and rows (pq) the same significantly different based DRMT 95%

Table 4 shows that the average weight of cob without husk is highest in the combined treatment of 40 g mycorrhiza fungi planting hole^{-1} and liquid fertilizer 50 ml L^{-1} and 40 g mycorrhiza fungi planting hole^{-1} each 0.32 kg. Independent influence of mycorrhiza on the cob with husk weight, the weight of cob without husk, cob length, cob diameter and number of lines of seeds cob^{-1} gives a significant influence. This shows that when the corn crop enters the generative phase, the number of mycorrhiza hyphae formed more so that the effectiveness of the absorption of nutrients to support plant growth becomes better. The treatment of mycorrhiza fungi as 40 g planting hole^{-1} gives a better effect on the growth and yield of sweet corn. This is because the mycorrhiza fungi infection on the roots of sweet corn more numerous so that the effectiveness of the absorption of nutrients to support the growth and yield of sweet corn to be better.

Application of mycorrhiza fungi and liquid organic fertilizer significantly affected the plant height at 42 DAP and weight cob without husk. It is thought to increase mycorrhiza fungi and liquid organic fertilizer can be complementary in supporting the growth and crop yield. Mycorrhiza fungi play a role in improving the tolerance of crops to the conditions of degraded land, helps the roots to increase the absorption of water and nutrients in the soil. While liquid organic fertilizer containing potassium, which are crucial elements in any process plant metabolism, namely in synthesis and plays a role in maintaining good turgor pressure so as to allow the smooth metabolic processes and ensure the sustainability of cell elongation [18].

3.3. Cob Length (cm), Cob Diameter (cm) and the Number of Seed Rows Cob[-1]

Application of mycorrhiza fungi significantly affect to cob length, cob diameter and number of seeds cob[-1] (Table 5). While the combination of application of mycorrhiza fungi and organic fertilizer no real effect.

Table 5. *Effect of the application of mycorrhiza fungi on the average length of cob (cm), cob diameter (cm) and the number of seed rows cob[-1].*

Treatment	variable		
	length of cob (cm)	cob diameter (cm)	number of seed rows cob[-1]
Without mycorrhiza fungi (M_0)	18.89 b	3.73 b	14.44 b
20 g mycorrhiza fungi (M_1)	20.89 a	3.96 a	16.00 a
40 g mycorrhiza fungi (M_2)	21.46 a	4.04 a	16.67 a
$DRMT_{0,05}$			
2=	1.197	0.117	1.267
3=	1.155	0.123	1.328

Note: The numbers are followed by the same letters in the same column, no significant based DRMT 95%

Table 5 shows that the average length of the highest cob sweet corn was obtained in treatment of M_2 as 21.46 cm different unreal with treatments of M_1, but significantly different with treatment of M_0. The cob diameter of sweet corn highest was obtained in treatment of M_2 as 4.04 cm were no significant with treatment of M_1 but significantly different with treatment of M_0. The number of seed row cobs[-1] highest obtained at treatment of M_2 as 16.67 cm different unreal with treatment of M_1, but significantly different with treatment of M_0. The size and number of seeds related to the uptake of nutrients accumulated in the form of sink as a result of mycorrhiza fungi infection on the roots of plants. Mycorrhiza fungi infection is related to metabolic processes such as photosynthesis that plants run optimally to ensure continuity of symbiosis between the plants with mycorrhiza fungi [5]. [2], suggests that the increase in Phosphorus uptake by plants due to volume expansion of the plant's roots and the acceleration of the movement of Phosphorus into hyphae of mycorrhiza fungi. Furthermore, Phosphorus is taken by the external hyphae transferred to arbuscular through internal hyphae that can increase crop Phosphorus uptake [23]. The effectiveness of mycorrhiza fungi thought to as a result of differences in the adaptation of species and nature of each mycorrhiza fungi [10]. [4] report that *Gigaspora* sp has a high tolerance to changes in soil chemistry compared to *Acaulospora* sp and *Glomus* sp. The different types of mycorrhiza fungi are contributing to the abundance populations of mycorrhiza fungi, further affect the ability to compete among types of mycorrhiza fungi in occupying space and obtain nutrients from the host plant root exudates as a source of energy [12:15].

3.4. Percentage of Mycorrhiza Infection on the Roots of Sweet Corn Plant

The average percentage of mycorrhizal infection on the roots of corn plants are listed in Table 6.

Table 6. *Average percentage of mycorrhiza infection (%) on the roots of corn plants.*

Treatment	Average percentage of mycorrhiza infection (%)	$DRMT_{0,05}$
Without mycorrhiza fungi (M_0)	0.00 c	2=4.662
20 g mycorrhiza fungi (M_1)	20.00 b	3=4.889
40 g mycorrhiza fungi (M_2)	27.78 a	

Note: The numbers are followed by the same letters in the same column, no significant based DRMT 95%

Table 6 shows that the average of percentage of mycorrhiza fungi infection on the roots of sweet corn plants is highest in the treatment of M_2 as 27.78% which is significantly different from the treatment of M_1 and M_0. The roots of sweet corn were infected by mycorrhiza fungi characterized by vesicles, internal and external hyphae. The vesicles was thick-walled structure and serves as a repository and exchange of food reserves. characterized external hyphae are outside the root cell growth, is an important structure as it continues to expand its distribution which aims to search for sources of food and organic and inorganic nutrients in the soil needed by the plant, so that the hyphae is better known as a helper, and the internal hyphae hyphae which serves as a liaison between vesicles vesicles with the other one. [20] report that an increase in uptake of N, P, K, Ca and Mg by plants when given the addition of mycorrhiza fungi.

Sweet corn plant roots more quickly associated with mycorrhiza fungi, converselly the mycorrhiza fungi otherwise would acquire the nutrient phosphorus from root exudates of plants that have been infected as a source of energy [13]. The result of this research shows that the higher dose given mycorrhiza fungi on sweet corn higher the infection on sweet corn roots. These results are in accordance with the statement of [3] that the surface area of the infected root mycorrhiza fungi 10 times greater than in uninfected mycorrhizae. The roots of which has an area greater absorption will have a chance to absorb nutrients larger, therefore the plant in association with mycorrhiza fungi will be able to improve its capacity to absorb nutrients and water. In additon, the roots of the plants have a metabolic rate 2-4 times higher when compared the root crops was not infected by mycorrhiza [21].

4. Conclusions

From the results of this reserach, the following conclusions can be drawn:

(a) The interaction between mycorrhiza fungi and liquid organic fertilizer has a significant effect on the plant height age 42 days after planting and cob weight without husk with the best results obtained in the treatment of mycorrhiza fungi as 40 g hole^{-1} with liquid organic fertilizer as 50 ml L^{-1} water.

(b) Application the mycorrhiza fungi as 40 g hole^{-1} gives the best results on the growth and yield of sweet corn.

Acknowledgements

The author would like to thank to the Ministry of Research, Technology and Higher Education of the Republic of Indonesia for the financial assistance through the scheme of National Priorities Research Grant Master Plan for the Acceleration and Expansion of Indonesian Economic Development 2011-2025 in 2014, 2015. The author also thank to the Rector of Halu Oleo University and the Chairman of the Research Institute of Halu Oleo University for the administrative of services.

References

[1] Aminuddin Mane Kandari, Sumbangan Baja, Ambo Ala, Kaimuddin, 2013. Agroecological zoning and land suitability assessment for maize (*Zea mays* L.) development in Buton regency Indonesia. J. International of Agriculture, Forestry and Fisheries. Vol.2. No. 6:202-211.

[2] Bolan. N. S., 1991. A Critical Review of the Role Mycorrhizal Fungi in the uptake of Phosphorus by Plants. J.of Plant and Soil. Vol.132. No.2:189-207.

[3] Brady, N.C., 2002. The Nature and Properties of Soils, Prentice Hall of India, New Delhi.

[4] Clark. R. B., 1997. Arbuscular Mycorrhizal Adaption, Spore Germination, Root Colonization and Host Plant Growth and Mineral Acquisition at Low pH. J.of Plant and Soil. Vol.192:15-22.

[5] Delvian, 2007. Penggunaan Asam Humik dalam Kultur Trapping Cendawan Mikoriza Arbuskula dari Ekosistem dengan Salinitas Tinggi. Jurnal Ilmu-Ilmu Pertanian Indonesia. Vol. 9. No. 2:124-129.

[6] Gonzalo.B.E. and A.Miguel, 2006. Mycorrhiza. An Ecological Alternative for Sustainable Agriculture. Melalui <http://www.micorrhizas.htm>.

[7] Hadisuwito, S., 2008. Membuat Pupuk Kompos Cair. PT Agromedia Pustaka. Jakarta.

[8] Halim dan F.S. Rembon, 2013. Peningkatan produksi tanaman jagung berbasis bioteknologi Mikoriza Indigenous Gulma. Laporan hasil penelitian MP3EI Dikti. Lembaga Penelitian Universitas Halu Oleo Kendari.

[9] Halim, 2012. Peran Mikoriza Indigen Terhadap Indeks Kompetisi antara Tanaman Jagung (*Zea mays* L.) dengan Gulma *Ageratum conyzoides*. Berkala Penelitian Agronomi. Vol.1.No. 1: 86-92.

[10] Halim, 2009. Peran Mikoriza Indigenous Gulma *Imperata cylindrica* (L.) Beauv dan *Eupatorium odorata* (L.) terhadap Kompetisi Gulma dan Tanaman Jagung. Disertasi Program Doktor Universitas Padjadjaran Bandung.

[11] Halim, Fransiscus S Rembon, Aminuddin Mane Kandari, Resman, Asrul Sani, 2014. Characteristics of indigenous mycorrhiza of weeds on marginal dry land in South Konawe, Indonesia J. International of Agriculture, Forestry and Fisheries. Vol.3. No.6:459-463.

[12] Janos. D. P., 1992. Heterogenity and Scale in Tropical Vesicular Arbuscular Mycorrhiza Formation. CAB International. Wallingford.UK.

[13] Juge. C., J.Samson, C.Bestien, H.Vierheilig, A.Coughlan and Y.Pieche, 2002. Breaking Dormancy in Spores of the Arbuscular Mycorrhizal Fungus *Glomus intraradies*. A Critical Cold Storage Period. J. of Mycorrhiza. 12:37-42.

[14] Mazi. A., 2004. Kebijakan Strategi Pengembangan Sistem Produksi Jagung di Provinsi Sulawesi Tenggara dalam Kerangka Program Celebes Corn Belt (CCB). Prosiding Seminar IX Budidaya Pertanian Olah Tanah Konservasi Kerjasama dengan Himpunan Ilmu Gulma Indonesia, Universitas Gorontalo dan Pemerintah Daerah Provisnsi Gorontalo. Gorontalo.

[15] Moutoglis.P and P.Widden, 1996. Vesicular Arbuscular Mycorrhizal Spore Population in Sugar Maple (*Acer saccharum marsh* L.) Forest. J. of Mycorrhiza. Vol.6:91-97.

[16] Prasetyo B.H. dan D.A. Suriadikarta, 2006. Karakteristik, potensi, dan teknologi Pengelolaan tanah ultisol untuk Pengembangan pertanian lahan Kering di indonesia. Balai Besar Penelitian dan Pengembangan Sumberdaya Lahan Pertanian. Bogor.

[17] Purwono dan R.Hartono, 2011. Bertanam Jagung Unggul. PT. Penebar Swadaya. Jakarta.

[18] Purwowidodo, 2007. Telaah Kesuburan Tanah. Penerbit Angkasa. Bandung.

[19] Rao S., 1994. Mikroorganisme dan Pertumbuhan Tanaman. Penerbit Universitas Indonesia. Jakarta.

[20] Sadaghiani, M.R, Hassana, A, Barin, M., Danesh, Y.R., & Sefidkon, F., 2010. Effects of arbuscular mycorrhizal (AM) fungi on growth, essential oil production and nutriends uptake in basil, Journal of Medicinal Plant Research. Vol.4.No.21: 2222-2228.

[21] Sieverding, E., 1991, Vesicular-Arbuscular Mychoryza Management in Tropical Agrosystems, Deutshe Gesellicaft Fur Tehnische, Germany.

[22] Sinwin, R.M., Mulyati dan Lolita, E.S., 2007. Peranan Kascing dan Inokulasi Jamur Mikoriza terhadap Serapan Hara Tanaman Jagung. Jurnal Ilmu Tanah. Faperta. Universitas Lampung.

[23] Smith. S. E., E.S.Dickon, F.A.Smith and V.P.Gianiazzi, 1993. Nutrient Transport between Fungus and Plant in Vesicular Arbuscular Mycorrhizal. Proceeding of Second Asian Conference on Mycorrhiza. Chiang Mai. Thailand. Biotrop Special Publication No.42 Seameo Biotrop. Bogor.

[24] Subagyo, 2000. Kriteria Kesesuaian Lahan untuk Komoditas Pertanian. Badan Penelitian dan Pengembangan Penelitian. Pusat Penelitian Tanah. Indonesia.

[25] Subowo dan Prihatin, T.A, Kentjanasari, 1996. Pemanfaatan Biofertilizer untuk Meningkatkan Produktivitas Lahan Pertanian. Jurnal Litbang Pertanian 15 (1), Medan.

[26] Suharta, N. dan B.H. Prasetyo, 1986. Karakteristik Tanah-Tanah Berkembang dari Batuan Granit di Kalimantan Barat. Pemberitaan Penelitian Tanah dan Pupuk 6:51-60.

[27] Talanca, H. A. dan Adnan. M., 2005. Mikoriza dan Manfaatnya Pada Tanaman. Balai Penelitian Tanaman Serealia. Prosiding Seminar Ilmiah dan Penemuan Tahunan PEI dan FKI xvi Komda. Sulawesi Selatan. ISBN: 979-9525-6-7.

[28] Waryanti, Sudarno, Sutrisno E., 2013. Studi Pengaruh Penambahan Sabut Kelapa pada Pembuatan Pupuk Cair dari Limbah Air Cucian Ikan Terhadap Kualitas Unsur Hara Makro (CNPK). FT UNDIP. Semarang.

18

Valuation of Environmental Services of Catchment Forests Within Baubau Wonco Watershed

Safril Kasim[1], Aminuddin Mane Kandari[2], La Ode Midi[3], Anita Indriasari[4]

[1]Spesifications Agroforestry, Department of Environmental Science, Faculty of Forestry and Environmental Science, Halu Oleo University, Southeast Sulawesi, Indonesia
[2]Spesifications Agroclimatology, Department of Environmental Science, Faculty of Forestry and Environmental Science, Halu Oleo University, Southeast Sulawesi, Indonesia
[3]Spesifications Agrohidrology, Department of Environmental Science, Faculty of Forestry and Environmental Science, Halu Oleo University, Southeast Sulawesi, Indonesia
[4]Spesifications Agriculture of Economic and Social, Department of Environmental Science, Faculty of Forestry and Environmental Science, Halu Oleo University, Southeast Sulawesi, Indonesia

Email address:
safrilkasim1970@gmail.com (S. Kasim), manekandaria@yahoo.com (A. M. Kandari), laodemidi@gmail.com (L. O. Midi),
anitayulardhi@gmail.com (A. Indriasari)

Abstract: Catchment forests cover a total of 2.750, 11 Ha. This is about 31, 85% of the total area of Baubau Wonco watershed. This forested land provides both tangible and intangible benefits of which some are perceived as environmental services. However, it has encountered high rate of deforestation and forests degradation [7]. A well managed catchment forests can bring about advantages to a wide range of stakeholders, normally far away from the forests in the form of water for domestic use, agriculture, industry, and preventing from flooding, erosion and landslide hazards. To this view, it is a logical assumption that these various stakeholders who are mostly living in the downstream area should provide costs for a good forest management as incentives to the local community who mostly occupy the upstream area. Therefore, the need of a model that regulates the upstream and downstream mechanism should be explored. The research is planned to be conducted for two years. The first year research has been carried out from July to October 2015. The study employed various methods of data analysis. Those are as follow: (i) Hedonic Price is used for estimating economic value of water for domestic and industrial use; (ii) Productivity approach used for analyzing economic value of water for agricultural use; (iii) Willingness to Pay (WTP) is used for analyzing economic value of catchment forests to preventing from erosion, flooding and landslide hazards. The results of the first year research show that the total volume of water domestic consumption reachs 6.163.488,50 m^3 year^{-1}, which is used by 18.950 households with the economic value obtains of Rp. 40.062.668.750 year^{-1}, while the economic value of water for agricultural use achieves Rp. 30.199.167/ha year^{-1}. This research will be continued to the second year study to (i) estimate the economic value of industrial water, (ii) to analyze the Total Economic Value of hydrological environmental services provided by catchment forests of the watershed area, and (iii) to develop a model that can facilitate downstream-upstream mechanism of a payment for the hydrological environmental services.

Keywords: Hydrological-Environmental Services, Valuation, Domestic Use, Agricultural Use, Industrial Use, Erosion, Flood and Landslide Hazard Control

1. Introduction

Catchment forests offer both tangible and intangible benefits of which some are perceived as environmental services. However, catchment forests in Baubau Wonco Watershed area have faced serious degradation due to infrastructure development, housing, mining activities, agricultural use and forest encroachment and illegal lodging activities. Forested land use changes within the watershed have brought about changes in river discharged, an increase of erosion hazard and sedimentation [7]. Integrated watershed management planning has been made as a guideline of water, land and forest resources management within the watershed area. However, there are still obstacles in the implementation phase including lack of funds [5].

Catchment forests within watershed area have significant roles to increase availability and quality of water, to minimize the rate of erosion and sedimentation as well as to decrease a potential risk of landslide hazard. These hydrological environmental services provided by the catchment forests are maintained by local communities. While the beneficiaries of these are vary and can be ranged from local actors to international stakeholders. To this point, there is a new believe proposed by neo-market natural resources economists that new methods and institutional frameworks need to be developed for facilitating a downstream-upstream incentive model or mechanism [17]. It is therefore, financial incentives have to be made available by international, national, regional and local stakeholders to compensate environmental services of the watershed area that are for long generation maintained by local people, To some extents, these are refer to as Payment for Environtal Services (PES). A downstrean and upstream model can become a policy umbrella to facilitate the PES as financial compensations to the local actors who have conducted forest conservation efforts. This financial subsidy can be used to sustain their livelihoods and to increase their capacity to managing natural forests. The PES can be well applied if Total Economic Value (TEV) of hydrological environmental services of the cathment forests is well determined. Furthermore, a need of institutional set-up is crucial. This study attempts to analyze the TEV and to develop an institutional model that can be implemented through downstream and upstream incentive mechanism.

An economic valuation of hydrological environmental services is a method to determine economic value as provided by cathcment forests. This is becoming more substantial since the need of clean water increase significantly over years. An increase of water demands seems to be parralel with an increase of population numbers. Nowadays, water resource is becoming a rare commodity because almost all activities of human beings need a clean water. A water resource is needed for domestic, agricultural, and industrial uses, etc. Moreover, a water resource can also be used for source of electricty. However, results of some researchs stated that within a last hundred years, number of world populations have increased three times. A ratio between water need and population numbers has led to water scarcity due to less of water supply in comparison with water demand.

Water scarcity has emerged as a global problem, especially in summer session. Water availabity that is increase and even overflows during the rainy/winter session has a low quality and if uncontrolable can lead to flooding whereas during the summer season water is becoming scarce commodity, thus water is getting more expensive and has high economic value. An increase trend of economic value of water resource from time to time needs to be systematically explored in order to get reliable data and information that can be used to determine incentive for PES.

The development of hydrological environmental services through a downstream-upstream model as one alternative to support local community efforts to conserving watershed area. Economic valuation of hydrological environmental services can describe a relationship between economic and environmental aspects that are needed on decision making process to determining alternative uses of water and proper water conservation programs in the future.

Baubau Wonco Watershed constitutes a total of 9999, 75 Ha in which consists of four Sub Watersheds namely : Wamoose Sub Watershed, Wasamparona, Sigari and Wancuawu. This watershed streams administeratively over two regencies: Baubau City and Buton Regency of which about 8.634,01 Ha is located in Baubau City [5].

The Baubau Wonco Watershed has a strategic location which is the downstream area streams to the center of

Baubau city and has an outlet in the Baubau strain. With this strategic position, the watershed has ecological, economic and social functions. Ecologically, Baubau Wonco Watershed has a significant role as a catchment forested area, to absorb and to stream rainy water to the outlet. Furthermore, catchment forest provide micro climate, oxygen supplies and carbon monooxyde absorber that is produced by industrial sectors, transportations and domestic sources. Besides, abundant and divers aquatic biota live in the river of Baubau Wonco Watershed.

The Baubau Wonco Watershed economically serves as agricultural land area for crop plantation and paddy field. Moreover, the downstream area that is located in the Central Business District of Baubau City has a great potency to be developed as an ecotourism destination.

Socially, the Baubau Wonco Watershed provides open space for the local community in which upstream-middle and downstream community has interacted to each other in doing their daily activities. This kind of social interaction that is using watershed area as media should be properly managed so that every actor can take benefits and conducts efforts to deal with the impacts resulted from watershed development programs. This Upstream-Downstream relationship can be seen, i.e. upstream community has to be aware that their activities in cultivating the upland area for agriculture or conserving forested land must bring about negative or positive impacts to the middle and downstream community. It is therefore, the need to establish a payment for environmental services by downstream community and other stakeholders is crucial. This is important to reenforce commitments of upstream community in order to protect catchment forested land and to implement agricultural practices that are in line with soil and water conservation principles. Besides, the funds will be very useful to saveguard their livelihood.

Based on description aboved, this research is intended to conduct economic valuation of hydrological environmental services of Baubau Wonco Watershed and to develop a proper model that can be implemented to facilitate the Upstream and Downstream Community as to provide Payment for Environmental Services. The logical framework of this research is presented in Figure 1.

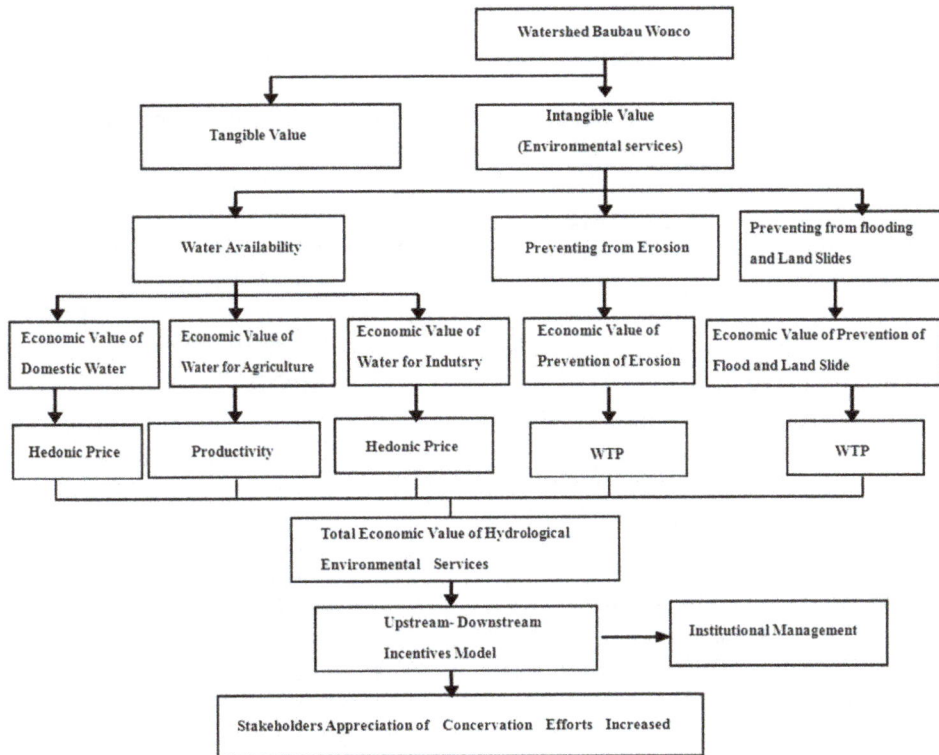

Figure 1. *The Logical Framework to A Valuation of Hydrological Environmental Services of Catchment Forests Within Baubau WoncoWatershed.*

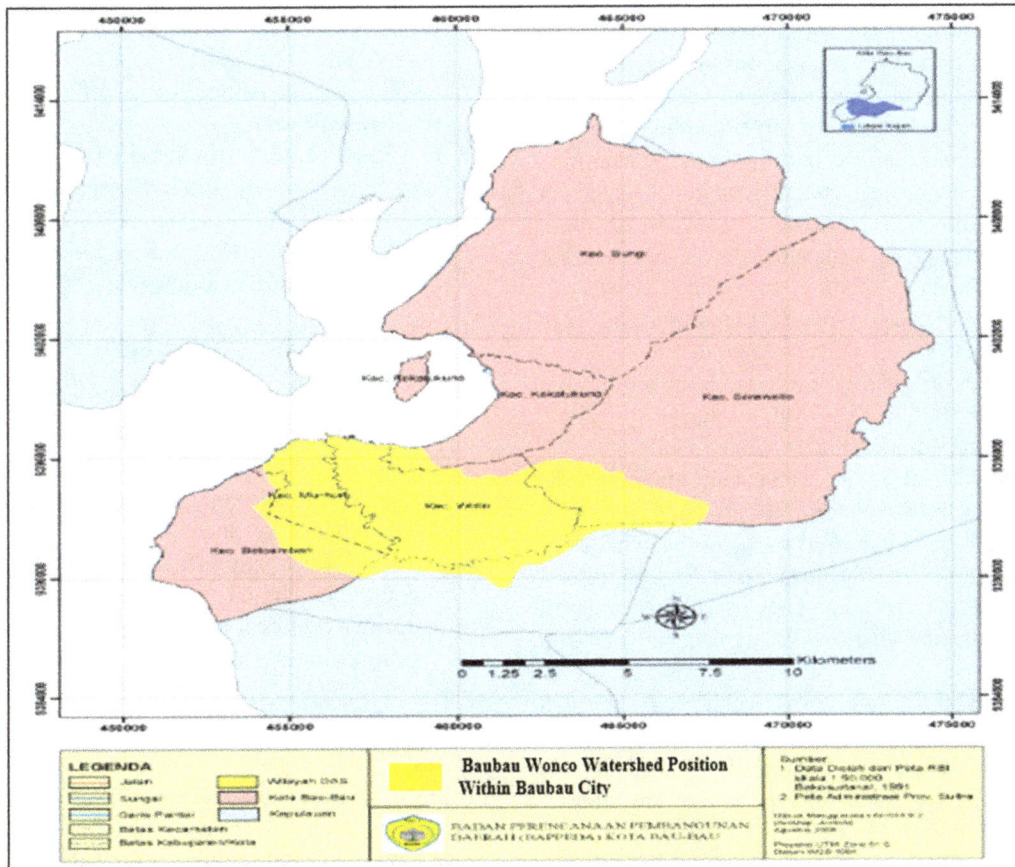

Figure 2. *Map of Baubau Wonco Watershed Position Within Baubau City.*

2. Study Area

This study is conducted within Baubau Watershed area. Total area of Baubau Watershed is 9999, 75 Ha and it consist of four subwatersheds namely Wamoose Subwatershed, Wasamporona, Sigari and Wancuawu., Baubau Watershed is administeratively located in two districts, Baubau City and Buton District, in which 8634,01 Ha is situated in Baubau City and 1365,74 ha in Buton District [5]. This watershed is geographically located in 5°27′8″ and 5°32′33″ South Latitude and 122°33′5″ and 122° 42′ 34′ East Longitude.

The soil types are dominated by litosol patch (51,30%) and Mediteran Types (41,77 %), while Latosol only occupy 6,93 of total watershed land. The watershed area receives rainfall 1785, 2 mm in average/year with 147 rainy days. The highest monthly averages of raifall happens in December (272,00 mm with 20 rainy days) whereas the lowest occurs in September (9,5 mm with 2 rainy days) [11].

There are four subdistricts belong to Baubau City and two others belong to Buton Regency. For the purposes of this study, data collected is only to be carried out in the four subdistricts of Baubau City namely : Sorawolio, Wolio, Murhum and Betoambari. Map of Location of Baubau Wonco Watershed can be seen in Figure 2.

3. Methodology

The first year research has been carried out from July to October 2015. The study employs various methods in valuation of hydrological environmental services of the watershed, namely : Hedonic Price for estimating economic value of water for domestic and industrial use, productivity approach used for analyzing economic value of water for agricultural use, while Willingness to Pay (WTP) approach using for the valuation of cathcment forest reserves for preventing from erosion, flooding and landslide hazards. Primary data was collected through observation and deep interview with respondents in each sub district. This is included : (i) number of households who are using clean water for domestic and agricultural uses from the catchment forests of the Baubau Wonco Watershed, (ii) number of industries that are using clean water from the cathcment forests of the watershed, (iii) The water volumes that are consumpted for domestic, agricultural and industrial uses, (iv) Spended costs for water consumption, (v) Spended costs for agricultural activities (paddys field), (vi) Paddys Fields Total Areas, (vii) Paddys Field Production/Year, (viii) An intensity of Crop Season/Year.

Whereas secondary data was collected through documenting such sources of data as : reports of previous study, government documents, especially related to the government policies and statistical reports that are relevant with the purposes of the study.

Data analysis was carried out using such formula as :

(a) Total volume of water consumption for domestic use using a formula as follows :

$$TVDC = \frac{WC \times 30}{n} \times \sum population \times 1\ year \qquad (1)$$

Notes :

- TVDC = Total volume of Water Consumption for Domestic Use (m^3 year^{-1})
- WC = Water Consumption for domestic use (m^3 HH^{-1} day^{-1})
- n = Number of Respondents

(b) Clean Water Economic Value is determined based on Hedonic Price that has been decided by Local Government of Baubau City, which is used a formula as follows:

$$EVDU = TVWC \times HP \qquad (2)$$

Notes:

- EVDU = Economic Value of Clean Water for Domestic Use (Rp/Year)
- TVWC = Total volume of Water Consumption for Domestic Use (m3/year)
- HP = Hedonic Price (Rp/m3)

(c) Economiv Value for Agriculture (Irrigated Paddy Field) is analyzed using a formula as follows :

$$EVAU = \frac{EVTP - C}{Xw} \qquad (3)$$

Notes:

- EVAU = Economic Value for Agriculture Used/Irrigated Paddy Field ((Rp/Year)
- EVTP = Economic Value of Total Productivity (Rp/Season/Year)
- C = Spended Costs (Rp/Season/Year)
- Xw = Volume of Water Consumption of Irrigation (m3/Second/Ha)

(d) Total volume of water consumption for industrial use using a formula as follows :

$$TVIC = \frac{WC \times 30}{n} \times \sum industries \times 1\ year \qquad (4)$$

Notes:

- TVIC = Total volume of Water Consumption for Industrial Use(m3/year)
- WC = Water Consumption for Industrial Use (m3/day)
- n = Number of Respondents

(e) Water Economic Value for Industrial Use is determined based on Hedonic Price that has been decided by Local Government of Baubau City, which using a formula as follows:

$$EVIU = TVWC \times HP \qquad (5)$$

Notes:

- EVIU = Economic Value of Clean Water for Industrial Use (Rp/year)
- TVWC = Total volume of Water Consumption for Industrial Use Use(m3/year)

- HP = Hedonic Price(Rp/m3)
- (f) Economic Value of Catchment Forests for Preventing from erosion using a formula as follows :

$$WTP = \sum_{i=1}^{21} AWPi \left(\frac{ni}{N}\right) \times Population \qquad (6)$$

Notes:
- AWPi = An average of Willingness to Pay
- ni = A number of respondents who are willing to pay for environmental services
- N = A number of Samples
- (g) Economic Value of Catchment Forests for Preventing Flooding and Landslide Hazards using the same formula as used to estimate economic valuation for preventing from erosion formula as follows :

$$WTP = \sum_{i=1}^{21} AWPi \left(\frac{ni}{N}\right) \times Population \qquad (7)$$

Notes:
- AWPi = An average of Willingness to Pay
- ni = A number of respondents who are willing to pay environmental services
- N = A number of Samples
- (h) Total Economic Value of Catchment Forests of Baubau Wonco Watershed is estimated using a formula as follows:

$$TEVES= EVDU + EVAU + EVIU+EVPE+EVPFL \qquad (8)$$

Notes:
- TEVES = Total Economic Value (Rp/year)
- EVDU = Economic Value for Domestic Use (Rp/year)
- EVAU = Economic Value for Agricultural Use (Rp/year)
- EVIU = Economic Value for Industrial Use (Rp/Year)
- EVPE = Economic Value for Preventing from Erosion (Rp/Year)
- EVPFL = Economic Value for Preventing from Flooding and Lanslides Hazards (Rp/Year)

4. Result and Discussion

4.1. Characteristics of Baubau Wonco Watershed

4.1.1. Slope

Slope is a basic element for analyzing and visualizing landform characteristics. It is important in studies of watershed units, landscape units, and morphometric measures [12];[13]. Slope combines with such other elements as soil texture, organic matters, existing types of vegetation, can be used to calculate runoff, forest inventory estimates, wild life habitat suitability and site analysis [16];[13]. Slope is a very crucial element for the Rate of Erosion Hazard Index.

A GIS analysis shows that topography classes vary ranges from 0-8 % until > 45 %. The highest gradient is > 45 % which occupie 2628, 15 Ha, whereas the lowest is gradient 15-25 % 392,98 Ha. Other topography classes present within the watershed land area are 0-8 % 2548,58 Ha, 8 – 15 % 2452, 01 Ha, 25 - 45 % 612,29 Ha respectively [7].

4.1.2. Soil

Soil is an important elemen of watershed ecosystem, as it serves as an anchorage for plant and source of nutrients. Thus soil is the fundamental raw material for plant growth. The knowledge of soil resources is essential for proper watershed development and planning.

Soil type analysis reveals that there are three types of soil : Litosol, Latosol and Mediteran. Litosol type administeratively spread over in Bukit Wolio Indah Village, Kadolokatapi, Kadolomoko, Lipu, Melai and Waborobo, in which occupies 4429 Ha. Mediteran type occurs in Bungi, a part of Kadolokatapi, Kaisabu Baru and Karya Baru Villages which occupies 3606, 17 Ha. While Latosol spread over Bungi, a part of Kadolokatapi, a part of Karya Baru and Waborobo Villages with 3606, 17 Ha wide.

Towards the North Eastern Part and along the river side, soil depth is deep as compare to southern western part of Wolio Regency which is shallow one. It might be a result of undulating and a rugged topography. Moderate soil depth has been observed along eastern part of Sorawolio Regency and plain areas of Wolio Regency which is lying in between shallow and moderate suggesting treatment measures within a watershed [13].

4.1.3. Land Use/Land Cover Analysis

Knowledge of land use/land cover is important element for watershed planning and management and the way to better understanding of earth as an ecosystem. The term land cover relates to the type of feature present on the surface of the earth and land use relates to the human activity [13].

Land use/cover analysis shows that there are five types of land uses : (a) Forest; (b) Settlement; (c) Mixed Cropping; (d) Plantation/ Unirrigated agricultural field; and (e) Bushes. Forest area is widest land use which occupies 2.750,11 Ha (31,85% of the total watershed area), following by Settlement 2.068, 31 Ha (23,96 %), Mixed Cropping 1551, 85 Ha (17,97 %), Plantation/ Unirrigated agricultural field 1.265, 21 Ha (14, 65%), and Bushes 11,57 Ha (17, 97 %) respectively (Figure 3).

Baubau Watershed has served many functions and such functions as ecology and hydrology should be properly managed. This is due to significant roles of the watershed ecosystem as a buffer zone of Baubau City, a clean water reservoir' for the citizen and erotion and flooding controlling in the rainy season. As it located in the City area, this watershed has been threatened by development activities of Baubau City, thus land use could be rapidly changed in the future. It is therefore important to promote sustainable watershed management.

4.1.4. Erosion Hazard Index

Potential erosion within watershed area can be analyzed using USLE equation. The result of USLE analysis is used to calculate Erosion Hazard Index. This index can be used to indicate a spread of critical land within watershed area. Erotion Hazard Index is a prediction of maximum soil loss in a piece of land, if land management has no improvement for the long term uses and including calculation of solum of soil.

Erosion Hazard Index analysis shows that more than 90 % of Baubau Watershed Area can be categorized as critical lands which Erosion Hazard Index ranged from very high to middle rate of erosion. This is due to forested area within Bau-Bau Watershed that has functions as buffer zone and recharge area has been changed to be settlement, mixed garden and dry land crop. Furthermore, a few area of forested land has been changed to be waste land. This change has resulted in high runoff and erosion rate, and can potentially cause flooding in rainy season and a decrease of water availability in summer season [6]. Since erosion hazard can affect many actors both within and outside of watershed area, significant efforts should be done, including payment for environmental services as one of budget source to conduct forest conservation efforts at the implemetation phase of the watershed planning and management.

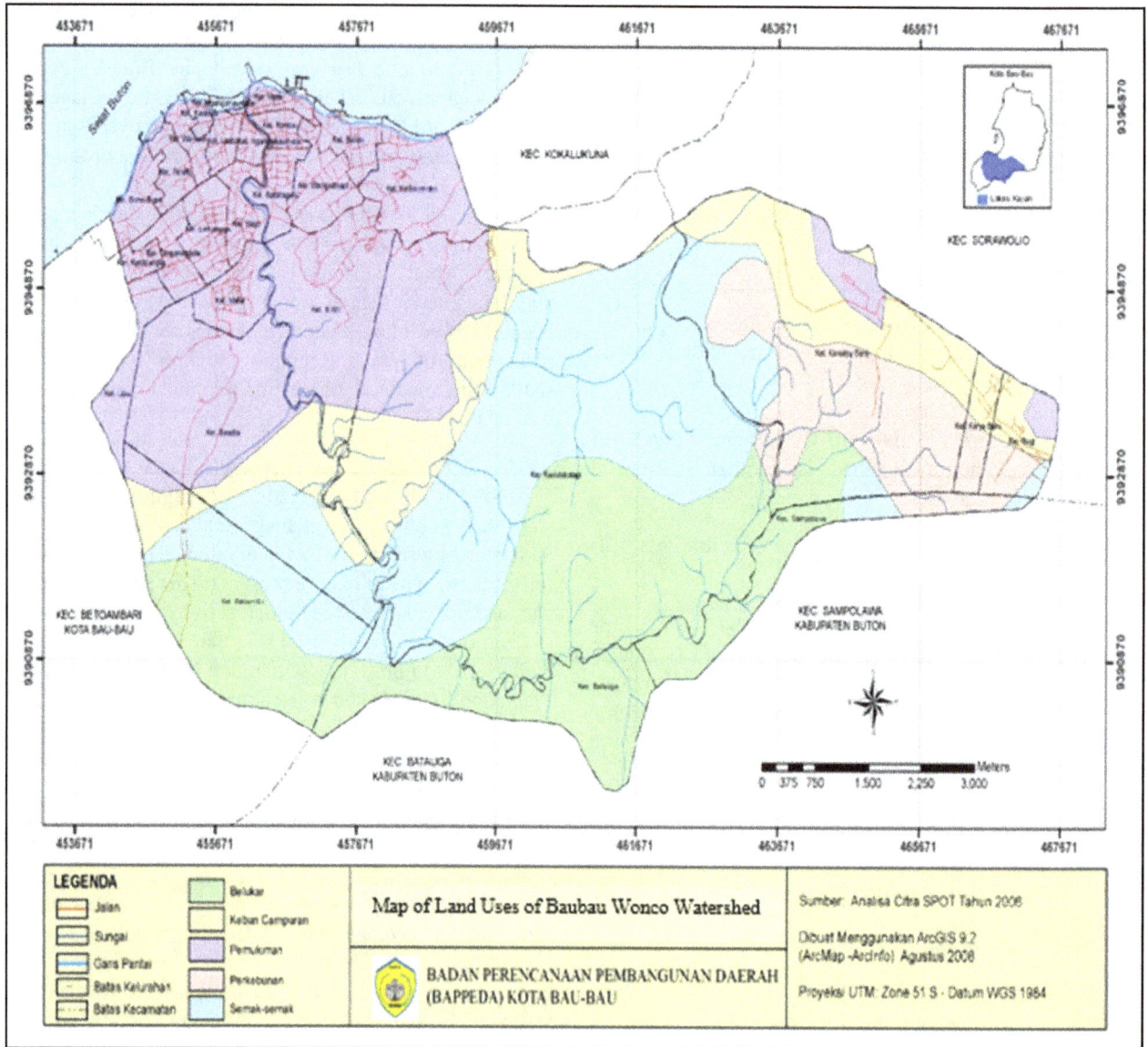

Figure 3. *The Map of Land Uses of Baubau Wonco Watershed.*

4.1.5. Sedimentation

Based on observation done in downstream of Baubau Watershed area, suspended load of sediment and sand has been found in ground floor of Baubau river. Sediment Delivery Ratio Analysis shows that total soil loss as a result of erosion which potentially deliver suspended sediment obtain 7.424,24 t year^{-1} [7].

4.1.6. River Discharge Fluctuation

River discharge analysis shows that maximum water dicharge is 10,245 m^3 sec^{-1} whereas minimum water discharged is 0,042 m^3 sec^{-1}. This means that ratio between maximum and minimum water dicharged is 243,929. This data indicated that overall, Baubau Watershed has been degraded because healhty watershed area has maximum and minimum ratio of water dicharge under 50.

There are several substreams within Baubau Watershed area namely Wakonti, Wamose and Matantolindu River. Current measurement of those rivers which conducted in last summer season shows that river dicharge ranges from 0,2 s.d 0,3 m^3

sec^{-1} [6]. This data indicates that Bau-Bau Watershed Area has been seriously degraded due to forested land use change. It is therefore important to take approriate conservation efforts to ensure that this watershed area will be used in sustainable manner. It is also important te explore possible fund resources through Payment for Environmental Services.

4.2. Economic Value of Domestic Use of Water

Clean water is a basic need for human beings. The use of water is different between one to others which depend on the need of the users. Thus, water consumption is diffrent between one region to others which relies on the extent of individual need of clean water in each. A region with high population numbers has a tendency to consume much more clean water than less population number [14]. A number of water consumption by local community of Baubau Wonco Watershed is a total need of clean water for domestic use. Baubau Wonco Watershed has served hydrological environmental services for local community of Baubau City both in a form of surface water and ground water as sources of clean water.

This research found that a total volume of water consumption for domestic use is 6.163.488,50 m^3 year^{-1} which consumpted by 18.950 households. Based on water consumption, using a formula 2, which clean water price is Rp.6.500 m^{-3}, resulting in the economic value of water consumption for domestic use as Rp 40.062.668.750 year^{-1}. This result is relatively high compared to consumption of domestic water in other regions, for example domestic water consumption of local people in surrounding catchment forests of Baini Village of Konawe District only achieved 278,55 m^3 with total population 445 people [9]. A relative high domestic water consumptions has a consequency to the hydrological conditions of Baubau Wonco Watershed as well as to the conservation efforts that should be made by watershed stakeholders. A high domestic water need can lead to water shortage, especially in the summer session. On the other hand, a lot of water is wasted through run off during the rainy season. It is therefore, source of available funds for catchment forest conservation efforts that will mostly be implemented by upstream community need to be explored. The available fund sources can be in a form of Payment for Environmental Services both from Baubau Wonco Watershed's community and local/ central government as well as from international community.

4.3. Economic Value of Water for Agricultural Use

Environmental economic primarily deals with a valuation of natural resources and environment. Before conducting valuation, there is a need to determine kinds of economic values that can be given to the those natural resources and environment. The analysis of economic value of water used for agricultural activity focusing on irrigated water for paddy field. Irrigated water is water from the river of Baubau Wonco Watershed which is streamed to paddy fields of local farmers in order to maintain water balancing of the paddy fields and to increase of its productivity.

Paddy fields within DAS Baubau Wonco cover about 100 Ha. There are 32 local farmers cultivate this annual crop for fulfilling their daily need and selling to the local market [1]. According to Central Government Decree by the year 1999 stated that local farmers can share budgets for maintaining facilities of irrigation in the tertier pipeline units through a payment for irrigation water by water user of local farmer association [14]. Furthermore, realocation of irrigated water for non agricultural use can bring about negative impacts to village economic conditions, a decrease of irrigated water volume, then a decline of a crop width and causing a decrease of food productivity and a loss of livelihood of local farmers (Rosegrant and Ringler, 1998) [14].

Nowadays, irrigated water users in Baubau City tend to increase and lead to an increase in water demands. In the other words, water resource is becoming more competitive resource. To this end, valuation of irrigated water is crucial in order to determine amount of budgets that can be shared by local farmers.

This study attemped to extrapolate the economic value of irrigated water using Productivity Function Method. This method is used through the analysis of costs that has been spended and revenues that has been obtained by local farmers from paddy product. There is also a need to calculate an average of irrigated water use. Using this method, local farmers can then calculate the economic value of irrigated water which has been used for increasing their paddy crop productivities.

The cost analysis shows that the average cost that has been spended by local farmers in producing paddy crop achieves Rp.9.080.000 year^{-1}. This value multiplies by 32 people (number of local farmers cultivating paddy crop) results in Rp 290.560.000 year^{-1}. Whereas the total revenue obtains Rp. 585.000.000 year^{-1}, with an average revenue of each farmer is Rp.18.281.250 year^{-1}. So, the total income of paddy crop farmers is Rp. 294.440.000 year^{-1}.

[15] explained that accounting irrigated water costs should be clearly determined the function of production which explains conseptual relationship between inputs and outputs, for example, in order to product paddy crop, there are four inputs that involved namely : capital, labour, such natural resources as land and irrigated water. To this point, economic value of water is a difference between the total economic value of paddy crop and the total spended costs of the three kinds of inputs.

The study found that an average water consumption for irrigation is 25,55 m^3 ha^{-1} dan the total volume of water consumption for irrigation is 2.555 m^3 ha^{-1}. Paddy crop is harvested once a year. The economic value of water achieves Rp. 30.199.167 ha^{-1} year^{-1}. This is relatively high compared to an economic value of irrigated water in Kampar Watershed that obtains 1.483.500 ha^{-1} year^{-1} [4], while an economic value of irrigated water in Van Der Wijce tirrigation, Sleman Regency obtains 1.388.742 ha^{-1} year^{-1} [15]. The study found that the difference level of economic value of irrigated water between the regions caused by both the difference acreage of the paddy fields and its productivity as well as the difference method of valuation.

Irrigated water is mostly assumed as free commodity by the farmers so that it can be freely used. This assumption can create problem with respect to the maintenance of irrigation facilities. This is due to lack of awareness and fund to maintain those facilities. The research has determined economic value of irrigated water within Baubau Wonco Watershed. This can lead to the arising awareness of local government and farmers to collaborate in developing a model of payment for environmental services that have been provided by catchments forests. The second year study will attempt to explore an appropriate model than can facilitate Baubau Wonco's stakeholders to share resources and funds for protecting the watershed from further degradation.

5. Conclusions

a. There are various major economic and environmental importance of Baubau Wonco Watershed supported by catchment forests namely: Water for domestic, industrial and agricultural use (irrigation); Erosion, flooding and land slide hazard prevention, wildlife habitat, and ecotourism.

b. As the first year study attempted to predict economic valuation of water for domestic and agricultural uses. Economic value of water provided by watershed area for domestic used obtains Rp. 40.062.668.750 year^{-1}, while for agricultural use achieves Rp. 30.199.167ha^{-1} year^{-1}.

c. These economic values are relatively high compare to other regions and it will bring a consequency to serious

d. conservations efforts. Otherwise, high environmental costs will be much more higher than determined economic values as found in this study to restore the ecosystem of the watershed.

e. The second year study will be intended to analyze the economic value of water for industrial use, and the economic value of catchment forest's functions to prevent both community and watershed ecosystem from erosion, flooding and landslide hazards. Thus, the Total Economic Value of the hydrological environmental services can be callculated. A proper model of downstream-upstream mechanism of a payment for hydrological environmental services will be explored and developed at the second year of the study.

Acknowledgment

The grateful thanks are delivering to the Head of Board Planning Agency of Baubau City that has provided supports for the planning and implementation of the research activities. We also thanks to the Directore of Central Higher Education Office of Ministry of Research and Technology and Higher Education of Republic Indonesia that has provided funds for the implementaion of this research.

In particular thanks goes to The Rector of Halu Oleo University for his legal and moral supports in implementation and compliance of this research. Many thanks also go to the head of Research and Community Service Center of Halu Oleo University for their supports in administeration matter and process. Last but not the least, a very grateful thank goes to the research team for their dedication and time management so that this research can be accomplished on schedule.

References

[1] Biro Pusat Statistik Kota Baubau, 2011. *Baubau City in Numbers*. BPS. Baubau.

[2] Brand, D. 2002. *Investing in the Environmental Services of Australian Forest*. In: Pagiola, S., Landell-Mills, N. And Bishop, J (Eds). (2002) Selling Forest Environmental Sevices; Market-Based Mechanism for Conservation and Development, London. Pp234-245.

[3] Isnin, M., Basri., H., and Romano, 2012. *Economic Value of Water availability of Krueng Jreu Sub Watershed Aceh Besa Regency*. Faculty of Agriculture, Syiahkuala University. Banda Aceh.

[4] Januaris, 2004. *The estimation of Economic Value of Irrigated Water to Support Agricultural Development in Kampar Watershed*. IPB. Bogor.

[5] Kasim, S., A.M..Kandari., Kahirun., 2007. *Study on Characteristics of Baubau Wonco Watershed*. Faculty of Agriculture, Halu Oleo University. Kendari.

[6] Kasim, S., Midi. L., 2012. *Impacts of Forested Land Use Changes on the Hydrological Functions of Baubau Wonco Watershed*. Proceeding of International Seminar on CRISHU Forum October 17th 2013.

[7] Kasim, S., Midi. L.,. 2013. *Agroforestry System as Vegetative Conservation Method for Land Use Development in Baubau Wonco Watershed*. Halu Oleo University. Kendari.

[8] Kasim, S., Midi, L., and Sarlina, 2014. *Valuation of Hydrological Environmental Services of Production Forest Area of Baini Village, Sampara Subdistrict, Konawe Regency*. Faculty of Forestry and Environmental Science, Halu Oleo University. Kendari.

[9] Kasim, S., Agustina, S., and Miduanto, 2014. *Valuation of Hydrological Environmental Services of Watu Mate Protected Forest, Waworaha Village, Lasolo Subdistrict of South Konawe Regency*. Faculty of Forestry and Environmental Science, Halu Oleo University, Kendari.

[10] Merryna, A. 2009. *Willingness To Pay Analysis of Local Community of Cirahab Sub Watershed*. Economy and Environmental Department of ITB. Bandung.

[11] Meteorology Station of Betoambari, 2013. *Recent Climatic Report of Baubau City*. Baubau.

[12] Moore I.D. and J.P. Wilson. 1992. *Length-slope factors for the revised universal soil loss equation: Simplified method of estimation*. Journal of Soil and Water Conservation. 47(5):423-428.

[13] Panhalkar S, 2011. *Land Capability Classification for Integrated Watershed Development by a Applying Remote Sensing and GIS techniques*. Journal of Agricultural and Biological Science. VOL. 6, NO. 4, APRIL 2011, Pages 46-55. ISSN 1990-6145. Asian Research Publishing Network (ARPN).

[14] Sumaryanto, 2006. *An Increase of Irrigated Water Efficiency Through an Implemetation of Environmental Tax based on Economic Value of Irrigation.* The Center of Social, Economy and Policy Analysis of Agriculture. Bogor.

[15] Syaukat, Y., dan Siwi, N., A., A, 2009. *An Estimation of Economic Value of Irrigated Water at Paddy Field Farm, Vab Der Wijce Irrigation Area of Sleman Regency, Yograkarta.* Jurnal Ilmu Pertanian Indonesia. Yogyakarta.

[16] Wilson J. P. and J. C. Gallant. 2000. *Terrain Analysis: Principles and Applications.* John Wiley and Sons, New York. pp. 87-131.

[17] Winrock International, 2004. *Financial Incentives to Communities for Stewardship of Environmental Resources; Feasibility Study. Winrock International, 621 North Kent Street, Suite 1200 Arlington, Virginia 22209 USA.* www.winrock.org, 50 pp.

[18] Zahabu, E., Malimbwi R.E., and Ngaga, Y.M. 2005. *Payment for Environmental Services as Incentive Opportunities for Catchment Forest Reserves Manaement in Tanzania.* Community Carbon Organization.

19

Growth Parameters of Onion (*Allium cepa* L. var. Cepa) as Affected by Nitrogen Fertilizer Rates and Intra-row Spacing Under Irrigation in Gode, South-Eastern Ethiopia

Weldemariam Seifu Gessesew[1, *], Kebede Woldetsadik[2], Wassu Mohammed[2]

[1]Department of Horticulture, College of Veterinary Medicine and Agriculture, Addis Ababa Univesity, Fiche, Ethiopia
[2]Department of Plant Sciences, Haramaya University, Dire-Dawa, Ethiopia

Email address:
whamlove@gmail.com (W. S. Gessesew)

Abstract: The productivity of onion crop is low due to poor agronomic and management practices in Gode district. Therefore, a field experiment was conducted at Gode Polytechnic College demonstration farm in 2013 under irrigation to assess the effect of N fertilizer rates and intra-row spacing on growth parameters of onion (Allium cepa L.). The treatments were consisting of six rates of N fertilizer (0, 46, 69, 92, 115, 138 kg ha[-1]) and four levels of intra-row spacing (7.5, 10, 12.5 and 15 cm) and the experiment was designed in RCBD with three replications. Results of the analysis revealed that the interaction effects of N rates and intra-row spacing showed highly significant ($P<0.01$) effect on leaf number. Plant maturity was delayed at higher N rates and wider intra-row spacings and vice-versa. The longest plant height was obtained from 15 cm intra-row spacing and 138 kg N ha[-1]where as the shortest was recorded from 7.5 cm intra-row spacing without N fertilizer application. Longest leaf length was obtained from plants spaced at 15 cm fertilized with 138 kg N ha[-1]. However, the shortest was recorded for plants grown in 7.5 cm intra-row spacings without N fertilizer. The overall result analysis showed thatgood growth performance of onion was obtained from 15 cm intra-row spacing combined with 138 kg N ha[-1] fertilizer applications.

Keywords: Growth, Intra-row, Nitrogen, Parameters, Rates, Spacing

1. Introduction

Onion (*Allium cepa* L.) is an important underground vegetable bulb crop of tropical and subtropical part of the world [1]. It belongs to the genus *Allium* of the family *Alleacea,* which is originated in southwest Asia and the Mediterranean regions. Onion is one of the oldest cultivated vegetables which traced back to at least 5000 years and has been in cultivation for more than 4000 years. Onion as food, medicine and religious object was known during the first Egyptian dynasty in 3200 B.C. Onion, which is different from the other edible species of *Alliums* for its single bulb, is one of the most important bulb crops cultivated commercially in most parts of the world. It is usually propagated by true botanical seed [2, 3].

Ethiopia has a great potential to produce onion throughout the year both for local consumption and export. It is grown primarily for its bulb which is used for flavoring the local stew, 'wet' which is considerably important in the daily diet, mostly used as seasonings or as vegetables in stews. It is one of the richest sources of flavonoid in the human diet and flavonoid consumption has been associated with a reduced risk of cancer, heart disease and diabetes. In addition it is known for anti bacterial, antiviral, anti-allergenic and anti-inflammatory potential [4].

Onions *(Allium cepa L.)* as bulb onion and/or shallot is probably cultivated in all countries of tropical Africa including Ethiopia [5]. Onion requires adequate soil moisture due to the relatively short and small root system. Onions are sensitive to photoperiod. Long days are favorable to onion production as this enhances leaf development and formation which, in turn, is directly related to bulb size. Early varieties require 13 hours for bulb initiation while late varieties require 16 hours for bulb initiation. In Ethiopia onion can grow between 500and 2400 meter above sea level, but the best growing altitude so far known is between 700and 1800 meter above sea level [6]. Thus, Onions are spread throughout the country being cultivated under both irrigated as well as rain fed conditions in different agro-climatic regions [7].

Onion is considered as one of the most important vegetable crops produced on large scale in Ethiopia. It also occupies economically important place among vegetables in the country. The area under onion is increasing from time to time mainly due to its high profitability per unit area and ease of production, and the increases in small scale irrigation areas. The crop is produced both under rain fed in the 'meher' season and under irrigation in the off season. In many areas of the country, the off season crop (under irrigation) constitutes much of the area under onion production. Despite the increase in cultivated areas, the productivity of onion is much lower than other African countries and the world average. The low productivity could be attributed to the limited availability of quality seeds and associated production technologies used [8]. According to [9] for private farmers' holdings in 'meher' season 2012/2013, the total area coverage by onion crop in the country was 21,865.4 ha, with total production of 219,188.6 t with average productivity of 10.02 t ha^{-1}. This is very low yield compared to the world average of 19.7 t ha^{-1}[10] due to low soil fertility, salinity effect and inappropriate cultural practice [11].

The use of appropriate agronomic management has an undoubted contribution in increasing crop yield. One of the important measures to be taken in increasing the productivity of onion is to determine the optimum amount of fertilizers rates and spacing in each agro-ecology. Among the fertilizers, nitrogen containing ones is the most important, since it is being a component of amino acids and chlorophyll, promotes rapid vegetative growth, protein content and yield of the crop. According to author [12], it is very difficult to give general recommendation for onion that can be applicable to different agro ecological zones of the country. Another author [13] reported also to optimize onion productivity; full package of information is required for each growing region of the country.

The Shebelle river basin along the Gode district is one of the most suitable areas for the production of onion. Recently onion growers in the study area started producing onion for home consumption as well as cash crop by irrigation. Farmers in the study area are mostly engaged in livestock production and few have recently started sedentary agriculture. Thus, productivity of most of the crops, including onion, is low due to poor agronomic and management practices. Moreover, lack of improved varieties and seed, absence of recommended nitrogen fertilizer rate and plant spacing are the pertinent problems of the study area. Currently the nationally recommended fertilizer rate of 100 kg DAP ha^{-1}(46 kgP$_2$O$_5$ ha^{-1}) and split application of 150 kg Urea ha^{-1}[14] are used along with 10 cm plant spacing [8] for onion production with no consideration of soil types. However, farmers in Gode area have no experience of applying the nationally recommended fertilizer rate and plant spacing rather they randomly practice undetermined fertilizer rate and plant spacing. In view of these, the present study was initiated to find out optimum and economic rates of fertilizer and intra-row spacing of onion crop for Gode province and the study was conducted with the

objective of determining the effect of nitrogen fertilizer rates and intra-row spacing on onion growth parameters.

2. Materials and Methods

2.1. Description of the Study Area

A field experiment was conducted in 2013 from January to June under irrigation at Gode Polytechnic College Demonstration farm in Somali National Regional State, South-eastern Ethiopia. The site is situated at latitude of 5°57' N and longitude of 43°27'E. The experimental site lies at an altitude of about 300 m.a.s.l. It is characterized by high temperature, erratic rainfall, and sandy clay loam soil texture. Gode district has a vast area of plain suitable for large scale irrigated agriculture and livestock production. The study site is typified as arid to semi-arid climatic zone with mean maximum and minimum annual temperatures of 35 and 23.6^{0}C respectively. The rainfall has a bi-modal distribution pattern with mean annual precipitation ranging from 150 to 340 mm which is not sufficient for rain fed production [15].

2.2. Description of the Experimental Materials

Onion cultivar called Seiyunn – Hadhramout – R.Y (Yemen F$_1$hybrid seed), locally named as Qalafo onion, was used as a test crop for the experiment. It is well adapted and widely cultivated in the study area and also has light red color, globe shaped bulb with pungent smell and can mature in 115 -130 days. Its yield potential is 35-46 t ha^{-1} (personal communication April, 2013). Urea (46% N) fertilizer was used as a source of nitrogen for the experiment. The national recommended rate of nitrogen fertilizer which was found adequate for dry bulb production in upper awash region was 92 kg N ha^{-1} and 10 cm plant spacing was investigated at Melkassa and Were Research centers [8] and they were used as the basis to set the N fertilizer rates and intra- row spacing in this study.

2.3. Treatments and Experimental Design

The treatments consists of factorial combination of six rates of N fertilizer (0, 46, 69, 92, 115 and 138 kg ha^{-1} N) and four levels of intra-row spacing (7.5, 10, 12.5, and 15 cm). There were a total of 24 treatment combinations. The experiment was laid out in randomized complete block design (RCBD) with three replications. The size of each plot was 2x3m^2 accommodating ten rows (five double rows planted on shoulders of 40 cm bed including furrow wide ridges) with 40, 30, 24 and 20 plants per row for the intra-row spacing of 7.5, 10, 12.5 and 15 cm, respectively. The recommended inter-row spacing of 40 cm was maintained for all plots. The distance between plots and blocks were 1 m and 1.5 m, respectively. The outer single rows at both sides of the plot and one plant at both ends of the rows were considered as border plants.

2.4. Experimental Procedure

Seedlings were raised on three sunken nursery beds with

size of 1.2 x 5m^2 from Yemen produced seed of locally named Qalafo onion. Seeds were obtained from shop of vegetable seed supplier and were sawn on January 01, 2013 at 10 cm distance between rows, lightly covered with soil and mulched with grass until seedlings are emerged (2-5 cm from the soil). Seedlings were managed under nursery for six weeks and then after transplanted to the experimental plots.

Before transplanting seedlings, the experimental field was ploughed and harrowed by tractor. Large clods were broken down in order to make the land fine tilth, and then a total of 72 plots each with size of 2x3m^2 were prepared in which 24 plots were allocated to each replication. Plots were leveled and furrows and ridges were prepared. The nursery beds were supplied with 100 g Urea 1.2 x 5 m^{-2} [14]. Then seedlings were transplanted when they reached 12-15 cm height or 3-4 true leaves stage by carefully uprooting from nursery bed. One day before transplanting of seedlings, the nursery was irrigated for safe uplifting of seedlings. During transplanting only healthy, vigorous and uniform seedlings grown at the center of seedbeds were transplanted and gap filling was done within a week after transplanting.

The experiment was conducted under furrow irrigation method. Four day irrigation interval was maintained for the 1st four weeks and then extended to seven days interval until 15 days to harvest, when irrigation was stopped completely. All other agronomic practices were applied as per the recommendation made for the crop for all plots throughout the experimental period. Harvesting of onion bulbs was done when 70% plants in each plot show neck fall. Harvested onion bulbs were cured for four days by windrowing on the ground before topping [14].

2.5. Soil Sampling

Soil sampling was done before transplanting of seedlings from five entire representative points of the experimental site from depth of 0-30 cm then mixed to form composite sample. This composite sample was sub-divided in to working samples for analysis and soil analysis for specific parameters was carried out at Addis Ababa city government environmental protection authority and water works design and supervision enterprise soil laboratories. The composite pre-planting soil samples were analyzed for soil EC and pH at 1:2.5 soils to water ratio using a glass electrode attached to pH digital meter, organic matter was determined by using [16], total N was determined using Kjeldhal method as described by [17], available P was determined by the methods of [18], exchangeable potassium and sodium determined by potentiometericaly with 1M ammonium acetate at pH 7.0, Soil cation exchange capacity (CEC) was determined by ammonium acetate method [19] and Soil texture was determined by Bouyocous hydrometer method [20].

2.6. Data Collection

Growth components of onion were collected from 10 randomly selected and pre-tagged plants from the six central rows of each plot. Data were collected as per the procedures mentioned as follows.

Days to maturity: was recorded as the actual number of days from the date of transplanting to the time when 70% of plants in each plot showed neck fall [14].

Plant height (cm): was measured in centimeters from the soil surface to the tip of matured leaf in the plant at maturity using a ruler.

Leaf number per plant: was determined by counting the total number of leaves produced by sampled plants and the mean leaf numbers per plant was obtained by calculating the average number of leaves.

Leaf length (cm): was recorded as the average leaf length of the longest leaves of sampled plants at maturity.

2.7. Data Analysis

The data were subjected to analysis of variance (ANOVA) using SAS version 9.1 GLM procedures and least significant difference (LSD) was used to separate means at $p < 0.05$ probability levels of significance.

3. Results and Discussion

3.1. Selected Soil Physico-Chemical Properties of the Study Area

Table 1. Soil physico-chemical properties of the experimental site before planting.

Soil properties	Results	Remark
Soil depth(cm)	0-30	
Particle size distribution (%)		
Clay (%)	23.08	
Silt (%)	25.84	
Sand (%)	51.08	
Soil textural class		Sandy clay loam
Bulk density (g/cm^3)	1.08	Satisfactory/ moderate
Organic carbon (%)	0.40	Low
Organic matter content (%)	0.70	Low
Total Nitrogen (%)	0.02	Very Low
Available Phosphorus (ppm)	29.34	High
CEC (c.mol/kg soil)	14.6	Moderate
Exchangeable Sodium c.mol kg^{-1}	0.70	Moderate
EC (dS m^{-1})	0.729	slightly saline
Soil pH	8.3	alkaline

Source: Addis Ababa city government environmental protection authority and water works design and supervision enterprise soil laboratories.

The results of the laboratory analysis of some selected physio-chemical properties of the soil of experimental site are presented below (Table 1). Results of the soil analysis before planting showed that the soil of the site is sandy clay loam in texture with alkaline (pH 8.3) reaction. The soil had a bulk density of 1.08 g cm^{-3}, and 0.02%, 29.34 ppm, 0.70% of total N, available phosphorous and, organic matter content, respectively. It had also 0.40%, 14.6 c.mol kg^{-1} soil, 0.729 dS m^{-1} and 0.70 c.mol kg^{-1} of organic carbon, CEC, EC, and exchangeable Sodium respectively. The rating under remark was done according to [21] and [22] suggestions.

According to the limit suggested by [16], the organic carbon (1.43%) or organic matter content (2.46%) of the soil was rated as very low before planting. According to the rating suggested by [23], the N content (0.15%) before planting was rated as low. According to the rating suggested by [24], the phosphorus content (15.5 ppm) before planting was rated as medium. The CEC (39, 13) before planting was rated as high as per the rating suggested by [23].

3.2. Days to Maturity

Days to maturity was significantly (P<0.05) affected by the main effect of fertilizer and intra-row spacing. However, fertilizer rates and intra-row spacing did not interact to influence days to maturity significantly. Days tomaturity of Qalafo onion variety was between 101and117 days.

Regardless of the intra-row spacing, days to maturity was prolonged in plants supplied with nitrogen levels, which were in statistical parity, as compared to plants which did not receive fertilizer. The highest N rate (138 kg ha^{-1}) delayed maturity of plants while plants which did not receive nitrogen fertilizer had early maturity. Plants supplied with 46, 69, 92, 115 and 138 kg N ha^{-1} extended days to maturity by about 2.61, 3.52, 8.02, 9.58 and 15.56%, respectively; as compared to the plants in the control plot (Table 2).

The delay maturity due to N fertilizer application is attributed to the prolonged canopy growth in response to nitrogen rate which maintain physiological activity for an extended period and thereby continuing photosynthesis [25,26]. Similar results were reported by [27] on days to maturity onion which was prolonged in response to increased levels of farmyard manure (FYM) and N. According to author [28], N rate above 100 kg ha^{-1} delayed maturity. Author [29] also observed that the N at the rate of 150 kg ha^{-1} recorded the highest number of days to reach bulb maturity. The result is also in line with the finding of [30] who reported that application of 180 kg ha^{-1} N prolonged the growing period of onion.

Irrespective of N fertilizer rates, closer plant spacing of 7.5 cm and 10 cm hastened by about 5 and 4 days, respectively, while wider plant spacing of 12.5 and 15cm showed slightly delayed maturity (Table 2). This might be attributed to competition for light and nutrient in closer spacing, causing early bulb maturity while wider spacing allowed plants to have access for more nitrogen which prolonged maturity. Maturity was shortened from 134 to 124 days as reported by [31] when seeding rate increased from 100,000 to 200,000 per acre. The result is in agreement with the findings of [32, 25] who noted that bulb maturity is advanced by higher density planting, which is associated with a high leaf area index and hence light interception by the leaf canopy that advance the date of bulb scale initiation.

3.3. Plant Height, Leaf Length and Leaf Number

Plant height and leaf length were significantly (P<0.01) influenced by the main effect of intra-row spacing and N rates but not by the interaction of the two. However, leaf number was significantly (P<0.01) affected by intra-row spacing and N rates and their interaction effect (P<0.05).

3.3.1. Plant Height

Considering the independent effect of N rates alone, the longest plants were obtained when N was supplied at the rate of 138 kg ha^{-1} and the shortest plants were observed in the control plot. The increase of plant height was 2.2, 4.4, 8.6, 13.17 and 25.17% as the increased N fertilizer rates of 46, 69, 92, 115 and 138 kg ha^{-1} N, respectively, over the control (Table 2). This is because of the fact that N is mainly concerned with the vegetative growth of plants through cell division and elongation [33, 26]. The result is in line with the research finding of [27] conducted at Gode who reported as increasing the rate of N supply from 0 kg ha^{-1} to 100 kg ha^{-1} significantly increased plant height by about 8%. This result is consistent also with the findings of [34, 35] who reported that as the N rates increased so did plant height of onion.

Disregarding the effect of N rates, significantly longest onion plants were obtained from intra-row spacing of 15 cm and comparatively shortest plants were observed at intra-row spacing of 7.5 cm which was statistically at par with 10 cm intra-row spacing (Table 2). In agreement to the current finding, author [36] reported tallest onion plant from 10 cm intra-row spacing than plants grown at 7.5 cm and 5 cm at Aksum, northern Ethiopia. Auther [37] also found tallest plant at 25 cm intra-row spacing than at 15 cm intra-row spacing. The reduction in plant height at increased plant density might be attributed to the possible competition for soil moisture and nutrients as indicated by [38]. In contrast to the present study result, [39] recorded significantly different and highest plant height on radish at 15 cm and 10 cm spacing.

Table 2. Main effects of N rates and intra-row spacing on days to maturity (days), plant height (cm) and leaf length (cm) of Qalafo onion variety grown at Gode under irrigated condition.

Treatments	Days to maturity (days)	Plant height (cm)	Leaf length (cm)
N (kg ha^{-1})			
0	101.75[e]	47.83[f]	42.19[e]
46	104.41[d]	48.88[e]	48.15[d]
69	105.33[d]	49.93[d]	52.63[c]
92	109.91[c]	51.93[c]	54.19[bc]
115	111.50[b]	54.13[b]	55.68[ab]
138	117.58[a]	59.87[a]	57.78[a]
LSD (5%)	1.21	0.76	2.47
Spacing (cm)			
7.5	106.28[d]	51.23[c]	50.16[b]
10	107.67[c]	51.64[bc]	51.06[b]
12.5	109.28[b]	52.22[b]	52.15[ab]
15	110.44[a]	53.69[a]	53.28[a]
LSD (5%)	0.99	0.62	2.02
CV (%)	1.36	1.79	5.81

LSD = least significant difference, CV= coefficient of variation in percent; Means within a column followed by the same letter are not significantly different at P< 0.05.

3.3.2. Leaf Length

Irrespective of the intra-row spacing treatments, the longest leaf was obtained by application of N at the rate of 138 kg ha[-1]while the shortest plant leaf was observed in the control plots. The increased leaf length observed with the increased N fertilizer rates of 46, 69, 92, 115 and 138 kg N ha[-1] was 14.13, 24.75, 28.44, 31.97 and 36.95%, respectively, over the control (Table 2). This is because of N which is mainly concerned with the vegetative growth of plants through cell division and elongation [33, 26]. The result is in line with the work of [27] who reported that a significant influence of leaf length was detected as the rate of N supply was increased from nil to 50 kg N ha[-1]. Other author [40] also reported that as the N rates increased so did leaf length.

Without considering the effect of N rates, significantly longest leaf was obtained from Intra-row spacing of 15 cm and comparatively shortest leaf length was observed at intra-row spacing of 7.5 cm which was statistically at par with 10 cm intra-row spacing (Table 2). This result is in agreement to [36] finding, who reported that highest onion plant leaf length obtained from 10 cm intra-row spacing than at 7.5 cm and 5 cm intra-row spacing at Aksum northern Ethiopia. In agreement with this result, many authors reported that maximum leaf length of onion was recorded in plants spaced at a wider spacing [41,42]. These authors also reported that shortest leaves were observed in plants with the closet plant spacing of 10 cm. The results showed that the wider spacing produced much longer leaves as compared to closer spacing, which might be due to relatively more nutrient and moisture availability and, hence, better growth of the plants.

3.3.3. Leaf Number

The leaf number of onion plant was significantly (P<0.01) affected by N fertilizer rates and intra-row spacing, and their interaction. The highest leaf numberwas obtained from the treatment combination of 138 kg N ha[-1]and 15 cm intra-row spacing which is statistically at par with all the other inra-row spacing combined with 138 kg N ha[-1]and 115 kg N ha[-1] combined with 15 cm intra-row spacing respectively.The lowest leaf numberwas recorded from the plots supplied with nil N fertilizer at all intra-row spacing, which was also statistically at parity with 46kg N ha[-1] combined with 7.5 cm intra-row spacing (Table 3). The increase in leaf number per plant at higher nitrogen and wider intra-row spacing could be attributed to the availability of enough growth factors that permit leaves to grow vigorously with less competition. This result is consistent with the findings of [43,44] who confirmed that increased N level ranging from 100 to 200 kg N ha[-1] resulted in increased number of leaves per plant of onion.

The result of the current study is in agreement with the result of [45] who reported that the number of leaves produced per plant was increased in response to decreasing plant population density. Similarly, auther [41] found more number of leaves per plant of onion at minimum plant population density (20 plants m[-2]) which decreased with increase in plant population. Auther [29] also reported that lower population density was best with regard to the number of leaves produced per plant.

Table 3. Interaction effect of N fertilizer rates and intra-row spacing on leaf number per plant of Qalafo onion variety grown at Gode under irrigated condition.

Intra- row spacing (cm)	N (kg ha[-1])					
	0	46	69	92	115	138
7.5	7.31[K]	7.52[k]	10.16[hij]	10.50[hij]	11.60[efg]	12.87[abc]
10	7.58[k]	10.10[ij]	9.73[j]	12.16[cde]	11.63[efg]	13.37[ab]
12.5	7.92[k]	9.96[ij]	11.06[fgh]	11.83[def]	12.66[bcd]	12.84[abc]
15	7.97[k]	9.86[ij]	10.76[hij]	12.03[cde]	12.82[abc]	13.76[a]
LSD (5%)	0.96					
CV (%)	5.42					

Means followed by the same letter are not significantly different at P< 0.05, LSD = least significant difference; CV= coefficient of variation in percent; ns= non-significant at p<0.05.

4. Conclusion

The field experiment was conducted to determine the effect of N fertilizer rates and intra-row spacing on onion growth. Results of the analysis revealed that the main effects of N rates and intra-row spacing showed a significant effect on days to maturity, plant height and leaf length. Besides, the interaction effect of N rates and intra-row spacing had significant effect on leaf number. Thus, according to this study, good growth performance of onion can be obtained from application of 138 kg N ha[-1] fertilizer combined with 15 cm intra-row spacing. However, this combination cannot be generalized for all onion cultivars and locations in areas under Shebelle river basin at Gode district. Therefore, the experiment should be repeated over locations and seasons by including intra-row spacing narrower than 7.5 cm as well as N rates higher than 138 kg ha[-1].

Acknowledgments

The author extends a special gratitude to the Federal Democratic Republic of Ethiopia (FDRE), Ministry of Agriculture (MoA), Agricultural Technical Vocational Educational Training (ATVET) coordination office for the financial support for this study. In addition, the administrative office of Gode Polytechnic College is also appreciated for facilitating the finance allocated for this research and for their cooperation to use facilities of the College for this research work.

References

[1]   I. J. Golani, M. A. Vaddoria, D. R. Mehta, M. V. Naliyadhara and K. L. Dobariya., 2006. Analysis of Yield Components in Onion. Vegetable Research Station, Gujarat Agricultural University, Junagadh - 362 001, India, *Indian J. Agric. Res.*, 40 (3): 224 – 227.

[2]   Hanelt P., 1990. "Taxonomy, Evolution, and History." In Onions and Allied Crops, edited by Haim D. Rabinowitch and James L. Brewster, 1-26. Boca Raton, Fla.: CRC Press.

[3]   Corgan, J., M. Wall, C. Cramer, T. Sammis, B. Lewis and J. Schroeder, 2000. Bulb Onion Culture and Management. NMSU Agricultural Experiment Station Circular 563: 1-16.

[4]   MoARD, 2009. Rural Capacity Building project. Course for Training of trainers on improved *horticultural crop technologies*. pp. 5-19.

[5]   Grubben JH, Denton DA., 2004. Plant resources of tropical Africa. PROTA Foundation, Wageningen; Back huys, Leiden; CTA, Wageningen.

[6]   Aklilu, S. 1997. Onion research and production in Ethiopia. *Acta Horticulturae* 433, 95-97.

[7]   Lemma Dessalegn and E. Herath. 1994. Agronomic Studies on *Allium*. Pp.139-145. In: Horticultural Research and Development in Ethiopia. 1-3 December, 1992. Institute of Agricultural research and food and Agricultural Organization. Addis Ababa, Ethiopia.

[8]   Lemma Desalegn and Shimeles Aklilu., 2003. Research experiences in onions production. Research report No. 55, EARO, Addis Ababa Ethiopia, P. 52.

[9]   CSA (Central Statistics Agency), 2013. Area and production of major crops. Agricultural sample survey 2012/2013, private peasant holdings, Meher season, Statistical Bulletin 532, Addis Ababa.

[10] FAO (Food and Agricultural Organization), 2012.Crop production Accessed on. http:/www.faostat.fao.org, September 10, 2012.

[11] MARC (Melkasa Agricultural Research Center), 2004.Progress report for 1995 – 2003. EARO, Ethiopia.

[12] Upper Awash Agro-Industry Enterprise 2001. Progress Report 1996-2002, Addis Ababa, Ethiopia. Agricultural product 2001/2002, Addis Ababa, Ethiopia. Van Eeden F. J., Myburgh J., 1971. Irrigation trials with onions. *Agroplantae*3: 57-62.

[13] Gupta RP, Srivastava KJ, Pandey UB, Midmore DJ. 1994. Diseases and insect pests of onion in India. *Acta Horticult.* 358:265-372.

[14] Ethiopian Agricultural research Organization (EARO), 2004. Directory of released crop varieties and their recommended cultural practice, Addis Ababa, Ethiopia.

[15] SCF-UK DPPA (Save the children fund –United Kingdom and disaster prevention and preparedness Agency), 2003.Managing risks and opportunities on understanding of livelihoods of Somali regional state, Ethiopia.

[16] Wakley, A. and Black, C. A., 1934. Determination of organic matter in the soil by chromic acid digesion. *Soil Sci.*, 63: 251-264.

[17] Dewis, J. and P, Freitas. 1975. Physical and chemical methods of soil and tissue analysis. FAO Bulletin No. 10. Rome. pp. 275.

[18] Olsen S. R. and L. A. Dean. 1965. Phosphorus. In: Methods of Soil Analysis. American Society of Agronomy. Madison, Wisconsin, 9: 920 – 926.

[19] Cottenie, I., 1980. Soil and plant testing as a base for fertilizer recommendation. FAO soils bulletin No.38/2, FAO, and Rome Italy.

[20] Moodie, C., Smith, D. and McCreerry, R., 1954. Laboratory manual for soil fertility. Washngton State College, Monograph, 31-39.

[21] Hazelton, P.A. and Murphy, B. W, 2007 *Interpreting soil test results* (2nded): what do all the numbers mean?

[22] Donald A., Dan M., Jim S., John M, 2011. *Soil testing interpretation guide*. Oregon State University, U.S.A.

[23] Landon, J. R., 1991. Booker tropical soil manual: A handbook for soil survey and agricultural land evaluation in the tropics and sub-tropics. Longman Scientific and Technical, Essex, New York. 474p.

[24] Olsen, S. R., C. V. Cole, F. S. Watanable and L. A. Dean, 1954. Estimation of Phosphorus in soils by extraction with sodium bicarbonate. USDA, circular, 939:1-19.

[25] Brewster, J. L., 1994. *Onions and other vegetable Alliums*.CAB, International, Wallingford, UK. 236p.

[26] Marschner, H., 1995. Mineral nutrition of higher plants.2nd Ed. Academic Press, New York, P.674.

[27] Girma Z. 2011. Response of Onion (*Allium Cepa* L. *Var. Cepa*) To Organic and Inorganic Fertilizers at Gode, South-Eastern Ethiopia. Thesis submitted to the school of Graduate stdies Haramaya University.

[28] Khan, H., Iqbal, M., Ghaffoor, A. and Waseem, K., 2002. Effect of various plant spacing and different nitrogen levels on growth and yield of onion (*Allium cepa* L.). Online *J. Biol. Sci.*, 2:545-547.

[29] Kumar, H., J. V. Singh, K. Ajay, S. Mahak, A. Kumar and M. Singh., 1998. Studies on the effect of spacing on growth and yield of onion (*Allium cepa* L.) cv Patna Red. Indian J. Agri. Res., 2: 134 -138.

[30] Islam, M. K., Alam, M. F., Islam, A. K. M. R., 2007. Growth and Yield Response of Onion (*Allium Cepa* L.) Genotypes to Different Levels of Fertilizers. *Bangladesh Journal of Botany* 36(1): 33-38.

[31] Hendrickson, P. and Swanson, M., 2003. Onion row spacing by seeding rate. NDSU, Crington research Extention Center.

[32] Brewster, J. L., 1990. Cultural systems and agronomic practices in temperate climates. pp. 2- 30. In: Rabinowitch, H. D. and Brewster J. L. (Eds.). Onions and allied crops. Vol. 2. Agronomy, biotic interactions, pathology, and crop protection. CRC Press, Boca Raton, Florida, USA.

[33] Brady, N. C. 1985. *The nature and properties of soils*. 9th Edition. New Delhi.

[34] Al-Fraihat, H. Ahmad, 2009. Effect of different nitrogen and sulphur fertilizer levels on growth, yield and quality of onion (Allium cepa L.). Jordan Journal of Agricultural Sciences, 5(2):155-165.

[35] Mozumnder, S. N., Moniruzzaman, M., Halim, G. M. A. 2007. Effect of N, K and S on the Yield and Storability of Transplanted Onion (*Allium cepa* L.) in the Hilly Region. Journal of Agriculture and Rural Development 5: 58-63.

[36] Kahsay Y. Fetien A. and Derbew B. 2013. Intra-row spacing effect on shelf life of onion varieties (*Allium cepa* L.) at Aksum, Northern Ethiopia. *J. Plant breeding and crop Sci.*, 5(6): 127-136.

[37] Abubaker, S., 2008. Effect of plant density on flowering date, yield and quality attribute of Bush Beans (*Phaseolus Vulgaris* L.) under Center Pivot Irrigation System. *Amer. J. Agri. and Biol. Sci.*, 3(4): 666-668.

[38] Karaye, A. K. and A. I. Yakubu, 2006. Influence of intra-row spacing and mulching on weed growth and bulb yield of garlic (*Allium sativum* L.) in Sokota, Nigeria. African Journal of Biotechnology., 5(3): 260-264.

[39] Perez MA, Ayub CM, Saleem BA, Virk NA, Mahmood N, 2004. Effect of Nitrogen Levels and Spacing on Growth and Yield of Radish (*Raphanussativus* L.). In *Int. J. Agric. Biol.* 6:3 publishing, UK.

[40] Muhammad, S. J., 2004. Studies on the management strategies for bulb and seed production of different cultivars of onions (Allium cepa L.). A dissertation submitted to Gomal University, Deraismil khan and Pakistan. 449p.

[41] Jilani M. S, Khan MQ, Rahman S., 2009. Planting densities effect on yield and yield components of onion (*Allium cepa* L.). *J. Agric. Res.* 47(4):397-404.

[42] Jilani, M. S., P. Ahmed, K. Waseem and M. Kiran, 2010. Effect of plant spacing on growth and yield of two varieties of onion (*Allium cepa* L.) Under the agro-climatic condition of D.I. Khan. *Pakistan Journal of Science*, 62: 37-41.

[43] Jilani M. S, and Ghaffoor A., 2004. Screening of Local Varieties of Onion for Bulb Formation. *Int. J. Agric. Biol.* 5(2):129-133. Pakistan.

[44] El-Tantawy, E. M. and A. K. El-Beik, 2009. Relationship between growth, yield and storability of onion (*Allium cepa* L.) with fertilization of nitrogen, sulphur and copper under calcareous Soil Conditions. Research Journal of Agriculture and Biological Sciences, 5(4): 361-371. http://www.insipub.com/rjabs/2009/361-371.pdf. Accessed June 2010.

[45] Tegbew. W. 2011. Yield and Yield Components of Onion (*Alliumcepa var.cepa*) Cultivarsas influenced by population density. Unpublished M.Sc. thesis submitted to the school of graduate studies of Haramaya University, Ethiopia.

20

Indigenous and Current Practices in Organic Agriculture in Nigeria

Ibeawuchi I. I.[1], Obiefuna J. C.[1], Tom C. T.[1], Ihejirika G. O.[1], Omobvude S. O.[2]

[1]Department of Crop Science and Technology, Federal University of Technology, Owerri, Nigeria
[2]Department of Crop and Soil Science, University of Port Harcourt, Rivers State, Nigeria

Email address:
ii_ibeawuchi@yahoo.co.uk (Ibeawuchi I. I.)

Abstract: The paper defines organic agriculture as holistic management system which promotes agro-ecosystem's health, biodiversity, biological cycle and biological activity without the external inputs of synthetic chemicals such as: fertilizers, pesticides, hormones and feed additives. Benefit include among others: high and comparable yield though could be supported by those receiving support from European community that provides monitory help to farmers as well as with conventional farming. Current practices of organic agriculture are a modification and continuation of indigenous practices that are more prominent in Nigeria. More research and funding by government and private sectors have been recommended.

Keywords: Organic Agriculture, Indigenous Crops, Current Practices, Nigeria

1. Introduction

Agriculture is major employer of workforce in Nigeria engaging about 70% of both the elderly and the young; male and female. Generally, man depends on plants for food, fibre, clothing, medicine, shelter etc. Thus, plant growth and yield in all ecosystems depend on the cycling and recycling of nutrients between the plant biomass and the organic and inorganic soil stores. Over the years, agriculture developed as a result of man's quest to feed himself, family and his animals. Through time man discovered that agriculture attracts various input from art, a science and also a business of producing plants and animal products for the beneficial use of mankind. Plants are primary producers in agriculture. During the process of photosynthesis, they take in carbon dioxide from the air, moisture and nutrients from the soil and by trapping the energy from sun light, convert these simple compounds into complex food materials.

Animals are secondary agricultural producers. They eat plants or parts of plants and convert the complex compounds they contain into animal products such as meat, milk, eggs, hides and wool. Primitive agriculture made use of the soil, rainfall and the local species of plants and animals as they naturally occur. Any plants and organisms which reduce the productivity of useful plants and animals were avoided or controlled. Productivity in the system was low and capable of feeding the farmers and his family alone. Due to population pressure and urbanization agricultural practices improved with use of high external inputs such as inorganic fertilizers, agro-chemicals for pest and weeds, these helped to increase productivity but with serious adverse effect on environment and man making the entire system very unsustainable. The trend in agricultural production system from indigenous to modern agriculture has developed by which agriculture is dynamic and improved overtime. Currently, there is a another strategy known as organic agriculture, which is a production system that maintains and sustains the health of soils, the ecosystem and the people living in that system and beyond. It completely relies on ecological processes, biodiversity and cycles adapted that are understood by local agricultural farmers in their conditions but exclude the use of external inputs with negative effects. Organic agricultural practices combine and interwoven in innovation, science, art, as well as business to benefit the environment and promote fair relationships and a good quality of life for all involved in it. Organic agricultural practices increases productivity by the slow release of nutrients to plants and habour as well as maintain soil biological stand. This brings about increase in productivity with little or no damage to the environment. There is need to review the indigenous and current practices of organic agriculture especially in Nigeria. A study of this nature is needed to further create awareness on indigenous practices of organic agriculture for proper comparison with the current practices. To further reposition agriculture for sustainable consumption and income, the knowledge of what is going on presently is needed to shape what will happen in future. This paper seeks to review the practicing of organic agriculture in order to create more awareness on the principles for which it is based, highlight the roles and problems of organic agriculture in our rural communities, review the indigenous practices and assess the current practices for proper recommendation.

2. Meaning of Organic Agriculture

Organic agriculture is based on production standards that require no synthetic inputs, instead using practices modeled on ecological processes to increase soil fertility and discourage pest infestation and disease infections in the fields, pens and ranges. This is to say that animals such as pigs lie in pens while cows feed on range lands. Prior the era of inorganic fertilizers, our forefather farmers usually restore fertility through the length of fallow period which may be 8 – 16 years and at times 20 years. They never used anything except may be the kitchen ash, poultry manure and dirt gathered after sweeping the compound. As population keep on increasing, unknowingly, the fallow period came down from 8-16 years to 8-12 years and to 3-4 years presently. The soil is no longer fertile as a result of many factors including soil erosion, nutrient mining by crops, exposure to rain and sunlight creating crusting etc. [5] viewed organic agriculture as providing a broad set of practices that increase resilience in farms. Thus: "Organic agriculture is a holistic production management system which promotes and enhances agro-ecosystem's health including increased functional biodiversity, biological cycles and soil biological activities. It emphasizes the use of management practices in preference to the use of off- farm inputs, taking into account that, regional conditions require locally adapted systems. This is accomplished by using where possible, cultural, biological and mechanical methods, as opposed to using synthetic materials, to fulfill any specific function within the system".

This definition serves in this review as a base line for organic agriculture around the world as well as in Nigeria [5]. Countries develop their own standards from those set out by the Codex Alimentarius Commission which is a commission of collection of internationally recognized standards, codes of practice, guidelines and other recommendations relating to foods, food production and food safety. Although organic agriculture adheres to certifiable standards, farmers can do much more to base their practices in ecology than standards required. The principles of organic agriculture concerns the way people interact with living landscapes, relate to one another and shape the legacy of future generation. The International Federation of Organic Agriculture Movement - IFOAM describes four principles of organic farming, on which its standards are based. Each of the principles contributes to the long term health of the farm, the surrounding environments, and the farming participants, which in turn builds resilience

for the long term success of the farm. These principles of this area of knowledge include:

- Health: Organic agriculture should enhance the health of soil, plant, animals, human and planet as one and indivisible.
- Ecology: Organic agriculture should be based on living ecological systems and cycles, work with them, emulate them and help sustain them.
- Fairness: Organic agriculture should build on relationships that ensure fairness with regards to common environment and life opportunities.
- Care: Organic agriculture should be managed in a precautionary and responsible manner to protect the health and well being of current and future generations and the environment[13].Presently organic agriculture has been found to play vital role in solving or reducing to barest minimum the following:

Environmental degradation:

Overtime high intensive agriculture has resulted to the global yield gap between best practices and farmers 'fields remains large, resulting to agricultural lands that continue to shrink and with corresponding global environmental threats: erosion of biodiversity, desertification, climate change and other trans-boundary pollution become a reality [25]. Intensive agriculture contributes to over 20% of global green house gas emissions - GHGs [25], Agricultural activities affects 70% of all threatened bird species and 49% of all plant species [25]) uniform cultures have dramatically reduced the number of plants and animals used in agriculture. Currently, 1,350 breeds face extinction, with two breeds being loss every week. Biodiversity erosion is exacerbated by the loss of forest cover, coastal wet lands and other wild relatives, important for development of biodiversity and essentials for food provision, particularly in times of food crisis.

Biodiversity Protection: Protecting biodiversity at genes, species and ecosystem levels through germplasm banks and protected areas is not sufficient. The maintenance cost of gene banks are high, up to half of the material collected is in need of regeneration and "freezing" genetic resources denies then-evolution. Biodiversity is best maintained through sustainable utilization and selection by food providers. Animals that move across boundaries and ecosystems are not immune to air and water pollution. The 12% of global land areas "fenced" for nature protection are located within or around a 40 percent land surface used by agriculture and forestry [24]. Food systems should be viewed as an integral part of the ecosystem. There is a need to manage agricultural land as part of a larger landscape that explicitly considers ecological functioning [25].

Self reliance, Rural Development and Nature Conservation:

Organic agriculture offers a means to address food self-reliance, rural development and nature conservation. The goal of organic agriculture is sustainable use of biodiversity in terms of both agriculture to biodiversity and biodiversity contribution to agriculture. Organic agriculture needs functional group of species and essential ecosystem processes as its main "input" to compensate for the restriction on or absolute lack of synthetic input use. [25] Maintained that a close relationship exists between organic agriculture and the maintenance of biodiversity. Rules and regulations that govern certified organic agriculture and by the practical experiences of organic farmers around the world justify the success expectation of organic agriculture.

Any form of agriculture means intervention and the alteration of the processes occurring in natural ecosystem by a human being [31]. Therefore such intervention always results in environmental damage. Organic agriculture has been found to operate with minimum environmental damage [15, 34 and 24]. Organic agriculture contributes to environmental protection and rational use of natural resources [30 and 22]. [21] made an update comparison between organic agriculture and conventional farming systems with regard to environmental performance and concluded that organic farming showed better environmental performance confidence interval than the conventional one. Crop yields obtained under organic agriculture compared with modern or conventional farming showed marked variations 20-40% in favor of conventional [15, 16, 17, 30]. However there are also studies that showed that organic agriculture showed far less significant differences [26].

In a 21-year study of the agronomic and ecological performance of biodynamic, organic and conventional farming systems in Switzerland, the crop yields in biodynamic and organic system found to be 20 percent lower than in conventional, although input of fertilizer and energy was reduced by 34 to 53 percent and pesticide input by 97 percent [22]. A 22-year trial study by Cornell University in USA proved that organic farming produces the same corn and soybean yields as conventional farms, but consume less energy and utilizes no pesticides [18]. [17] made an assumption that the yield reductions associated with organic farming in Western Europe will also apply in other parts of the world. Evidence from the less developed countries including Nigeria, indicates that organic yield levels are often very similar to those achieved in conventional systems [29, 17, 36].

Several authors suggest that in the so called less developed countries and Eastern Europe, Organic yields can even be higher than conventional [15, 6]. Organic agriculture also involves stockless organic systems. Although organic farming systems are often perceived to require both livestock and crop production enterprises to form a viable agronomic and economic unit, there may be economic and other circumstances leading to stockless farming production [23]. The stockless organic farming systems have proved to be capable of supporting soil fertility in several European regions [23]. In terms of crop yield, [23, 11] reported that the yield obtained are comparable with those achieved by other methods of organic farms. Thus, stockless organic agriculture seems to be economically viable. Another benefit of organic stockless systems is improved sanitary condition of the farm and produce especially in Europe where agri-environment has been introduced [11], In the mid nineties the proportion of stockless farms in Germany varied between 20 and 50

percent [32] while France and U.K also recorded a sustainable portion of stockless organic farmers [7].

Profitability of Organic Agriculture: Most available studies report similar or better economic performance than conventional farm, even without taking into account external costs of production [19, 16, 30]. This is mainly due to lower production cost. Organic products attract on average a 20-50% higher price than conventional products.

Promising for less developed countries: Several authors suggested that organic farming is a promising solution for central and Eastern Europe both from the economic and environmental point of view [6, 33, 35, 37]. Options and implication of converting to organic agriculture in the less developed countries have been positively assessed. Besides, [30] observed that organic agriculture can well feed the populations of less developed countries including Nigeria.

Problems of organic agriculture:
The problems as reported in wikipedia freeencyclopedia are as follows:

Soil Management:
Plants need nitrogen, phosphorus and potassium as well as micro-nutrients, but getting enough nitrogen, and particularly synchronization so that plants get enough nitrogen at the right time (when plants need it most), is likely the greatest challenge for organic farmers.

Weed Control:
Techniques for controlling weeds have varying levels of effectiveness and including hand weeding, mulch, a natural pre-emergence herbicide, flame, garlic and clove oil, borax, pelargonicacid, solarization (which involves spreading clear plastic across the ground in hot weather for 4-5 weeks), vinegar, and various other homemade remedies. One recent innovation in rice farming is to introduce ducks and fish to wet paddy fields, which eat both weeds and insects. Note:(Oil of cloves - wikipedia, the free encyclopedia) Clove oil: it is a natural analgaesic and antiseptic used primarily in dentistry for its main ingredient eugenol. It can be purchased in pharmacies over the counter, as home remedy. There are three types of clove oil:

- Bud Oil: derived from flower-buds of Syzyguim aromaticum. Itconsists of 60-90% eugenol, eugenyl acetate, caryophyllene andother minor constituents.
- Leaf Oil: is derived from the leaves of Syzyguim aromaticum. Itconsists of 82-88% eugenol with little or no eugenyl acetate, andminor constituents.
- Stem Oil: is derive from the twigs of Syzyguim aromaticum. Itconsists of 90-95% eugenol, with other minor constituents.

Controlling other organisms:
Organisms aside from weeds which cause problems include arthropods (e.g. insects, mites) and nematodes. Fungi and bacteria can cause diseases. Insect pests are a common problem and insecticides both non-organic and organic, are controversial due to their environmental and health effects. It may be argued in a way that to manage insect is to ignore them and focus on plant health, since plants can survive the loss of about a third of leaf area before suffering severe

growth consequences though this assertion may not be acceptable in insects that feed on fruits to a larger extent.

Indigenous Practices in Organic Agriculture:

Development and practice or organic agriculture started in the tropics when most of the world's presently unused but arable lands were farmed [14 and 8]. In Nigeria and India more than 80% of the people work on land [8]. Due to availability of arable land then, people practiced the type of agriculture that was basically free from use of synthetic inputs.

Indigenous organic agriculture based on the farming systems involving:

Shifting cultivation or slash and burn, mixed cropping or multiple cropping, and Crop rotation were practiced.

Shifting cultivation or slash and burn:

This indigenous farming system depends on natural cycles to sustain it. Therefore the farmer farmed on a piece of land until fertility declined before moving to another fertile plot. In the beginning, the farmer had no intention to go back to the already exhausted lands. Production was mainly based on naturally accumulated nutrients resulting from fallowing. Crude implement such as hoe, matchets, axe, etc were used while the farmer and his family provided the labour. Productivity in this system was low. Due to scarcity of land resulting from increased population and urbanization fallow periods were reduced. The situation ushered in other modified forms of farming systems.

Shifting cultivation was practiced by the ancient people that are now considered a type of organic agriculture that is in practice today because the use of synthetic inputs: fertilizers, insecticides, plant growth hormones, etc were not involved. Rather the use of organic manures such as: poultry droppings, compost, farmyard and green manure were emphasized for improved crop production and soil condition [3].

Mixed Cropping or Multiple Cropping: This is the growing of two or more crops on the same piece of land. The advantages of these systems include reduction of diseases and pests, maintenance of soil conditions, facilitates vertical and horizontal variations thus, allowing cultivation of crops adapted to light and shade. Also permit phased harvesting – that is in crops with different maturity period that can be phase the harvesting by harvesting the early maturing varieties before the late maturing ones. Permit phased harvesting serves as insurance as when a farmer plants two or three crops on a piece of land in an intercropping or mixed cropping system, the failure of one crop does not mean the failure of all crops, the remaining one or two crops out of the two or three planted that remained serves as an insurance to the farmer. At organic level multiple cropping received no off farm input of fertilizers while land preparation is mechanically done.

Crop Rotation:

This involves a carefully considered cropping sequence with or without a fallow period. The main feature of crop rotation is that a given combination of crops is grown in a particular sequence on the same piece of land for several years without loss of soil fertility of reduced yield. This is done to; maintain and improve soil fertility, prevent build-up of pest, weeds and soil borne diseases, control soil erosion, reduce the period of peak requirement of irrigation water and conserve soil moisture from one season to the next. In crop rotation, crops were selected on the basis of their relationship to one another to ensure complementary or supplementary relationship rather than competitive. Again, the practice depended solely on organically produced input such as: compost, farmyard and poultry manures.

Current practices:

Organic agriculture is not totally new in Nigeria. It has been practiced age long in its rudimentary form by our forefathers before the advent of inorganic fertilizations. Our forefathers practiced shifting cultivation which involved the total use of natural fertility. However, engaging in shifting cultivation, crop rotation and mixed cropping systems as means of restoring soil fertility and sustainable pest and disease control without the use of agrochemicals [4]. Current practices in Nigeria for instance are based on and are still rooted in the indigenous systems since organic agriculture is not yet certified [4]. However, current practices of organic agriculture is a modification of the already existing practices with more attention on creation of integrated, humane, environmentally and economically viable agricultural systems.

Agroforestry:

This is an integrated approach of using the interactive benefits from combining trees and shrubs with crops and/or livestock Agroforestry was introduced to Nigeria in 1926 by International Council for Research in Agroforestry [9]. Practical applications of agroforestry evolve from:

- Traditional (through experience by generation of farmers).
- Modern (with the help of agricultural science) such as: shelter belt and hedgerow barriers. Recent/current (evolved with the help of research in agroforestry)
- Alley cropping and improved fallows.

Major products in this system include:

Timber, Fuel wood and Fodder [2] reported the practical applications of agroforestry as following:

- Live fences to control animal movement in the farmstead. This couldbe (a) live stakes serving as fences (b) hedges, example: Gliricidia sepuim (extracts obtained from its leaves is used to remove external parasites in animals, farmers in Latin America often wash their livestock with a paste made of crushed G. sepuim leaves to ward off torsalos) (http://en.wikipedia.org/wiki/Gliricidia_sepuim).
- Hedgerow barriers: Are trees planted along contours of sloping lands to check erosion and provide organic matter through leave fall. Extensive root system improves soil structure, example: Pterocarpus specie (oha), tender leaves also used as vegetable and trunk used as timber.
- Windbreaks and shelter belt: planted to protect community and their lands from strong wind, example: Casuarina species, Azadirachta indica, Leucaena and Senna species.

- Parklands (scattered trees): common type of agro forestry system in the tropics [2]. Trees are pruned and used as fodder.
- Alley cropping: it is a system in which trips (alleys) of annual crops are grown between rows of trees or shrubs. The trees used are mainly Cassia siamea, Gliricidia sepuim and Leucaena leucocephala and are capable of fixing nitrogen.
- Improved fallows: Current practices involving the use of fast growing leguminous species to shorten the long period inherent in the indigenous shifting cultivation practice.

Furthermore, agroforestry systems can be classified based on the basic components of woody perennials, herbaceous plants and animal components as: (a) agrosilvicultural system (production of woody plants and seasonal plants) (b) silvopastoral system involving production of woody plant species and livestock (c) agrosilvopastoral system (production of woody plant species, seasonal plants and livestock).

Intercropping/Multiple Cropping/Mixed Cropping:

Involve planting of two or more crops on the same piece of land in such a way that compatibility is maintained within the crop communities. In the system soil fertility is maintained by use of leguminous crops with non-leguminous ones [12, 28].

Bush Fallow System:

This is still in practice in areas where land for agriculture is available. However, production under fallow system is at a subsistence level even in the current practices. One feature note-worthy is absolute, elimination of synthetic input especially in the rural farming communities. One cannot conclusively divorce the use of synthetic input with large farms owned by cooperative societies under fallow systems.

3. Conclusion and Recommendation

From literature reviewed, organic agriculture appears to be a system of agriculture that preserve ecosystem, biodiversity and maintains sustainable food and health of humans. However, information on current practices in the Africa especially in Nigeria is not enough to compare the profitability of organic farming and conventional farming as is the case in advance countries: Europe and America. It is therefore recommended that more research work on organic agriculture should be encouraged through funding by government at all levels - federal, state and local government. Organic agriculturist should be given certificate and grant awards, and their organic product certified at subsidized rates for their sustained income generation through sales of such certified products at premium prices.

Refernces

[1] Agroforestry Wikipedia, the free encyclopedia. (http://en.wikipedia.org/wiki/Agroforestry).

[2] Agromisa (2003). Agroforestry CTA. Wageningen Netherlands.

[3] Asiegbu I.E. and Oikeh S.O. (1993). Growth and yield of tomato to sources andrates of organic manure in ferralitic soils. Bioresource Technology. 45: 21-25.

[4] Atungwu, J.J., Aiyelaagbe, I.O.O., Sobowale, P.A.S., Oni, A.O. and Garba, S.H., (2009). Integrating African Traditional Farming and Organic Agriculture. In: Organic Agriculture for Health, Wealth and Environmental Conservation. Proceedings of the fifth national conference of organic agricultural project intertiary institutions in Nigeria. FUTO. November, 15-19, 2009.

[5] Boron, S. (2006) Building resilience for an unpredictable future: How organic agriculture can help fanners adapt to climate change, FAO Rome, August2006. P6.

[6] Buys, J. (1993) "Conversion Towards organic agriculture in Russia - A preliminary study" Biological Codex Alimentarius commission (2001). Organically produced foods. Rome, Italy: FAO AND who 77pp.

[7] David, C., B.; (1996). Towards modeling the conversion of stockless farming to organic fanning. On-farm research in south East of France. In New Research in organic Agriculture. N. H. Kristensen and H. Hogh-Jensen pp 23-27. Tholey-Theley., IFOAM: 23-27.

[8] Evans, J. (1992). Plantation forestry in the tropics (second edition). Oxford University press Inc. New York. P6.

[9] Gholz, H. L., (1987). Agroforesty: realities, possibilities, and potentials.(http://books.google.com/books7id-YlYflgzthEsC&p^PA137&Ipg=PA137&dq^Year+agrofores1r y+came+mto+mgeria%3F&l&sig=pLhySxcjWsl sHueqfvd6vOsji3g&hlen).

[10] Gliricidia sepuim Wikipedia, the free encyclopedia. (http://en.wikipedia.org/wiki/Gliricidiasepuim).

[11] Huxham, S. K., (2005). The effect of conversion strategy on theyield of the first organic crop. Agriculture Ecosystems and Environment 106(4) 345-357.

[12] Ibeawuchi, I.I. and Ofoh M.C. (2000). Productivity of maize cassava/food legume mixtures in south-eastern Nigeria. Nig. J. of Agric. Rural Development 1(1): 19.

[13] International federation of organic agriculture movement (2006). The IFOAM normsfor organic production and processing. Version 2005 Bonn, Germany.

[14] Kellogg, C.E. and Orvedai, A.C. (1996). Potentially arable soils of the world andcritical measures for their use. Advances in Agronomy. 21, 109-70.

[15] Lampkin, N. (1990). Organic farming. Ipswich, farming press Books.

[16] Lampkin, N. (1992). The economic implication of conversion from conventional to organic fanning systems. Ph.D. Thesis, Dept of economics and agricultural economics, university of Wales, aberystwyth.

[17] Lampkim, N. (1999). Converting Europe- The potential for organic farming as mainstream. Paper presented to 1 1th National Organic fanning conference. Cirencester.

[18] Lang, S.S. (2005) Organic farming produces same corn and soybean yields as conventional farms but consumes less energy and no pesticide. Cornell university News Service.

[19] LEI (1990). Productie en atzet Van BD- en Eko-Produkten (production end economics or bio-dynamic food). National agro-economic institute of the whether lands. Band 2. LEI medelingen 425, Den Haeag. 89p.

[20] Mader, P. (2002). Soil fertility and biodiversity in organic fanning. Science 296(5573): 169-1697.

[21] Mader, P. (2004). Soil fertility in sustainable farming system. KSALT, Journal of the Royal Swedish Academy of Agriculture and Forestry 143 (1): 37-40. Oil of cloves wikipedia,   the   free   encyclopaedia, (http://en.wikipedia.org/wiki/oil_of_cloves).

[22] Pretty, J.N. (1995). Regeneration agriculture: policies and practices for sustainability and self-Reliance London, Washington. National academy press, Action Aidand Vikas.

[23] Philipps L. and Welsh J.P. (1999). Ten years experience of all-arable rotations. Designing and testing crop rotations for organic farming. Foulum, Danish Research Centre for organic fanning.

[24] Scialabba, E.N and C. Hattam, Eds. (2002). Organic agriculture, environment andfood security. Environmental and natural.

[25] Scialabba, E. N. (2003). Organic agriculture: The challenges of sustaining food production while enhancing biodiversity. United Nations. The matic Group meeting on wild life, biodiversity and organic agriculture. Anakara, Turkey, 15-16 April 2003, FAO Corporate documents repository.

[26] Stanhill, G. (1990). The comparative productivity of organic agriculture. Agriculture, Ecosystems and Environment. Vol30:1:1-26.

[27] The encyclopedia of organic gardening (1978) organic matter. Rodale press, inc. U.S.A. 814-819.

[28] Tom, C.T. and Asiegbu, I.E. (2002). Evaluation of short duration pigeon peacultivars in intercrop with FARZ-7 maize

in derived savanna of Nigeria. International journals of agricultural science. Vol.2.

[29] Van Elzakker, B. R. Witte et al. (1992). Benefits of diversity. New York. UNDP United Nations Development Programme.

[30] VanMansvelt, J. D. and J. A. Mulder (1993). European feature for sustainable development: a contribution to the dialogue. Landscape and urban planning: 27: 67-90.

[31] Van Mansvelt, J.D. and D. Znaor (1999) Criteria for the abiotic and biotic realm: environment and ecology. Checklist for sustainable landscape management. J.D. Van Mansvelt and M.J. Lubbe. Amsterdam-Lausanne-New York, Oxford, Shannon, Scingapore, Tokio, Elsevier.

[32] Von Fragstein P. (1966). Organic arable farming a contradiction? Fourth congress of the ESA-Book of Abstracts, Colmar Cedex, European Society Agronomy.

[33] Znaor, D. (1994) Ecological Agriculture: Analysis of the most commonly criticized aspects. Why shall Estonian Agriculture be Ecological? Proceedings of the international conference heed in tartu June 13, 1994; centre of Ecological Engineering tartu.

[34] Znaor, D. (1995) Ekoloska Poljoprivreda suitrasnjice Ecological agriculture of tomorrow. Zagreb, Nakladin Zavod Globus.

[35] Znaor, D. (1997) what future for sustainable agriculture? Danube watch 3(2): 2-3.

[36] Znaor, D. and H. Kieft (2000). Environmental impact and macro-economic feasibility of organic agriculture in the Danube River Basin. The world grows organic: Proceedings of the 13th international IFO AM Scientific conference Basel, Switzerland.

[37] Znaor, D. (2002): "Contribution of organic agriculture to macroeconomic and environmental performance of the countries with economic in transition". Vagos research papers 53(6) Lithuanias University of agriculture, Akademija.

21

Evaluation of Plant Growth Regulator, Immunity and DNA Fingerprinting of Biofield Energy Treated Mustard Seeds (*Brassica juncea*)

Mahendra Kumar Trivedi[1], Alice Branton[1], Dahryn Trivedi[1], Gopal Nayak[1], Sambhu Charan Mondal[2], Snehasis Jana[2, *]

[1]Trivedi Global Inc., Henderson, USA
[2]Trivedi Science Research Laboratory Pvt. Ltd., Bhopal, Madhya Pradesh, India

Email address:
publication@trivedisrl.com (S. Jana)

Abstract: Among the oilseeds grown around the world, mustard is one of the important crop worldwide due to its wide adaptability and high yielding capacity. Owing to the importance of its utilities as condiment, cooking oil and some medical aids, the demand for its seed production is too high. The present study was carried out to evaluate the impact of Mr. Trivedi's biofield energy treatment on mustard (*Brassica juncea*) for its growth-germination of seedling, glutathione (GSH) content in leaves, indole acetic acid (IAA) content in shoots and roots and DNA polymorphism by random amplified polymorphic-DNA (RAPD). The sample of *B. juncea* was divided into two groups. One group was remained as untreated and coded as control, while the other group was subjected to Mr. Trivedi's biofield energy treatment and referred as the treated sample. The growth-germination of *B. juncea* seedling data exhibited that the biofield treated seeds were germinated faster on day 5 as compared to the control (on day between 7-10). The shoot and root length of seedling were slightly increased in the treated seeds of 10 days old with respect to untreated seedling. Moreover, the major plant antioxidant *i.e.* GSH content in mustard leaves was significantly increased by 206.72% ($p<0.001$) as compared to the untreated sample. Additionally, the plant growth regulatory constituent *i.e.* IAA level in root and shoot was increased by 15.81% and 12.99%, respectively with respect to the control. Besides, the DNA fingerprinting data using RAPD revealed that the treated sample showed an average 26% of DNA polymorphism as compared to the control. The overall results envisaged that the biofield energy treatment on mustard seeds showed a significant improvement in germination, growth of roots and shoots, GSH and IAA content in the treated sample. In conclusion, the biofield energy treatment of mustard seeds could be used as an alternative way to increase the production of mustard.

Keywords: Mustard, Biofield Energy Treatment, Seedling, RAPD, Glutathione, Indole Acetic Acid

1. Introduction

Indian mustard (*Brassica juncea*) is a winter oilseed crop grown across the Northern Indian plains. Among the various oilseed crops, it is one of the important because of its potential utilities in the growing biofuels industries [1]. Mustard seed is widely used as a condiment and as an edible oil. The pungency of mustard oil is due to the presence of allyl-isothiocyanate. The low pungency of mustard oil can be obtained after inactivating the myrosinase enzyme present in it and used as a filler component in various processed meat products [2].

Glucosinolates are the major class of bioactive phytocontituents mainly rich in mustard. The hydrolytic product of glucosinolates plays an important role in plant defense against microorganisms and insects. However, it itself may act as nutrients as an essential component of nitrogen and sulfur [3]. Based on the literature, it was reported that the mustard seeds extract have the potential chemo-preventive and chemotherapeutic activities *in vitro* by scavenging the hydroxyl radicals, it also induces apoptosis of cancer cells [4]. Several studies have reported the antioxidant activities of mustard seeds extract [1, 5]. Rance and Morisset *et al.* reported the allergic reactions of mustard such as atopic dermatitis, urticaria and angioedema accounts 1.1% of total food allergy in children [6, 7]. The agricultural productivity depends on the most vital abiotic stress factor; salinity *i.e.* dissolved salts in water. The metabolic impairment in the plant cell occurs due to the osmotic and toxic effects of salt concentration in water. The different levels of salinity have affected the lipid components of mustard seeds. With increasing the salinity, phospholipids and glycolipids content were increased, while total and neutral lipids content were declined [8]. Generation of reactive oxygen species (ROS) is the main output of such metabolic impairment during salinity stress [9, 10]. The ROS such as superoxide radical (O^{2-}), hydrogen peroxide (H_2O_2), and hydroxyl radical (OH^-) are produced through the reduction of molecular O_2 during aerobic metabolism in mitochondria. Apart from the metabolic derived ROS, plant cell also produces singlet oxygen ($^1O_2$) in the chloroplast during photosynthesis [11, 12]. Among the various antioxidant pathways, the ascorbate–glutathione (ASC–GSH) cycle has been played an important role [13]. In plants, GSH is crucial for biotic and abiotic stress management. It is a pivotal component of the ASC–GSH cycle, a system that reduces poisonous hydrogen peroxide produced during photorespiration in peroxisomes. GSH and GSH-dependent enzymes represent a regulated defense against oxidative stress not only against ROS but also against their toxic products. Recent advances in molecular biology, development of polymerase chain reaction (PCR), and DNA sequencing have resulted in a powerful technique that can be used for the characterization of genetic diversity. Besides, the genetic diversity can also be assessed by the study of morpho-agronomic variability for plant breeders. For characterization of genetic profile a powerful tool has been developed as the molecular marker, so called DNA fingerprinting [14].

The National Center for Complementary and Integrative Health (NCCIH), allows the use of Complementary and Alternative Medicine (CAM) therapies like biofield energy as an alternative in the healthcare field. About 36% of US citizens regularly use some form of CAM [15], in their daily activities. CAM embraces numerous energy-healing therapies; biofield therapy is one of the energy medicine used worldwide to improve the overall human health. Mr. Trivedi's unique biofield treatment (The Trivedi effect®) has been extensively contributed in scientific communities in the field of agricultural science [16-19] and chemical science [20]. Due to the necessity of mustard as the food resource, and to improve the overall productivity of mustard plants an effective control measure need to be established. Under these circumstances, the present work was undertaken to evaluate the effect of biofield energy treatment on mustard in relation to germination growth in seedlings, level of GSH and IAA and the molecular analysis using DNA fingerprinting.

2. Materials and Methods

The *Brassica juncea* (*B. juncea*) seeds were distributed into two parts. One part was remained as control, no treatment was provided. The other part was subjected to Mr.

Trivedi's biofield energy treatment and considered as treated. The random amplified polymorphic DNA (RAPD) analysis was done by Ultrapure Genomic DNA Prep Kit; Cat KT 83 (Bangalore Genei, India).

2.1. Biofield Energy Treatment Strategy

The treated sample of mustard seeds was subjected to Mr. Trivedi's biofield treatment under ambient conditions. Mr. Trivedi provided the treatment to the seeds through his inherent unique energy transmission process without touch. The treated sample was assessed for growth germination of seedlings, glutathione (GSH) level and indole acetic acid (IAA) content in roots and shoots of the mustard plant.

2.2. Growth Germination of Mustard Seedlings

Control and treated mustard seeds (*B. juncea*) were soaked in distilled water separately. The water soaked seeds were wrapped with moist filter paper and kept in darkness at 20°C for germination. The percent of germinated seeds and length of roots and shoots in seedling were observed.

2.3. Measurement of Glutathione in Mustard Leaves

The extraction and estimation of GSH levels in plant leaves were followed as per standard method by Moron *et al.* 1979. For the extraction of GSH, approximately 5 gm of mustard leaves were crushed and mixed with 5 mL of 80% chilled methanol (as a solvent). Then the extract was sonicated for about 10 minutes. Then 1 mL of 5% trichloroacetic acid (TCA) was added to the extract. This sample was used for the analysis of GSH content. The GSH levels were estimated as per Moron *et al.* and TCA was taken as blank [21].

2.4. Measurement of Indole Acetic Acid (IAA) Content in Shoots and Roots of Mustard Seedlings

The extraction and analysis of IAA were done using Tang and Bonner's method. Freshly prepared Salkowski's reagent was used for the detection of IAA content in shoots and roots of mustard seedlings. For the extraction of IAA, approximate 200 mg plant tissue (shoots and roots) was grinded with 5 mL of 80% chilled methanol. The extract was filtered through Whatman filter paper (No. 1). After filtration the final volume of the extract was made upto 10 mL using 80% cold methanol. Optical density was measured after 30 minutes at 530 nm using ultra-violet visible spectrophotometer [22].

2.5. Isolation of Plant Genomic DNA Using CTAB Method

After germination when the plants were reached the appropriate stage, leaves disc were harvested from each plant. Genomic DNA was isolated according to standard cetyl-trimethyl-ammonium bromide (CTAB) method [23]. Approximate 200 mg of plant tissues (seeds) were grinded to a fine paste in approximately 500 μL of CTAB buffer. The mixture (CTAB/plant extract) was transferred to a microcentrifuge tube, and incubated for about 15 min at 55oC in a recirculating water bath. After incubation, the mixture was centrifuged at 12000g for 5 min and the supernatant was transferred to a clean microcentrifuge tube. After mixing with chloroform and iso-amyl alcohol followed by centrifugation the aqueous layers were isolated which contains DNA. Then, ammonium acetate followed by chilled absolute ethanol were added, to precipitate the DNA content and stored at -20°C. The RNase treatment was provided to remove any RNA material followed by washing with DNA free sterile solution. The quantity of genomic DNA was measured at 260 nm using spectrophotometer [24].

2.6. Random Amplified Polymorphic DNA (RAPD) Analysis

DNA concentration was considered about 25 ng/μL using distilled deionized water for the polymerase chain reaction (PCR) experiment. The RAPD analysis was performed on the treated mustard seeds using five RAPD primers, which were labelled as RPL 4A, RPL 5A, RPL 6A, RPL 13A, and RPL 19A. The PCR mixture including 2.5 μL each of buffer, 4.0 mM each of dNTP, 2.5 μM each of primer, 5.0 μL (approximately 20 ng) of each genomic DNA, 2U each of *Thermus aquaticus (Taq)* polymerase, 1.5 μL of MgCl₂ and 9.5 μL of water in a total of 25 μL with the following PCR amplification protocol; initial denaturation at 94°C for 5 min, followed by 40 cycles of annealing at 94°C for 1 min, annealing at 36°C for 1 min, and extension at 72°C for 2 min. Final extension cycle was carried out at 72°C for 10 min. Amplified PCR products (12 μL of each) from control and treated samples were loaded on to 1.5% agarose gel and resolved by electrophoresis at 75 volts. Each fragment was estimated using 100 bp ladder (Genei™; Cat # RMBD19S). The gel was subsequently stained with ethidium bromide and viewed under UV-light [25]. Photographs were documented subsequently. The following formula was used for calculation of percentage of polymorphism.

$$\text{Percent polymorphism} = A/B \times 100$$

Where, A = number of polymorphic bands in treated plant; and B = number of polymorphic bands in control plant.

3. Statistical Analysis

Data from GSH content in mustard leaves were expressed as Mean ± S.E.M. and analyzed through a Student's t-test to ascertain statistical differences between control and treated mustard seeds at the end of the experiment. A probability level of *p<0.05* was considered as statistically significant as compared to the control.

4. Results and Discussion

4.1. Growth Germination of Mustard Seedlings

Allelopathy is the process of plant communication system through the direct or indirect, detrimental or advantageous effects of one plant to another. They communicate through the release of allelochemicals *i.e.* the secondary metabolites or waste products of plants into the environment through

leaching, root exudation, volatilization and decomposition of plant residues. The mustard plant is belongs to *Brassicaceae* family cited as allelopathic crop [26]. The growth germination of mustard seedling data of control and treated samples are shown in Table 1.

Table 1. *Growth-germination of mustard (Brassica juncea) seedlings.*

Group	Germination (Day)	Germination (%)	Length (cm) (Mean ± S.E.M.)	
			Shoot	Root
Control	7-10th	65	8.88 ± 0.16	5.73 ± 0.05
Treated	5th	100	8.91 ± 0.25	5.78 ± 0.10

n = 100 (Shoot); n = 65 (Root); S.E.M.: Standard error of mean

It was found that the control seeds of *B. juncea* were germinated by 65% between days 7-10, while the biofield treated seeds were germinated on day 5 with 100% germination. After germination, on day 10th the mustard plants shoot and roots were measured. The shoot length in control sample was 8.88 cm and in treated sample it was 8.91 cm (n =100). The shoot length in the treated sample was slightly increased as compared to the control. Moreover, the length of root in control sample was 5.73 cm and in the treated sample it was 5.78 cm (n = 65). The root length in the treated sample was also slightly increased with respect to the control. The seeds of the majority of plant species have failed to germinate due to salt and osmotic stresses that may attribute either osmotic effects or specific ion toxicities towards seedling development. Establishment of the seedling is a critical stage in crop production that depends on the biochemical and physiological structures of seed. For faster and good development of seedling it is necessary a good physical strength and health of seeds to utilized essential nutrient during germination and early growth of seedling [8, 27]. Based on the findings, it is assumed that the early germination in biofield energy treated sample may be due to the increase in the ability of oxygen mediated metabolism to fight against stress that ultimately shortens the germination time as compared with the untreated seeds. Moreover, may be the biofield treated seeds protect themselves from allelopathic harmful effect from the environment. Hence, the treated seeds were germinated faster and the growth of roots and shoots were increased as compared to the untreated seeds.

4.2. Measurement of Glutathione in Mustard Leaves

The level of endogenous GSH content in leaves of both control and treated samples are illustrated in Table 2.

Sulfur is an essential component of all living organisms for protein synthesis. It is the integral constituent of various amino acids and cellular endogenous components like GSH. GSH (γ-L-glutamyl-L-cysteinylglycine) is sulfur containing thiol tripeptide, found in most of the organisms including plants. Deficiency of sulfur retard the growth of shoot, while did not affect the growth of the roots [13, 28]. The concentration of endogenous GSH was 0.29 ± 0.006 mM in control group, whereas it was increased to 0.89 ± 0.009 mM (n = 4) in the treated group. The result indicated that the GSH level in mustard leaves were increased significantly ($p<0.001$) by 206.72% in the biofield energy treated sample as compared to the naive (Table 2).

Table 2. *The glutathione (GSH) level in the leaves of mustard (Brassica juncea).*

Group	Endogenous GSH (mM)				Mean GSH (mM) (Mean ± S.E.M.)	Increased (%)
	Sample 1	Sample 2	Sample 3	Sample 4		
Control	0.29	0.29	0.28	0.30	0.29 ± 0.006	206.72
Treated	0.92	0.89	0.88	0.90	0.89* ± 0.009	

S.E.M.: Standard error of mean; n = 4; *$p \leq 0.001$

As literature reported that mustard seeds extract have the chemo-preventive and chemotherapeutic activity by scavenging free radicals and induced death of cancerous cells by apoptosis [4]. This study revealed, the significant improvement of GSH in mustard leaves after biofield energy treatment. Hence, it is assumed that the increase in GSH content in the biofield treated sample might accelerate the rate of free radical scavenging activity. Besides, the chemo-preventive activity of mustard seeds extract on cervix cancer was reported by Gagandeep and their co-researcher [29]. It is expected that the biofield treated mustard seeds in the diet may contribute better ability to reduce the risk of cancer incidence due to its chemo-preventive effect in the human population. Germination of seeds is the rate limiting factor in oxidative stress condition in plant cells. GSH is one of vital antioxidant in plant cells and plays a crucial role to plants defense system. Additionally, this data was well supported with early germination of mustard seedling in the biofield treated group (Table 1). Soil and water are polluted day by day due to heavy-metals derived by mining and from burning of fossil fuels. Plants like mustard have an important mechanistic pathway to detoxify heavy-metals through sequestration with heavy-metal with GSH [30]. This polluted soil and water cannot be chemically degraded or biodegraded by microorganisms. An alternative biological approach to deal with this problem is called phytoremediation, *i.e.* cultivation of mustard plants to clean up polluted waters and soils. Many studies have reported that live mustard plant tissues such as seeds and roots, contain a compound that acts as soil biofumigants by killing nematodes and pathogenic fungi [31].

4.3. Measurement of Indole Acetic Acid (IAA) Content in Shoots and Roots of Mustard Seedlings

IAA is one of the main constituents of plant growth substance *i.e.* auxin produced by several plant-associated commensal bacteria. The production of auxin is the key factor for determination of plant pathogenicity. Fewer chances of plant infection due to more level of IAA content in plants [32]. The IAA content in mustard shoots and roots of both control and treated samples are shown in Fig. 1.

In this experiment, the IAA content in mustard roots was 25.3 μg/g in the control sample, whereas it was increased to 29.3 μg/g in the treated sample. There was 15.81% increased of IAA content in mustard roots after biofield energy treatment. Furthermore, the level of IAA in mustard shoots was 17.7 μg/g in control sample, whereas it was slightly increased to 20.0 μg/g in the treated group. The data showed 12.99% increase in the IAA content in the shoots of biofield treated group. The growth of roots and shoots of mustard seedling was well increased due to over production of plants growth substance IAA. These findings were well corroborated with the literature [33, 34]. Based on the findings, it is assumed that the increased IAA content after biofield energy treatment in both roots and shoots might be helpful for their growth and overall development of plants.

4.4. Random Amplified Polymorphic DNA (RAPD) Analysis

The polymorphic DNA is responsible for giving the information about genetic marker due to its selectively of neutral nucleotide sequence and distinct genomes pattern [35, 36]. Here, RAPD was used as a DNA fingerprinting technique for evaluation of mustard seeds. The control and treated samples were evaluated based on their various RAPD patterns. It is very simple to detect because there is no need of DNA sequence information or synthesis of specific primers. It is a preferred tool being used nowadays to correlate the genetic similarity or mutations between species. The simplicity and wide field acceptability of RAPD technique due to short nucleotide primers, which were unrelated to known DNA sequences of the target organism [37]. The DNA fingerprinting by RAPD method was performed using five primers in the control and treated samples. The polymorphic bands are marked by arrows as shown in Fig. 2.

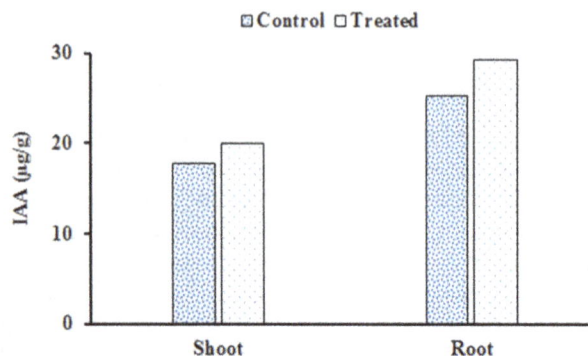

Figure 1. *The indole acetic acid (IAA) content in shoots and roots of mustard (Brassica juncea).*

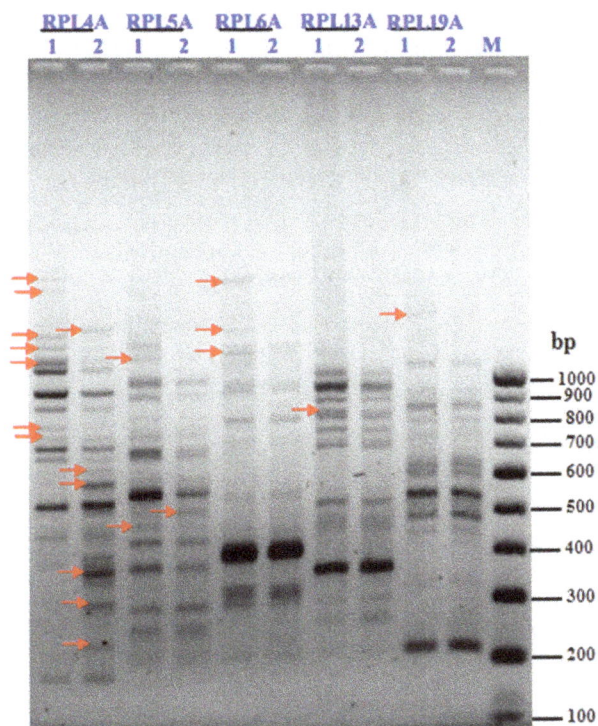

Figure 2. Random amplified polymorphic-DNA (RAPD) profile of mustard seeds (Brassica juncea) generated using Genei five RAPD primers, RPL 4A, RPL 5A, RPL 6A, RPL 13A and RPL 19A. 1: Control; 2: Treated; M: 100 bp DNA Ladder.

The RAPD patterns of treated sample showed some unique and dissimilar patterns. DNA polymorphism analyzed by the RAPD analysis was presented in Table 3. The level of true polymorphism in treated sample was observed at 59% with RPL 4A primer, at 21% with RPL 5A primer, at 30% with RPL 6A primer, at 10% with RPL 13A primer and at 11% with RPL 19A primer with respect to the control sample. The polymorphism was detected between control and treated samples. The percentage of true polymorphism was observed between control and treated samples of mustard seed an average 26% after the biofield energy treatment.

Table 3. DNA polymorphism analyzed by random amplified polymorphic-DNA (RAPD) analysis of mustard seeds (Brassica juncea).

S. No.	Primer	Band Scored	Common Band in Control and Treated	Unique Band	
				Control	Treated
1	RPL 4A	21	8	7	6
2	RPL 5A	13	8	2	1
3	RPL 6A	8	4	3	–
4	RPL 13A	8	8	1	–
5	RPL 19A	10	8	1	–

–, No band

The highest change in DNA sequence (46%) was observed in the treated group with RPL 4A primer as compared to the control. The lowest change was found in the treated group with RPL 13A primer as compared to the control. Biofield energy treatment could be responsible to improve the GSH and IAA content in mustard shoots and roots. Based on the findings of

growth germination pattern of seedling, GSH level, and IAA content followed by RAPD analysis there was positive impact of Mr. Trivedi's biofield energy treatment on the seeds of *B. juncea*. Based on these results, it is expected that biofield energy treatment has the scope to be an alternative approach related to improve the plant growth, development and simultaneously could reduce the pathogenicity.

5. Conclusions

Based on the study outcome, the biofield energy treated *B. juncea* showed faster (on day 5) and 100% germination as compared to the control (between day 7-10, 65%). Moreover, the GSH content in treated sample was increased significantly ($p<0.001$) by 206.72% of *B. juncea* leaves as compared with their respective control. Apart from this, the plants growth regulating constituent IAA was also increased by 15.81% in mustard roots, while 12.99% increased in mustard shoots as compared to the control. RAPD analysis data of the treated sample showed an average 26% of polymorphism among the primers as compared to the control. In conclusion, the present investigation demonstrates that Mr. Trivedi's unique biofield treatment could be utilized as an alternate therapeutic approach concurrent with other existing therapy to improve the productivity of mustard in the field of agriculture in the near future.

Abbreviations

ROS: Reactive oxygen species; ASC–GSH: Ascorbate–glutathione; PCR: Polymerase chain reaction; NCCIH: National Center for Complementary and Integrative Health; CAM: Complementary and Alternative Medicine; RAPD: Random amplified polymorphic DNA; TCA: Trichloroacetic acid; CTAB: Cetyl-trimethyl-ammonium bromide; IAA: Indole acetic acid.

Acknowledgements

Financial assistance from Trivedi science, Trivedi testimonials and Trivedi master wellness is gratefully acknowledged. Authors thank Monad Nanotech Pvt. Ltd., Mumbai, for their support.

References

[1] Dubie J, Stancik A, Morra M, Nindo C (2013) Antioxidant extraction from mustard (*Brassica juncea*) seed meal using high-intensity ultrasound. J Food Sci 78: E542-E548.

[2] Tsuruo I, Yoshida M, Hata T (1967) Studies on the myrosinase in mustard seed part I. the chromatographic behaviors of the myrosinase and some of its characteristics. Agr Biol Chem 31: 18-26.

[3] Bones AM, Rossiter JT (1996) The myrosinase-glucosinolate system, its organization and biochemistry. Physiol Plant 97: 194-208.

[4]   Bassan P, Sharma S, Arora S, Vig AP (2013) Antioxidant and *in vitro* anti-cancer activities of *Brassica juncea* (L.) Czern. seeds and sprouts Int J Pharma Sci 3: 343-349.

[5]   Amarowicz R, Wanasundara UN, Karamac M, Shahidi F (1996) Antioxidant activity of ethanolic extract of mustard seed. Nahrung 40: 261-263.

[6]   Rance F, Dutau G, Abbal M (2000) Mustard allergy in children. Allergy 55: 496-500.

[7]   Morisset M, Moneret-Vautrin DA, Maadi F, Fremont S, Guenard L, et al. (2003) Prospective study of mustard allergy: first study with double-blind placebo-controlled food challenge trials (24 cases). Allergy 58: 295-299.

[8]   Parti RS, Deep V, Gupta SK (2003) Effect of salinity on lipid components of mustard seeds (*Brassica juncea* L.) Plant Food Hum Nutr 58: 1-10.

[9]   Munns R, Tester M (2008) Mechanisms of salinity tolerance. Ann Rev Plant Biol 59: 651-681.

[10]   Dionisio-Sese ML, Tobita S (1998) Antioxidant responses of rice seedlings to salinity stress. Plant Sci 135: 1-9.

[11]   Mittler R, Vanderauwera S, Gollery M, Van Breusegem F (2004) Reactive oxygen gene network of plants. Trends Plant Sci 9: 490-498.

[12]   Asada K (2006) Production and scavenging of reactive oxygen species in chloroplasts and their functions. Plant Physiol 141: 391-396.

[13]   Noctor G, Foyer CH (1998) Ascorbate and glutathione: Keeping active oxygen under control. Annu Rev Plant Physiol Plant Mol Biol 49: 249-729.

[14]   Khan MA, Rabbani MA, Munir M, Ajmal SK, Malik MA (2008) Assessment of genetic variation within Indian mustard (*Brassica juncea*) germplasm using random amplified polymorphic DNA markers. J Integr Plant Biol 50: 385-392.

[15]   Barnes PM, Powell-Griner E, McFann K, Nahin RL (2004) Complementary and alternative medicine use among adults: United States, 2002. Adv Data 343: 1-19.

[16]   Shinde V, Sances F, Patil S, Spence A (2012) Impact of biofield treatment on growth and yield of lettuce and tomato. Aust J Basic Appl Sci 6: 100-105.

[17]   Sances F, Flora E, Patil S, Spence A, Shinde V (2013) Impact of biofield treatment on ginseng and organic blueberry yield. Agrivita J Agric Sci 35.

[18]   Lenssen AW (2013) Biofield and fungicide seed treatment influences on soybean productivity, seed quality and weed community. Agricultural Journal 8: 138-143.

[19]   Nayak G, Altekar N (2015) Effect of biofield treatment on plant growth and adaptation. J Environ Health Sci 1: 1-9.

[20]   Trivedi MK, Tallapragada RM, Branton A, Trivedi D, Nayak G, et al. (2015) Characterization of physical, spectral and thermal properties of biofield treated 1,2,4-Triazole. J Mol Pharm Org Process Res 3: 128.

[21]   Moron MS, Depierre JW, Mannervik B (1979) Levels of glutathione, glutathione reductase and glutathione S-transferase activities in rat lung and liver. Biochim Biophys Acta 582: 67-78.

[22]   Tang YW, Bonner J (1947) The enzymatic inactivation of indoleacetic acid. I. Some charasteristics of the enzyme contained in pea seedlings. Arch Biochem 13: 11-25.

[23]   Green MR, Sambrook J (2012) Molecular cloning: A laboratory manual. (3rdedn), Cold Spring Harbor, N.Y. Cold Spring Harbor Laboratory Press.

[24]   Borges A, Rosa MS, Recchia GH, QueirozSilva JRD, Bressan EDA, et al. (2009) CTAB methods for DNA extraction of sweet potato for microsatellite analysis. Sci Agric (Piracicaba Braz) 66: 529-534.

[25]   Welsh JW, McClelland M (1990) Fingerprinting genomes using PCR with arbitrary primers. Nucleic Acids Res 18: 7213-7218.

[26]   Turk MA, Tawaha AM (2002) Inhibitory effects of aqueous extracts of barley on germination and growth of lentil. Pak J Agron 1: 28-30.

[27]   Sharma P, Sardana V, Banga SS (2013) Salt tolerance of Indian mustard (*Brassica juncea*) at germination and early seedling growth. Environ Exp Biol 11: 39-46.

[28]   Rausch T, Wachter A (2005) Sulfur metabolism: A versatile platform for launching defense operations. Trend Plant Sci 10: 503-509.

[29]   Gagandeep, Dhiman M, Mendiz E, Rao AR, Kale RK (2005) Chemopreventive effects of mustard (*Brassica compestris*) on chemically induced tumorigenesis in murine forestomach and uterine cervix. Hum Exp Toxicol 24: 303-312.

[30]   Zhu YL, Pilon-Smits EAH, Jouanin L, Terry N (1999) Overexpression of glutathione synthetase in Indian mustard enhances cadmium accumulation and tolerance. Plant Physiol 119: 73-80.

[31]   Black H (1995) Absorbing possibilities: Phytoremediation. Environ Health Perspect 103: 1106-1108.

[32]   Yamada T (1993) The role of auxin in plant-disease development. Annu Rev Phytopathol 31: 253-273.

[33]   Morris RO (1986) Genes specifying auxin and cytokinin biosynthesis in phytopathogens. Annu Rev Plant Physiol 37: 509-538.

[34]   Chen Q, Qi WB, Reiter RJ, Wei W, Wang BM (2009) Exogenously applied melatonin stimulates root growth and raises endogenous indoleacetic acid in roots of etiolated seedlings of *Brassica juncea*. J Plant Physiol 166: 324-328.

[35]   Kimura M (1983) The neutral theory of molecular evolution. Cambridge Univ. Press, Cambridge.

[36]   Bretting PK, Widrlechner MP (1995) Genetic markers and plant genetic resource management. John Wiley & Son Inc. Canada.

[37]   Williams JG, Kubelik AR, Livak KJ, Rafalski JA, Tingey SV (1990) DNA polymorphisms amplified by arbitrary primers are useful as genetic markers. Nucleic Acids Res 18: 6531-6535.

22

Beekeeping Practice and Forest Conservation in Gwer-West Local Government Area of Benue State, Nigeria

Francis Sarwuan Agbidye[*]**, Thompson Orya Hyamber**

Department of Forest Production and Products, University of Agriculture, Makurdi, Nigeria

Email address:

fagbidye@yahoo.com (F. S. Agbidye)

Abstract: The research was conducted in Gwer West Local Government Area of Benue State, Nigeria as to ascertain the number of beekeepers in the area, identify the beekeeping methods used, the quantity of honey produced, the income generated, bee plants in the area, conservation methods used, and the challenges faced by beekeepers in the study area. Multi-stage and purposive sampling techniques were used to administer one hundred (100) copies of questionnaire in the study area as follows: 20 in Tsambe/Mbesev, 20 in Tyouhatiee/Injah, 15 in Gbaange/Tongov, 10 in Tijime, 12 in Avihijime, 15 in Nyamshii, and 8 in Meeikyeh respectively. The results showed that beekeepers exist in the study area with majority of them falling between the ages of 20 and 40 years and they use traditional methods of beekeeping. The study also revealed that the main reason for keeping bees in the study area was for income generation while the major challenge of beekeeping in the study area was pests. Majority (78%) of the beekeepers produce about 40 litres of honey while a few (18%) produce above 80 litres of honey annually generating between ₦28, 000.00 and ₦100, 000.00 (about 160 - 500USD) annually per beekeeper. The major bee plants in the study area were *Daniellia oliveri, Citrus sinensis, and Mangifera indica* while the major conservation methods used include deliberate retention of bee plants on farmlands, planting of bee plants in home gardens, fire tracing among others. It was recommended that training on modern beekeeping methods should be carried out in the study area to improve beekeeping practice in the study area.

Keywords: Beekeeping Practice, Beekeepers, Bee Plants, Forest Conservation, Home Gardens

1. Introduction

It has been stated that beekeeping is concerned with the practical management of the social species of honey bees, which live in large colonies of up to 100,000 individuals [1]. Beekeeping is usually carried out for the purpose of honey production and other hive products such as bees wax, pollen, propolis, royal jelly and bee venom and for the pollination of agriculture and other plants. The oriental or the African bee, *Apis mellifera adansonii* (Lestis) is the most widely used species in Africa for honey production [2]. According to [3], honey is the name given to the sweet, thick, liquid substance composed mainly of simple sugars produced by honey bees from the nectar of flowers. Honey bees collect nectar, transform and combine enzymatically with specific substances of their own and deposit in the cells of the honeycomb to mature. Honey has a taste that is distinctive from sugar and a nutritional value that is much higher than the latter [3]. Traditional beekeeping is practiced by various tribes all over the world, especially in the savannah regions. The practice relies on the wild honey bee colonies which are baited to the hives at the beginning or end of the raining season. Beekeeping is environmentally friendly and closely linked with other human activities [4]. It is an established fact that community forestry projects are strengthened by the involvement of beekeepers. This is so because beekeepers are experienced craftsmen and women with much knowledge and interest in what the forest and the trees produce. Beekeepers have knowledge of all the trees which are beneficial to bees and to humans. They know which of the trees are most preferred for nesting by the honey bees in the locality [5]. Bee plants are plants that are visited by bees either for nectar or pollen or both. Some plants provide only nectar while others provide only pollen and some provide both nectar and pollen. Some of the bee plants in Nigeria include the following: *Ancardium occidentale* (cashew), *Mangifera indica* (mango), *Citrus spp* (Orange), *Elaeis guineensis* (oil palm), *Zea mays* (maize or corn), *Parkia biglobosa* (Locust bean tree), *Prosopis africana* (Iron tree), *Gmelina arborea, Azadirachta indica* (Dogonyaro), *Daniellia oliveri* (African balsam), *Chromolaena odorata* (Siam weed, *Vitellaria paradoxa* (Shea butter tree), *Tectona grandis* (Teak), *Khaya* spp, *Dacryodes edulis* (African pear), *Irvingia gabonensis* (Bush mango) among others [5] and [6]. Forest conservation is recognized as the wise utilization, preservation and or renewal of forests, waters, lands and minerals for the greatest good of the people over a very long and endless period of time [7]. All over the world people have been known to conserve forest tree species that are of importance to them. Some of the ways that rural people conserve biodiversity is the deliberate retention of some trees [8]. Farmers in Benue State have been reported to deliberately maintain multipurpose forest trees on their farms for economic, social and ecological uses and thus encourage their conservation [9]. Some of these trees include *Vitellaria paradoxa, Afzelia africana, Prosopis africana* and *Parkia biglobosa*. Furthermore, [10] asserted that in Ethiopia, forest tree species like *Schefflera abyssinica,* and *Croton macrostachyus* are deliberately conserved by the rural people for beekeeping.

However, beekeeping practice is accessible to everybody and does not require serious academic knowledge and therefore anyone can engage in beekeeping irrespective of age [6]. Honey yield however, depends on the skills of beekeepers, climate, vegetation, and bee species [1]. Beekeeping is a sound economic activity which can be used for wealth creation and harnessed to tackle poverty problems in rural extension communities [5]. Honey production has not been appreciably exploited as to realize its perceived potential in Nigeria [5]. There is paucity of information on beekeeping activities in developing countries of the world especially Nigeria [5]. For instance; the amount of honey and other important hive products produced per annum as well as the number of beekeepers is hard to come by. This study was therefore carried out with the following objectives: (a) To determine the number of beekeepers in the study area (b) To identify the beekeeping methods used in the study area (c) To determine the quantity of honey produced and the income generated in the study area (d) To identify the major bee plants in the area and the conservation methods used for these plants and (e) To identify the challenges faced by beekeepers in the study area. This information is important in knowing the intervention measures to improve the beekeeping practice in the study area. The study was carried out in only seven of the fifteen council wards of the local government area.

2. Methodology

2.1. The Study Area

This study was conducted in Gwer West Local Government area (LGA) of Benue State. Gwer West LGA which was carved out of Gwer LGA in 1991 is located between latitudes 9 and 12^{0}N and longitudes 6 and 9^{0}E. It is bounded by Makurdi and Doma LGAs to the north, Gwer East LGA to east, Otukpo LGA to the South and Apa and Agatu LGAs to the West. The main inhabitants of the LGA area are the Tiv people. Gwer LGA has fifteen council wards namely; Sengev/Yengev, Saghev/Ukusu, Tsambe/Mbesev, Tyouhatiee/ Injah, Mbapa,

Tijime, Gambe/Ushin, Avihijime, Meeikyegh, Nyamshi, Ikyaghev, Gbaange/Tongov, Mbachohon, Mbabuande, Sengev. The headquarters of the LGA is Naka which is strategically located at killometre 40 along the Makurdi - Ankpa interstate road [11]. Gwer West LGA has annual rainfall of between 1500mm and 2000mm. The rainy season starts in April and continues through October, with the highest peak in September. The LGA occupies a landmass of about 456.45sq km. According to the 2006 census (projected figures for 2015), Gwer West LGA has a population of 154, 942 [12]. The soils of the study area are generally ferruginous in nature dominated by clay, loamy, and sandy with lateritic soils also occurring in nature. The vegetation of the study area is guinea savanna characterized by dense grass cover consisting of dominant species of trees like, *Burkea africana, Daniellia oliveri, Khaya senegalensis, Prosopis africana, Azadirachta indica, Parkia* biglobosa, *Vitellaria paradoxa, Citrus* spp, *Tamarindus indica* among others.

2.2. Data Collection and Analyses

The study was carried out in seven of the fifteen council wards of the LGA namely: Tsambe/Mbesev, Tyouhatiee/injah, Tijime, Avihijime, Meeikyegh, Nyamshi, Gbaange/Tongon. The study population comprised some of the beekeepers in the LGA. The sample size of 100 was drawn from 135 beekeepers identified in the selected council wards. Following [13] with slight modifications, multi-stage sampling technique was adopted to select seven council wards out of fifteen. From each of the council wards selected, respondents (beekeepers) were purposely sampled for the study. Data for the study were collected using semi-structured questionnaire. The questionnaire was validated through a pre-test. The validated questionnaire was administered on 100 respondents in the seven selected Council wards as follows: 20 in Tsamba/Mbesev, 20 in Tyouhatiee/Injah, 15 in Gbaange/Tongov, 10 in Tijime, 12 in Avihijime, 15 in Nyamshii, and 8 in Meeikyeh. Data collected were analyzed using descriptive statistics (frequencies and percentages).

3. Results

3.1. Socio - Economic Characteristics of Beekeepers in Gwer West LGA, Benue State, Nigeria

The socio-economic characteristics of beekeepers (Table 1) showed that 98% percent of the beekeepers are male while only 2% are female. Majority of the beekeepers (85%) are married are between 20 to 40 years (youths). A good percentage of the elderly (34%) are also engaged in beekeeping. Most of the beekeepers are adherents of the Christian faith. The results also showed that majority of the beekeepers (45%) have only primary education followed by those with no formal education (30%). Majority of the beekeepers engage in farming as their primary occupation while 21% of the beekeepers do not engage in any other activity apart from beekeeping.

3.2. Reason for Practice Beekeeping

The results obtained (Table 2) showed that respondents engaged in beekeeping mainly for income generation (86.0%) while some do so because it was the profession of their forefathers (11.0%) with and as a hobby (3.0%).

3.3. Duration of Beekeeping Practice in Gwer West LGA, Benue State, Nigeria

The results obtained (Table 3) showed that majority of the beekeepers in the study area have been in the business for about 10 years (56%) followed by those who have practiced beekeeping for about 5 years (27%) while only 4% of the beekeepers have been in the practice for over 20 years (4%).

3.4. Training on Modern Beekeeping

Information on Table 4 below showed that 98.0% of beekeepers are not trained into beekeeping while 2.0% are trained in beekeeping.

Table 1. *Socio-economic characteristics of Beekeepers in Gwer West LGA, Benue State, Nigeria.*

Variable	Frequency	Percentage
Sex		
Male	98	98.0
Female	2	2.0
Marital Status		
Single	9	9.0
Married	85	85.0
Divorced	6	6.0
Age		
Below 20 years	6	6.0
20 – 40 years	60	60.0
41- 60 years	34	34.0
Religion		
Islam	0	0.0
Christianity	74	74.0
Traditional	26	26.0
Non adherents	0	0.0
Educational Status		
No Formal Education	30	30.0
Primary	45	45.0
Secondary	21	21.0
Tertiary	4	4.0
Primary Occupation		
Farming	78	78.0
Beekeeping	21	21.0
Trading	1	1.0
Civil Service	0	0.0

Table 2. *Reason for practicing beekeeping in Gwer West LGA, Benue State, Nigeria.*

Reason	Frequency (F)	Percentage (%)
Hobby	3	3.0
Source of income	86	86.0
Inheritance	11	11.0
Total	100	100

Table 3. *Duration of Beekeeping Activity by Beekeepers in Gwer West LGA, Benue State, Nigeria.*

Beekeeping practice (period)	Frequency(F)	Percentage (%)
1-5 years	27	27.0
6-10 years	56	56.0
11-15 years	13	13.0
16-20 years	0	0.0
21 above	4	4.0
Total	100	100

Table 4. *Training on Modern Beekeeping Practice in Gwer West LGA, Benue State, Nigeria.*

Response	Frequency(F)	Percentage (%)
Yes	2	2.0
No	98	98.0
Total	100	100

Types of hives used by beekeepers in Gwer west LGA, Benue state, Nigeria.

The results of the study showed that all the beekeepers in the study area (100%) used hollowed tree trunk hives for their beekeeping practice (Table 5, Plate 1).

Table 5. *Types of Hives used by beekeepers in Gwer West LGA, Benue State, Nigeria.*

Hive Used	Frequency(F)	Percentage (%)
Basket	0	0.0
Bamboo	0	0.0
Hollowed Tree Trunks	100	100
Clay Pots	0	0.0
Langstroth	0	0.0
Kenya top bar	0	0.0
Total	100	100

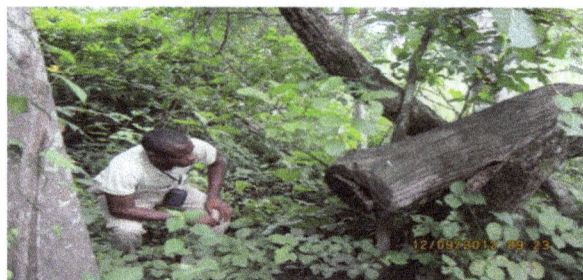

Plate 1. *Hives used by beekeepers in Gwer West LGA, Benue State, Nigeria.*

3.5. Baiting Materials

The result obtained (Table 6) showed that all the beekeepers drum the hollowed tree trunks to attract bees to the hives.

Table 6. *Baiting materials used by beekeepers in the study area.*

Baiting Materials	Frequency (F)	Percentage (%)
Tobacco leaf	0	0.0
Cow dung	0	0.0
Lemon grass	0	0.0
Drumming the hive	100	100.0
Total	100	100

3.6. Location of Hives in the Study Area

Results obtained (Table 7, Plate 1) showed that majority of the beekeepers (90%) place their hives under trees while some do so on trees (10%).

Table 7. *Placement of Hives in the Study Area.*

Placement of hive	Frequency(F)	Percentage (%)
On trees	10	10.0
Under Tree Shades	90	90.0
on the rock	0	0.0
Other	0	0.0
Total	100	100

3.7. Obtaining Bee Colonies in the Study Area

The results obtained (Table 8) showed that all the beekeepers in the study area obtained their bee colonies from the wild (nature).

Table 8. *Obtaining Bee Colonies in the Study Area.*

Colonizing Method	Frequency(F)	Percentage (%)
Buying	0	0.0
Nature	100	100
Total	100	100

3.8. Major Bee Plants in the Study Area

The results obtained (Table 9) showed that the major bee plant in the study area was *Daniellia oliveri* (31.2%) followed by *Citrus sinensis* (25%).

Table 9. *Major Bee Plants in the Study Area.*

Bee Plant	Frequency(F)	Percentage (%)
Daniellia oliveri	50	31.2
Mangifera indica	30	18.8
Elaeis guineensis	20	12.5
Citrus sinensis	40	25.0
Parkia biglobosa	20	12.5
Total	160	100

3.9. Forest Conservation Measures Adopted in the Study Area

The results obtained showed that the beekeepers in Gwer-West LGA deliberately retain bee plants on their farm lands (40%) and also plant bee plants in home gardens (23.3%) among others measures to ensure their conservation (Table 10).

Table 10. *Conservation measures adopted by beekeepers for bee plants in the Study area.*

Conservation Method	Frequency(F)	Percentage (%)
Deliberate retention of bee plants on farmlands	60	40.0
Fire tracing around bee plants	30	20.0
Planting of bee plants in Home gardens	35	23.3
Enlightenment campaigns against bush burning	15	10.0
Establishing plantations of bee plants (*Citrus*)	10	6.7
Total	150	100

3.10. Marketing Honey in the Study Area

The result obtained (Table 11) showed that all the beekeepers (100%) in the study area sell their honey in the local market (Naka Market).

Table 11. *Marketing Channels for honey in the Study Area.*

Marketing Channel	Frequency(F)	Percentage (%)
Honey Hawking	0	0.0
Local Market	100	100
Cooperative	0	0.0
Total	100	100

3.11. Quantity of Honey Produced/Income Realized in the Study Area

Results obtained (Table 12) showed that majority of beekeepers (78.0%) in the study area produce below 40 litres followed by those who produce above 80 litres of honey in a year (18.0%). Beekeepers make from 28,000.00 to 560,000 annually from selling honey.

Table 12. Quantity of Honey Produced by Beekeepers in the Study Area.

Quantity (Litres)	Frequency (F)	Percentage (%)
<40	78	78.0
40-80	4	4.0
Above 80	18	18.0
Total	100	100

3.12. Problems Encountered by Beekeepers in the Study Area

Results obtained from the study area on the problems encountered by beekeepers showed that the major problem is pest attack (85%) followed by abscondment (10%) and then low colonization (5%).

Table 13. Problems Encountered by Beekeepers in the Study Area.

Problems	Frequency(F)	Percentage (%)
Pest attack	85	85.0
Disease	0	0.0
Fire Outbreak	0	0.0
Low Colonization	5	5.0
Abscondment	10	10.0
Theft	0	0.0
Total	100	100

4. Discussion

The results of this study have revealed that beekeeping is carried out in Gwer West LGA and that all categories of people; male, female, married, unmarried, divorced, young, middle aged, elderly, Christians, and adherents of traditional religion are all involved in beekeeping activities. Even though males were mostly involved, few women were also involved in the practice. Majority of the beekeepers were not involved in beekeeping as their primary occupation. This result agrees with [6] and [4] who stated that beekeeping practice can be combined with other human activities and the practice is accessible to everybody as it does not require serious academic knowledge and therefore anyone can engage in beekeeping irrespective of profession, age, sex, marital status, religion and educational background. It was also clear from the study that many people engage in beekeeping for income generation. This is in line with the results of [4], [5] and [6] who reported that beekeeping is a lucrative business and can bring extra income. It is very clear from the study that the beekeepers in Gwer West LGA have not received any form of training on modern beekeeping and hence engage only in traditional beekeeping methods with low outputs and hence low incomes. It had earlier been reported [6] that modern beekeeping training programmes have rarely been done by the government, non-governmental organizations (NGOs) and institutions. They advocated that if training programmes are carried out, beekeeping skills would be enhanced, number of beekeepers and productivity would increase leading to higher incomes from beekeeping. [1] had also reported that honey yield depends on the skills of beekeepers among other factors. The study also showed that

Gwer-West LGA has abundance of bee plants like *Daniellia oliveri, Citrus sinensis, Mangifera indica, Parkia biglobosa, Elaeis guineensis* among others which makes the area suitable for beekeeping practice. These bee plants found in Gwer-West LGA are similar to the ones reported by [6] and [5]. The beekeepers in the study area adopt various methods to conserve the bee plants in the area to ensure sustainability of their practice. These forest conservation methods include: deliberate retention of bee plants on farm lands, planting of bee plants in home gardens, fire tracing around bee plants, enlightenment campaigns against bush burning and establishment of plantations of bee plants especially *Citrus sinensis* and *Elaeis guineensis*. [9] and [8] had also reported deliberate retention of multipurpose forest trees on farmlands and also planting of in home gardens by farmers in Benue State for economic, social and ecological reasons thus ensuring their conservation. In Ethiopia, [10] reported that forest tree species like *Schefflera abyssinica*, and *Croton macrostachyus* are deliberately conserved by the rural people for beekeeping. The study also revealed that the major challenge facing beekeeping in Gwer West LGA is pest attack (ants). This problem could be solved by training and provision of relevant inputs.

5. Conclusion and Recommendations

The study has shown that a good number of the male population of Gwer West who have not received any form of training in modern beekeeping, engage in traditional beekeeping mainly for income generation. Beekeeping in the study area is mostly done by males as the two females involved inherited it from their late husbands. The major bee plant in Gwer-West LGA *Daniellia oliveri* while the major method of conserving bee plants is deliberate retention the plants on farmers' farmlands. Honey yield in Gwer West is low and all that is produced is sold at the local market Naka. The major challenge of beekeeping in the study area was pest attack (ants). Beekeeping practice in the study area has a lot of potential for income generation which can be harnessed to bring about significant improvement in the livelihood of the people, creation of job opportunities, and self empowerment scheme to the rural dwellers of Gwer-West Local Government area of Benue State. Since the people of Gwer West LGA do not engage in modern beekeeping practices, Government, NGOs, institutions and individuals should aid in providing an enabling environment to substantially increase level of honey production by training and providing the local beekeepers with modern beekeeping equipment for increased production.

Acknowledgements

The authors wish to acknowledge the contribution of the research assistants who helped with the administration of the questionnaires and the beekeepers in Gwer-West LGA for their cooperation in volunteering information for this research.

References

[1] V. T. Leen, J. B. Willem, M. Marieke, S. Piet, and V. Hayo. Beekeeping in the tropics. Digigrafi Netherlands. 2005. pp7-49.

[2] C. D. Michener. The bees of the world. The John Hopkins University Press. Baltimore, MD, USA. 2000. pp. 913.

[3] R. T. Wilson. Current status and possibilities for improvement of traditional apiculture in sub-saharan Africa. Livestock Research for Rural Development. 2006. Vol. 13 (8) : 1-11.

[4] H. D. Usman. "The role of beekeeping in poverty alleviation". Magazine of the National Park Service, Nigeria. 2000. Pp. 43-44.

[5] A. A. Kareem, O. A, Iroko, A. F. Adio, O. C. Jegede, A. O. Olaitan and A. A. Jayeola. Role of non-timber forest products (NTFPS) in creating wealth: A case of honey production. In: L. Popoola, F. O. Idumah, V. A. J. Adekunle and I. O. Azeez (eds). The global economic crisis and sustainable renewable resources management. *Proceedings of the 33rd Annual Conference of the Forestry Association of Nigeria held in Benin City, Edo State, Nigeria, 25th – 29th October, 2010 pp 429 - 435.*

[6] A. A. Ukoima and U. F. Edeki. Apiculture: A panacea for poverty alleviation in the Niger delta, Nigeria. In: L. Popoola, F. O. Idumah, V. A. J. Adekunle and I. O. Azeez (eds). The global economic crisis and sustainable renewable resources management. *Proceedings of the 33rd Annual Conference of the Forestry Association of Nigeria held in Benin City, Edo State, Nigeria, 25th – 29th October, 2010, pp 567- 572.*

[7] International Union for the Conservation of Nature (IUCN). Indigenous people and climate change/REDD-An overview of current discussions and main issues. 2010. Available at: *http//cmsdata.iucn.org/downloads/iucn_briefing_ips_and_red d_march_2010.pdf.* Retrieved October 15, 2011.

[8] T. N. Tee and A. Ageende. Community woodlot production as a sustainable forest management option in Benue State, Nigeria. In: Popoola, L and Oni, P. I. (eds.). Sustainable Forest Management in Nigeria- lessons and prospects. *Proceeding of the 30th Annual* FAN Conference held at Kaduna. 2005. pp 371-388.

[9] T. N. Tee and J. Amonum. Domestication of non-timber forest tree products for sustainable livelihood. *Journal of Research in Agriculture* 2008. 5 (4):76-81.

[10] K. Hundera. Traditional forest management practices in Jimma Zone, South West Ethiopia. *Ethiopia Journal of Education and Science.* 2007, 2 (2):1-9.

[11] Benue State Diary. Benue State of Nigeria Diary. Produced by the Ministry of Information and Culture Makurdi, Benue State, Nigeria. 2013.

[12] A. Tsee. The Dynamics of Benue State Population (1963-2016). Micro Teacher and Associates, Makurdi. 2013. 91pp.

[13] F. S. Agbidye. Some Aspects of the Ecology, Nutritive Value and Marketability of Edible Forest Insects in Benue state, Nigeria. PhD. Thesis, Department of Forestry and Wood Technology, Federal University of Technology Akure, Nigeria. 2008. 118pp.

23

Variations in Stem Borer Infestation and Damage in Three Maize (*Zea mays* L.) Types in Southern Guinea Savanna and Rainforest Zones of Nigeria

Edache Ernest Ekoja[1, *], Olufemi Richard Pitan[2], Folashade Temitope Olaosebikan[2]

[1]Department of Crop and Environmental Protection, University of Agriculture, Makurdi, Nigeria
[2]Department of Crop Protection, Federal University of Agriculture, Abeokuta, Nigeria

Email address:
ernestekoja@yahoo.com (E. E. Ekoja)

Abstract: The effects of location, maize types and borer control with carbofuran (Furadan 3G®) on the severity of maize stem borer infestation and damage was investigated in the late maize planting season of 2011. Treatments were laid out in randomized complete block design using a split-slip-plot factorial arrangement. Whole plot factor consisted of two locations (Southern Guinea Savanna and Rainforest agro-ecological zones of Nigeria), subplot factor consisted of 1.5 kg a.i.ha^{-1} and 0.0 kg a.i.ha^{-1} of carbofuran, while the sub-sub-plot factor comprised of three endosperm types of maize (flint, pop and sweet corn). Stem borer infestation (quantified by dead heart count and larval population per plant) and damage (quantified by %lodged stem, %bored internodes, %bored ears, number of exit holes, number of stem borer cavities and number of damaged seeds per plant) as well as yield were compared. Results revealed that borer infestation and damage were significantly higher ($P < 0.05$) in the Rainforest compared with the Savanna. Single dose application of carbofuran (1.5 kg a.i. ha^{-1}) also significantly ($P < 0.05$) increased grain yield in all the maize types at both locations. For all parameters, no significant ($P > 0.05$) location × carbofuran × maize type and location × maize type effect was detected. However, significant ($P < 0.05$) location × carbofuran and carbofuran × maize type interaction effects were observed. We conclude that in both agro ecologies, flint corn was more tolerant of borer attack while sweet corn was more susceptible compared to either flint or popcorn. In addition, carbofuran at 1.5 kg a.i.ha^{-1} can significantly reduce stem borer population in the three maize types.

Keywords: Borer, Flint, Pop, Sweet, Corn, Control

1. Introduction

Expansion in the cultivation of flint, pop and sweet endosperm types of maize in Nigeria is increasingly becoming inevitable given government's ban on continued importation of food grains. The demand for maize in Nigeria is high and it comprises high demand for it as food for the teeming human population, as feed for livestock, and as raw material for the production of some food and non-food products in agro-allied industries [14, 15]. About 42% of the 785 million tons of maize produced worldwide annually are from the United States of America, while Africa accounts for only for 6.5% [16]. The 2011 estimated figure for maize yield in Nigeria shows a 17.5% increase in yield between 2000 and 2011 [7], yet indications are that production of maize inadequately meets demand quantitatively and qualitatively.

Insect pests are the most limiting factor to the production of maize and research efforts have shown that maize stem borers are the major insect pest of maize [5, 18, 15, 17]. [19] listed 21 species considered being of economic importance; however, [18] reported that only a subset is damaging within any region/crop combination. The common species infesting maize in Nigeria include: *Busseola fusca* Fuller (Noctuidae), *Eldana saccharina* Walker (Pyralidae), *Sesamia calam*istis Hampson (Noctuidae), *Chilo partellus* Swinehoe (Cambridae) and *Acigona ignefusalis* Hampson (Pyralidae) [11, 22]. Depending on the species, the larval stage may last 25-58 days and may have 6-8 instars [20]. Pupal stage normally takes 5-14 days after which adult moths emerge [12, 13, and 19]. Maize stem borers usually pupate close to the tunnel exit or even partly outside the stem [28]. The larvae occupy and feed on different parts of maize whorl, leaves, stalk, tassel and ears. At the early stages of plant growth, damage to the growing point causes 'dead heart'. Borer damage also causes early leaf senescence, reduced translocation, lodging, direct damage to ears and increment in the incidence off stalk rot and ear rot diseases, sometimes resulting to significant yield losses [25, 4]. The severity and nature of stem borer damage depend upon the borer species, the plant growth stage, the number of larvae feeding on the plant, and the plant's reaction to borer feeding [4, 17]. Yield losses caused by maize borers in Africa have been estimated to range from 10 - 100 % [4].

Manipulating time of sowing to avoid severe borer infestation, removal of damaged cobs and stems from the field, selection of resistant varieties, biological controls with naturally occurring biotic agents are relied upon for stem borer control. However, where control failure exists or where they are pest outbreak, control with synthetic insecticides has been recommended. Carbofuran (Furadan 3G) applied at the rate of 1.5 kg a.i.ha^{-1} as well as Carbaryl (Vetox 85) applied at the rate of 0.75 kg a.i.ha^{-1} has been recommended [6, 23]. Most of these recommendations for maize stem borer control emanated for researches with flint corn.

As regular pests of maize, the knowledge of yield–loss relationships between maize types grown in any locality and the stem borers is important in planning effective management strategies for the pest [10]. We therefore set up this experiment to assess the impact of maize types (flint, pop and sweet corn) and borer control with carbofuran on the severity of maize stem borer infestation and damage in two agro ecological zones of Nigeria.

2. Materials and Method

2.1. Description of the Planting Materials and Study Areas

Three maize types [flint (SUWAN-1-SR(DMR)), pop (Kaduna pop corn) and sweet corn (Oba Super 2] obtained from the Department of Crop Production, University of Agriculture, Makurdi were used for field experiment carried out between the month of August and November, 2011 at the Teaching and Research Farm of the University (coordinates: 07°41'N 05°40'E) and Federal University of Agriculture, Abeokuta (coordinates: 07°15'N, 03°25'E). The maize type varieties used are open pollinated and medium maturing. The flint corn was also resistant to Downy mildew disease of corn.

2.2. Research Design

The plots were laid out in randomized complete block design made up of six plots replicated four times giving a total of 24 plots in each location. The experiment was set up in split-slip-plot factorial arrangement with the whole plot factor consisting of two locations (Southern Guinea Savanna and Rainforest agro ecological zone of Nigeria), while the subplot factor consisted of 1.5 kg a.i.ha^{-1} and 0.0 kg a.i.ha^{-1}

rates of carbofuran (Furadan 3G®). The sub-sub-plot factor was the three endosperm types of maize (flint, pop and sweet corn). Each plot was 5 m long and six rows wide. An inter row spacing of 0.75 m was maintained within plots while a 1 m alley way existed both within and between blocks. The total area of the field in each location was 736 m^2. Paraquat at 3.0 kg a.i.ha^{-1} and pendimethaline at 2.5 kg a.i.ha^{-1} were applied immediately after sowing for weed control. This was supplemented with manual hoe weeding 7 weeks after planting (WAP). Four seeds per hole were sown on ridges and seedlings were thinned down to 2 plants per stand at 2 WAP to give a population density of approximately 53,333 plants ha^{-1}. Four weeks after planting, NPK (20:10:10) fertilizer was applied.

2.3. Data Collection

Sampling for stem borer damage assessment was carried out at harvest (12 WAP). Data on dead heart count, % lodged stem, % bored stem, % bored internodes, % bore ears, mean number of exit holes per plant, mean number of damaged seeds per plant and mean number of filled cobs per plant were collected from 10 plants selected at random from plants in the four middle rows of each plots. At 4 WAP, plants showing "dead heart" symptoms (destruction of the growing point in the whorl as a result of the stem borer's feeding activities on young maize plants) were counted per plot. Furthermore, the stem of each sample was split and the number of larvae recovered was recorded. The length of borer tunnel on one half of the stem was measured and cumulated. The number of borer cavities was derived as a quotient of total tunnel length and mean length (32 mm) of mature larvae of B. fusca, the predominant borer species recovered in both locations. Ears from the sample of 10 plants were picked, dehusked and sorted into grain-filled and unfilled categories, thereafter; the ears were dried. Grain weight was recorded after threshing.

2.4. Statistical Analysis

All data in percentage were transformed to arcsine before the analysis of variance. Mean larval population per plant was analyzed as $\sqrt{x + 0.5}$ before the preliminary F-test which was carried out for each parameter using GENSTAT Discovery Edition 4, [8]. Significantly different mean values were separated using least significant difference (LSD) at 5% level of probability.

3. Results

Busseola fusca was consistently the most abundant borer species in the three types of maize in both locations (51.7% of total collection in flint corn, 58.3% in popcorn and 56.7% in sweet corn at Makurdi, Southern Guinea Savanna agro ecological zone and 48.3%, 54.2% and 50.0% of total collection in flint, pop and sweet corn respectively in the Rainforest agro ecological zone) (Figures 1 and 2). Higher populations of B. fusca were recovered from popcorn in both locations compared to the other maize types. *S. calamistis* was next in abundance with a higher population of the borer was recovered in the Rainforest zone compared to the Southern Guinea Savanna zone. Next in relative abundance was *E. saccharina* followed by *A. ignefusalis* and *C.*

partellus. The later was not found in the Rainforest zone. In addition, *A. ignefusalis* was not encountered in flint corn in the Southern Guinea Savanna zone and in all the maize types evaluated in the Rainforest zone.

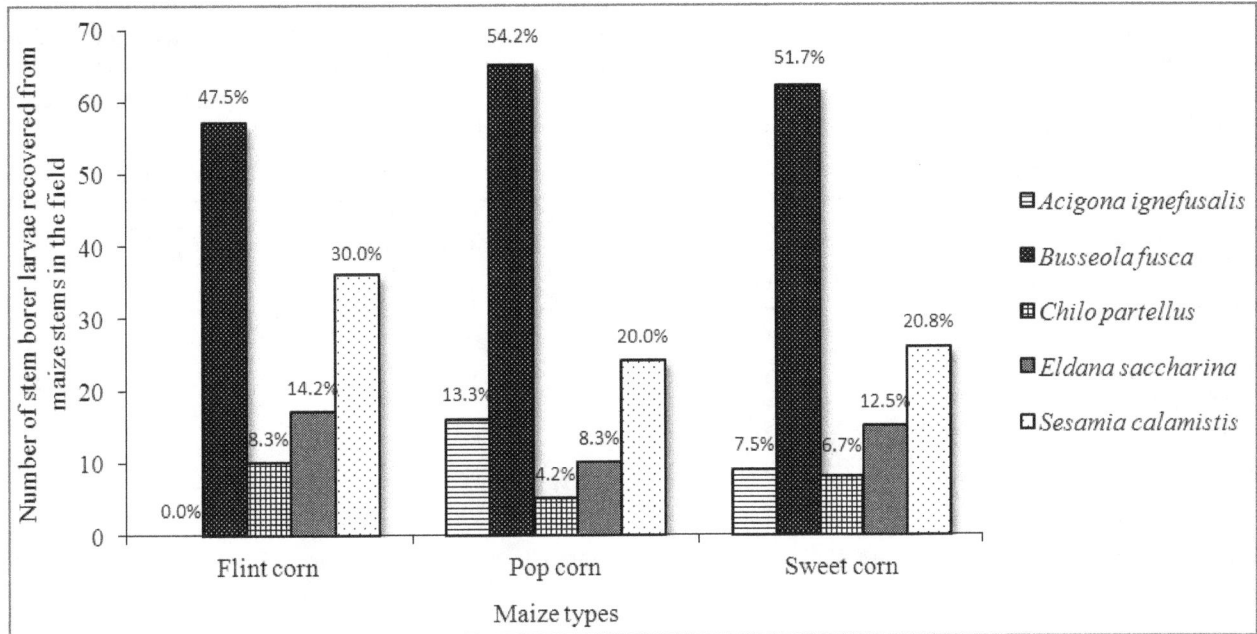

Figure 1. *Species composition and relative abundance (%) of borer recovered from the stems of maize at harvest in Makurdi, Southern Guinea Savanna agro-ecological zone of Nigeria. (n = 120).*

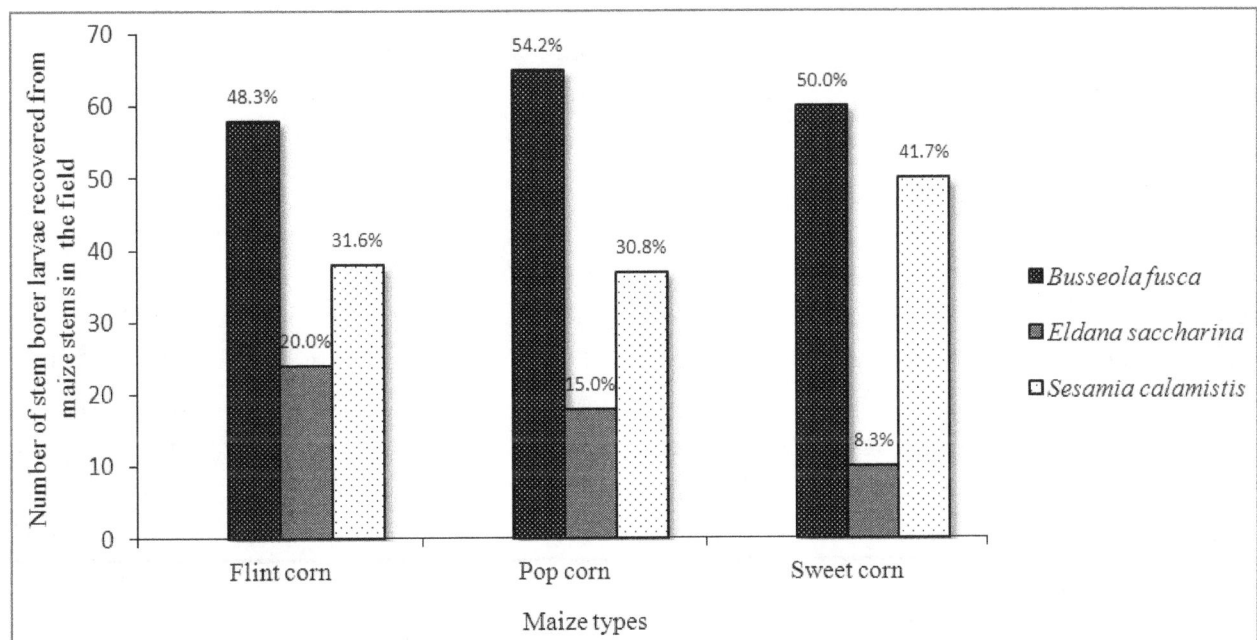

Figure 2. *Species composition and relative abundance (%) of borer recovered from the stems of maize at harvest in Abeokuta, Rainforest agro-ecological zone of Nigeria. (n = 120).*

For all parameters, stem borer infestation and damage were significantly higher ($P < 0.05$) in the Rainforest zone compared to the Southern Guinea Savanna zone and this resulted in a significantly lower ($P < 0.05$) number of filled cobs and total grain yield of the crops grown in that environment (Table 1). Maize grain yield was about 22.8%

lower in the Rainforest compared with the Savanna zone. However, the mean number of exit holes per plant and mean number of stem borer cavities per plant in both locations were not significantly different ($P > 0.05$) from each other.

Single dose soil application of carbofuran at the rate 1.5 kg a.i.ha^{-1} significantly suppressed ($P < 0.05$) borer infestation and damage in the three maize types in both locations (Table 1). This increased the number of filled cobs per plant and grain yield from 0.91 to 1.48 and 1.36 to 1.96 tha^{-1} respectively. About 44.9% increase in grain yield was recorded from the pesticide application in all the maize types at both locations

The three maize types were susceptible to borer attacks, given the relatively high borer activities observed in them in both locations (Table 1). Apparently, sweet corn was more susceptible to the attacks given the significant ($P < 0.05$) loss of seedlings due to dead heart and other damages caused by the borer which resulted in low yield values from sweet corn plots. Marginal increase in mean larval population per plant, number of exit holes per plant and number of stem borer cavities per plant observed in sweet corn plots over pop and flint corn plots, but the differences were not significant ($P > 0.05$). Flint corn was more tolerant of the borer attack, in spite of the relatively high borer activity observed in the maize type; a significantly higher yield value of 2.26 tha^{-1} was obtained.

Table 1. Main effects of location and stem borer control with carbofuran on infestation, damage and yield in three endosperm types of maize (Zea mays L.).

Treatments	Dead heart count (3WAP)	% lodged stem	% bored stem	% bored internodes	% bored ears	Larval population per plant	Number of exit holes per plant	Number of stem borer cavities per plant	Number of damaged seeds per plant (n= 100)	Number of filled cobs per plant	Grain yield (tha^{-1})
Location											
Guinea Savanna	1.38	20.65	54.40	23.90	49.20	1.70	1.00	1.96	36.71	1.35	1.83
Rain forest	2.50	31.44	59.40	26.99	53.46	1.98	1.01	2.47	40.46	1.05	1.49
LSD$_{0.05}$	0.45	5.12	2.29	1.66	2.74	0.08	Ns	Ns	1.94	0.17	0.11
F-value	62.49	45.01	45.27	34.98	24.09	116.92	0.15	4.89	37.73	32.04	30.99
P-value	0.004	0.007	0.007	0.010	0.016	0.002	0.723	0.114	0.009	0.011	0.002
Df = 1, 3											
Carbofuran											
0.0 kg a.i.ha^{-1}	2.54	15.09	75.09	32.68	66.50	2.56	1.32	3.08	62.17	0.91	1.36
1.5 kg a.i.ha^{-1}	1.33	37.01	38.89	18.21	36.20	1.12	0.67	1.36	15.00	1.48	1.97
LSD$_{0.05}$	0.67	1.69	8.35	2.31	8.27	0.11	0.32	0.39	2.10	0.13	0.21
F-value	19.26	1011.59	112.70	67.52	80.16	1096.67	24.05	115.10	1479.70	116.16	51.79
P-value	0.005	<0.001	<0.001	<0.001	<0.001	<0.001	0.003	<0.001	<0.001	<0.001	<0.001
Df = 1, 6											
Maize types											
Flint corn	1.63	21.19	49.93	23.52	42.82	1.75	0.79	1.91	29.69	1.44	2.26
Popcorn	1.75	24.14	57.23	25.18	51.02	1.90	1.08	2.34	40.06	1.24	1.70
Sweet corn	2.44	32.82	63.81	27.64	60.19	1.90	1.14	2.41	46.00	0.91	1.03
LSD$_{0.05}$	0.47	3.35	4.65	1.79	4.78	Ns	Ns	Ns	2.82	0.15	0.16
F-value	7.47	27.76	19.03	16.10	28.11	1.82	10.71	2.59	97.31	28.00	121.99
P-value	0.003	<0.001	<0.001	0.001	<0.001	0.122	0.184	0.096	<0.001	<0.001	<0.001
Df = 2, 24											

Means are values of four replicates
LSD$_{0.05}$ = Fisher's Least Significant Difference at 5% level of probability
Data in percentage were transformed to arcsine values and insect count was analysed as $\sqrt{x + 0.5}$ before the F-test

There was no significant (P > 0.05) location × carbofuran × maize type and location × maize type interaction effect was detected for all parameters (Table 2). But significant (P < 0.05) location × carbofuran and carbofuran × maize type interaction effect in some parameters were observed. Percentage bored stem, mean larval population per plant and number of exit holes in carbofuran-untreated plots in the rainforest zone were significantly higher (P < 0.05) compared with the treated plots and treated and untreated plots in the southern Guinea Savanna zone. The untreated plots of sweet corn had significantly higher (P < 0.05) %bored stem, %bored ear, mean number of exit holes per plant, mean number of stem borer cavities per plant, number of damaged seeds per plant and a significantly lower (P < 0.05) number of filled cobs per plant compared with the treated and untreated plots of the other maize types.

Table 2. Interactions effects of location stem borer control with carbofuran and maize types on infestation, damage and yield parameters.

Treatments	Dead heart count (3WAP)	% lodged stem	% bored stem	% bored internodes	% bored ears	Larval population per plant	Number of exit holes per plant	Number of stem borer cavities per plant	Number of damaged seeds per plant (n=100)	Number of filled cobs per plant	Grain yield (tha⁻¹)
Location × Carbofuran											
Guinea Savanna × 1.5 kg a.i.ha⁻¹	0.66	8.26	37.16	17.64	35.41	1.04	0.66	1.09	13.50	1.70	2.39
Guinea Savanna × 0.0 kg a.i.ha⁻¹	2.08	32.95	71.98	30.17	63.05	2.36	1.33	2.85	59.92	1.00	1.36
Rain forest × 1.5 kg a.i.ha⁻¹	2.00	21.82	40.61	18.78	37.00	1.21	0.69	1.63	16.50	1.28	1.85
Rain forest × 0.0 kg a.i.ha⁻¹	3.00	41.06	78.20	35.19	69.91	2.76	1.33	3.31	64.42	0.82	1.78
LSD$_{0.05}$	Ns	4.80	Ns	Ns	Ns	0.12	0.32	Ns	Ns	Ns	Ns
F-value	0.57	15.10	0.16	1.22	0.61	7.02	0.01	0.07	0.37	5.06	2.31
P-value	0.478	0.008	0.699	0.312	0.465	0.038	0.909	0.801	0.563	0.066	0.180
Df = 1, 6											
Carbofuran × Maize types											
1.5 kg a.i.ha⁻¹ × Flint	1.00	10.23	36.88	16.70	33.45	1.01	0.58	1.43	12.13	1.850	2.73
1.5 kg a.i.ha⁻¹ × Pop	1.25	11.52	39.86	17.88	36.94	1.18	0.66	1.34	16.00	1.54	2.13
1.5 kg a.i.ha⁻¹ × Sweet	1.75	23.52	39.92	20.06	38.24	1.18	0.77	1.32	16.88	1.08	1.51
0.0 kg a.i.ha⁻¹ × Flint	2.25	32.15	62.98	30.34	52.19	2.48	0.99	2.39	47.25	1.03	1.90
0.0 kg a.i.ha × Pop	2.25	36.75	74.60	32.47	65.10	2.61	1.51	3.35	64.12	0.95	1.13
0.0 kg a.i.ha × Sweet	3.13	42.12	87.70	35.23	82.14	2.57	1.49	3.50	75.13	0.75	0.77
LSD$_{0.05}$	Ns	Ns	9.21	Ns	9.22	Ns	0.35	0.66	3.81	0.20	Ns
F-value	0.36	2.09	11.75	0.56	15.05	0.09	3.80	3.74	47.96	6.21	1.49
P-value	0.704	0.145	<0.001	0.578	<0.001	0.911	0.037	0.038	<0.001	0.007	0.246
Df = 2, 24											
Location × Maize type											
LSD$_{0.05}$	Ns	Ns	Ns	Ns	Ns	Ns	Ns	Ns	Ns	Ns	Ns
F-value	0.46	3.32	0.20	0.84	0.64	0.01	0.38	0.16	0.45	2.58	0.72
P-value	0.638	0.056	0.816	0.446	0.534	0.994	0.964	0.849	0.642	0.097	0.495
Df = 2, 24											
Location × Carbofuran × Maize type											
LSD$_{0.05}$	Ns	Ns	Ns	Ns	Ns	Ns	Ns	Ns	Ns	Ns	Ns
F-value	0.05	1.98	0.12	2.88	0.19	0.05	0.03	0.55	0.17	1.77	0.61
P-value	0.951	0.161	0.886	0.075	0.826	0.951	0.974	0.547	0.842	0.191	0.550
Df = 2, 24											

Means are values of four replicates

LSD$_{0.05}$ = Fisher's Least Significant Difference at 5% level of probability

Data in percentage were transformed to arcsine values and insect count was analysed as $\sqrt{x + 0.5}$ before the F-test.

4. Discussion

Population density and damage activities of stem borers observed in the study confirm previous reports which implicated the insect as a major pest of maize. The higher density of *B. fusca* in the maize types observed in both locations is similar to the findings of Nigeria by [2] in southwestern part of Nigeria, but contrary to the findings of [3] and [26] who reported that *E. saccharina* and *S. calamistis* as the most important stemborers on maize in the

forest zone of eastern Nigeria, the forest/savanna transition zone of south western Nigeria. Though not investigated, the nature of the vascular tissues of popcorn may have contributed to the relatively higher preference of *B. fusca* for it. The absence of *A. ignefusalis* in flint corn in the Guinea Savanna zone is yet to be determined whether or not it is a case of escape of infestation. But its absence in the Rainforest zone maybe due to location effects and absence of millet in neighboring plots. *A. ignefusalis* is a millet stem borer, but could infest other cereals [29]. In addition, *C. partellus* was absent in larval population recovered from both location. But [27] reported that factors such as temperature, rainfall and humidity influences the distributions of *B. fusca* and *C. partellus*, with temperature being the most important. He indicated that *C. partellus* was found in warmer regions and *B. fusca* in cooler areas. This generalization may be the possible reason for the absence of *C. partellus* in cooler environment of the Rainforest zone of Nigeria.

The significant reduction in borer population by single dose application of carbofuran at the rate of 1.5 kg a.i.ha^{-1} further confirms the effectiveness of the chemical in the control of stem borers. Similar results have been reported by [6, 4]. In spite of health and environmental risks associated with the use of chemicals for pest control, they continues to play the important role of minimizing crop loss associated with increase in insect pest population in farmer's field. It is important to note that carbofuran is a systemic insecticide which is effective even after the larvae penetrate into the stem [4]. Their use in late season maize could leave residues in the grains; hence, non-systemic alternative is suggested.

The three maize types used in the experiment were susceptible to stem borer infestation and damage. The high population of larvae observed in their stems may be the reason for high level of borer damage observed in the study. This agrees an earlier observation by [9] who found significant increase in infestation and damage as well as yield reduction with increase larval population.

The higher grain yield observed in flint corn compared with either pop or sweet corn may be due to the lower damage observed in the maize type in both locations as well as differences in grain morphometrics and weight. Noteworthy, infested plants also yielded cobs filled with grains in spite of borer tunneling activities which resulted in stem lodging and cavities in the stems of the three maize types. This result was not unexpected because maize has scattered vascular bundle typical of a monocotyledonous plant, which enables the plant to translocate nutrients and water through undamaged tissues to the yield bearing sink(s) without significant reduction in yield. In addition, adequate rainfall as well as the use of fertilizers has been reported to exert significant influence on maize yield and stem borer damage [24, 21].

The no significant location × carbofuran × maize type and location × maize type interaction effect detected for all parameters show that infestation, damage and yield were not significantly affected by the combinations of these factors. However, the significantly higher damage values observed in carbofuran-untreated plots in the Rainforest zone compared with the treated plots and treated and untreated plots in the Southern Guinea Savanna zone reveals the impact of location and borer control with carbofuran on bored stem, larval population per plant and number of exit holes in the maize types. Generally, herbivore activities can vary with habitat type depending on the prevailing conditions in a given ecosystem [1, 20]. Furthermore, the combination of borer control with carbofuran and maize types had significant effect on bored stem, bored ear, number of exit holes per plant, number of stem borer cavities per plant, number of damaged seeds per plant and number of filled cobs per plant. This may also have resulted from differences in maize type's ability to absorb carbofuran from the soil.

5. Conclusion

The study was carried out to assess the effect of three maize types (flint, pop and sweet corn) and borer control with carbofuran on the severity of maize stem borer infestation and damage in Southern Guinea Savanna and Rainforest agro-ecological zones of Nigeria. The results revealed that yield loss due to stem borer infestation and damage could be severe if the pest is not managed during the late maize cropping season (August to November) in both locations. Furthermore, the lower infestation and damage levels in flint corn compared with pop and sweet corn in carbofuran-treated and untreated plot in the Southern Guinea Savanna and Rainforest zone of Nigeria, suggests that flint corn may be more productive when grown by resource–poor farmers in both locations. However, infestation and damage by stem borer can be suppressed with single dose soil application of carbofuran at the rate of 1.5 kg a.i.ha^{-1}. Further studies still should be conducted to assess the extent of damage caused by individual borer species and their implications on the yield of each maize type in controlled environment. This will provide additional empirical evidence of the economic status of the pest.

Acknowledgements

The authors are grateful to Professors O. E. Ogunwolu and M. O. Adeyemo for their invaluable inputs during the planning and execution of the field work at the Makurdi, Nigeria.

References

[1] Altieri, M. A. and Nicholls C. I. 2004. *Biodiversity and pest management in agroecosystems*, 2nd editors. The Haworth Press Inc. URL: http://www.scribd.com/doc/12591031/Biodiversity-and-Pest-Management-in-Agroecosystems

[2] Balogun, O. S. and Tanimola, O. S. 2001. *Preliminary studies on the occurrence of stem borer and incidence of stalk rot under varying plant population densities in maize*. Journal of Agricultural Research and Development 1: 67-73. http://dx.doi.org/10.4314/jard.v1i1.42191

[3] Bosque-pérez N A. and Mareck J. H. 1990. *Distribution and composition of lepidopterous maize borers in southern Nigeria*. Bulletin of Entomological Research 80: 363–368. DOI: http://dx.doi.org/10.1017/S0007485300050604

[4] Bosque-pérez, N. A. 1995. *Major insect pests of maize in Africa: biology and control*. IITA Research Guide 30. Second edition. Training Program, International Institute of Tropical Agriculture (IITA), Ibadan, Nigeria. 30 p. URL: http://www.fao.org/sd/erp/toolkit/books/majorinsectpestsofmaize.doc

[5] Bosque-perez, N. A. and Schulthess, F. 1998. *Maize: West and Central Africa*. In Polaszek, (ed.). African Cereal Stem borers: Economic Importance, Taxonomy, Natural Enemies and Control. CAB International, Wallingford, UK, pp. 11-24. URL: http://www.cabi.org/bookshop/book/2399

[6] Egwuatu, R. I. and Ita, C. B. 1982. *Some effects of single and split application of Carbofuran on the incidence of damage by Locris maculata, Busseola fusca and Sesamia calamistis on maize*. Tropical Pest Management: 28:227-283. DOI: 10.1080/09670878209370721

[7] FAOSTAT 2013. *Maize yield estimate for Nigeria*. FAO Statistics Division. Retrieved 10:00 GMT 18 April, 2013.

[8] Genstat Discovery Edition 4 (2011). VSN International Ltd., Rothamsted Experimental Station.

[9] Girling, D. J. 1980. *Eldana saccharina as a crop pest in Ghana*. Tropical Pest Management. 26: 156-156. DOI: http://dx.doi.org/10.1080/09670878009414386

[10] Hammond, R. B. and Pedigo, L. P. 1982. *Determination of yield-loss relationships for soyabeans defoliators by using simulated insect-defoliation techniques*. Journal of Economic Entomology 75: 102-107. DOI: http://www.ingentaconnect.com/content/esa/jee/1982/0000007 5/00000001/art00026

[11] Harris, K. M. 1962. *Lepidopterous stem borers of cereals in Nigeria*. Bulletin of Entomological Research. 53:139-171. DOI: http://dx.doi.org/10.1017/S0007485300048021

[12] Harris K. M. 1990. *Bioecology of Chilo species*. Insect Science and its Application 11: 467-477. DOI: http://dx.doi.org/10.1017/S1742758400021044

[13] Holloway J. D. 1998. *Noctuidae: Introduction*. In: Polaszek A, editor. African cereal stem borers: economic importance, taxonomy, natural enemies and control, pp. 79-86. CTA/CABI International. URL: http://www.cabi.org/bookshop/book/2399

[14] Iken J. E and N. A. Amusa 2010. Consumer acceptability of seventeen popcorn maize *(Zea mays L.)* varieties in Nigeria *African Journal of Agricultural Research Vol.* 5(5), 405-407. URL: http://www.academicjournals.org/journal/AJAR/article-abstract/2B5B9D627885

[15] Iken, J. E. and Amusa, N. A. 2004. *Maize Research and Production in Nigeria*. African Journal of Biotechnology. 3 (6): 306. DOI: 10.5897/AJB2004.000-2056

[16] IITA (International Institute of Tropical Agriculture) 2014. *Maize*. URL: http://www.iita.org/maize 22nd June, 2014

[17] Kfir, R. 2013. Maize Stem Borers in Africa: Ecology and Manage-ment. *Encyclopedia of Pest Management*. Taylor and Francis: New York, Published online http://www.tandfonline.com/doi/abs/10.1081/E-EPM-120048597

[18] Kfir, R. 2002. *Increase in cereal; stem borer populations through partial elimination of natural enemies*. Entomologia Experimentalis et Applicata. 104: 299-306. http://dx.doi.org/10.1046/j.1570-7458.2002.01016.x

[19] Maes K. 1998. Pyraloidea: Crambidae, Pyralidae. pp. 87–98 In. Polaszek A. 1998. *African Cereal Stem Borers: Economic Importance, Taxonomy, Natural Enemies and Control*. Wallingford, UK: CABI. 530 pp. DOI: http://www.cabi.org/bookshop/book/2399

[20] Mailafiya D.M., Le Ru B.P., Kairu E. W., Dupas S, Calatayud P.A. 2011. *Parasitism of lepidopterous stem borers in cultivated and natural habitats*. Journal of Insect Science 11:15. http://dx.doi.org/10.1673/031.011.0115

[21] Moyal, P. 1995. *Borer infestation and damage in relation to maize stand density and water stress in the Ivory Coast*. International Journal of Pest Management. 41:114-121. http://dx.doi.org/10.1080/09670879509371934

[22] NRI (Natural Resource Institute) 1996. *A Guide to Insect Pests of Nigerian Crops: Identification, Biology and Control*. Federal Ministry of Agriculture and Natural Resources and the Oversea Development Administration. UK.

[23] Okrikata, E., and C. Anaso 2008. *Influence of some inert diluents of neem kernel powder on protection of sorghum against pink stalk borer (Sesemia Calamistis, Homps) in Nigerian Sudan Savanna*. Journal of Plant Protection Research. 48 (2):161-168. DOI: 10.2478/v10045-008-0019-4.

[24] Polaszek, A. 1998. *African Cereal Stem borers: Economic Importance, Taxonomy, Natural Enemies and Control*. CAB international, Wallingford, UK, 345p. DOI: http://www.cabi.org/bookshop/book/2399

[25] Schulthess, F., Bosque-perez, N.A. and Gonuou, S. 1991. *Sampling lepidopterous pests on maize in West Africa*. Bulletin of Entomological Research. 8: 297-301. http://dx.doi.org/10.1017/S0007485300033575

[26] Schulthess, F., Bosque-Pe´rez, N.A., Chabi-Olaye, A., Gounou, S., Ndemah, R. & Goergen, G. 1997. Exchanging natural enemies species of lepidopterous cereal stemborers between African regions. *Insect Science and its Application*, 17, 97–108.

[27] Sithole, S. Z. 1987. *The effect of date of planting on shootfly and stem borer infestations on sorghum*, pp. 174 - 183. In Proceedings of the Third Regional Workshop on Sorghum and Millets for Southern Africa, 6 - 10 October 1986, Lusaka, Zambia, ICRISAT, Patancheru.

[28] Smith, J. W., Wiedenmann R. N., Overholt W. A. 1993. *Parasites of lepidopteran stem borers of tropical gramineous plants*. ICIPE Science Press. URL: http://www.abebooks.com/Parasites-Lepidopteran-Stemborers-Tropical-Gramineous-Plants/10814893292/bd

[29] Youm, O., Harris, K. M., and Nwanze, K. F. 1996. *Coniesta ignefusalis (Hompson), the millet stem borer: a handbook of information*. (In En. Summaries in En, Fr, Es.) Information Bulletin no. 46. Patancheru 502 324, Andhra Pradesh, India: International Crops Research Institute for the Semi-Arid Tropics. 60 pp. URL: http://trove.nla.gov.au/version/26607464

24

A Comparative Study on the Effect of Organic and Inorganic Fertilizers on Agronomic Performance of Faba Bean (*Vicia faba* L.) and Pea (*Pisum sativum* L.)

Bhaskarrao Chinthapalli[*]**, Dagne Tafa Dibar, D. S. Vijaya Chitra, Melaku Bedaso Leta**

Department of Biology, College of Natural Sciences, Arba Minch University, Arba Minch, P.O. Box No. 21, Ethiopia

Email address:

chinthapalli.bhaskar@amu.edu.et (B. Chinthapalli)

Abstract: The study was conducted to observe the comparative effect of organic fertilizer (cow dung) and inorganic fertilizers like urea and potassium chloride on the growth, biomass and biochemical parameters of two legumes of pea (*Pisum sativum*) and faba bean (*Vicia faba*). Experiments were done using two plant species of legume family. Organic fertilizer like cow dung (15t/ha) and inorganic fertilizer was applied at rate of urea (120kg/ha) and potassium chloride (125kg/ha). The application of cow dung at 15t/ha showed significant growth over the inorganic fertilizer urea and potassium chloride in terms of germination percentage, fresh weight and dry weight, plant height, shoot length, and root length as well as number of leaves in both the legume plants. Similarly, biochemical parameters have also shown significant differences from organic fertilizer over the inorganic fertilizers and control. Thus our study provides the evidence for using organic fertilizer like cow dung by farmers to have better yield to produce quality grains as cow dung is easy available, environmentally safe and cost effective in pea and faba bean plants.

Keywords: Growth, Cow Dung, Urea, Potassium Chloride, Biomass, Pea (*Pisum sativum*), Faba Bean (*Vicia faba*)

1. Introduction

Less soil fertility is considered to be one of the most important constraints on improved agricultural production [1]. Fertilizers are used to improve fertility and are indispensable for sustained food production, but excessive use of mineral fertilizers has aroused environmental concerns. Organic fertilizers coming from fermented and decomposed organic materials are generally nutritious and safe. Microbial fertilizers are apparently environment friendly, low cost and non-bulky agricultural inputs which play a significant role in plant nutrition as a supplementary and complementary factor to mineral nutrition [2]. Therefore fertilizer and plant nutrition research should establish a workable relation between environment preservation and fertilizers [3].

Legumes are an important source of protein for humans and live stock. They provide nutritionally rich crop residues for animal feed and also humans and play a key role in maintaining the productivity of soil. They also play a unique role due to their ability of fixing atmospheric nitrogen [4]. In Ethiopia, faba bean (*Vicia faba* L.) and pea (*Pisum sativum* L.) are mainly grown for human consumption and one of the most important food crops [5]

Fertilizer is very important for crop growth and productivity. One example, of fertilizer is cow dung which is obtained from cow, which is environmental friendly, is easily used and compared with chemical fertilizer which increases the environmental problems. Organic fertilizers are used easily from locality products and livestock wastes and cost effective than chemical fertilizer [6].

The most useful factor is nutrient supply; organic farming is one of the fastest growing sectors of the agriculture worldwide. Its main objective of to create a balance between inter connected system of the soil organism, plants, animals and humans [7]. The choices of suitable forms of fertilizer of the crop growth of the plant are governed by local, natural condition and variation in soil and climate with regard to their suitability for crops cultivation [8]. Numerous test and experience have shown that farm yard manure with its long time effective nutrients is an ideal fertilizer for crop growth [9].

According to [8], fertilizers are resource of plant nutrient that can be added to supplement soil natural fertility. Crop plants have great demand for nitrogen, phosphorous and potassium as a whole main significance in maintenance of normal physiological function of the cell. In similar way according to [10] lack of nitrogen can result in slow growth and poor yield, but excess nitrogen results in deleted maturity and low quality of leaf, phosphorus deficient crop plants are turned abnormally dark green and late maturing, excessive phosphorus shorten leaf, burn duration, lack potassium as common cause of low quality, the source of potassium also important. Additional studies on corn crop plants on cow dung mixed soil is a good source of different plant nutrients particularly N.P.Ks [11]

The present study was conducted to compare the effects of organic and inorganic fertilizers on two legume species of faba bean and pea to determine the germination, seedling growth and biomass of crop plants and also to compare the physical characters or growth parameters with biochemical parameters treated with organic and inorganic chemical fertilizers.

2. Materials and Methods

2.1. Experimental Site

Arba Minch is found in SNNPRS. It located at $30°56'N$ of the equator and $37°44'E$ with surface area of 2184 hectares of land with altitude ranged from 1200-1400 masl with average temperature 30.6°C, annual rainfall 575mm, situated in 505 km away towards south of Addis Ababa in Great Rift Valley of lake Abaya and Chamo. The study was conducted in Abaya Campus, sikela 5 Km away from Arba Minch University.

2.2. Experimental Design and Layout

The experiment was laid out in a randomized complete block design was used with three replicates. Faba bean and pea seeds were procured from the Arba Minch Agriculture College. The treatments of organic (cow dung) fertilizer and inorganic fertilizer like Urea (120kg/ha having 46% N) and Potassium Chloride (125kg/ha having 60% K_2O) calculated according to the application of fertilizer per hectare soil and added to the prepared plots with the bean and pea seeds separately [12]. The cow dung that was used for these experiments were collected from the farm yard. The cow dung collected was subjected to different management practices like mixing thoroughly with a shovel with the aim of reconciling it and applied 15t/ha [13].

The size of the experimental plots was 3.2 m x 4.5m (14.4m²) with 1.3 m spacing between blocks and 0.8 m between plots. Each plot had 7 rows. The inter-row and intra-row spacing were respectively 0.7 m and 0.3 m. The central five rows were used for data collection. All agronomic practices such as land preparation and weeding were performed as per the local farmers' practices.

2.3. Physical Parameters

Measurements of percentage of germination, shoot length, root length, shoot Fresh weight and dry weight and root fresh weight and dry weight of both legume plant species were taken at 35 days after-planting (vigorous vegetative growth).

2.4. Biochemical Parameters

Leaf samples from two plant species in the net rows were harvested for chlorophyll estimation following the methods [14-15], in which 100 mg fresh leaf was crushed in 20 ml of 80% acetone and the extract centrifuged for 10 min at 1000 rpm. Absorbance of the supernatant was recorded at 645 and 663 nm in a UV-Spectrophotometer (Hitachi). Chlorophyll content (expressed as mg/g-1 of each sample) was estimated according to [15] as follow:

Chlorophyll a (mg/g-1) = 12.7 (A663) – 2.69 (A645) x VW

Chlorophyll b (mg/g-1) = 22.9 (A645) – 4.86 (A663) x VW

Total Chlorophyll t (mg/g-1) = [20.2 (A645) – 8.02 (A663) x VW]/1000

Where A = absorbance at the given wavelength, W = weight of fresh leaf sample, V = final volume of chlorophyll solution.

The total soluble protein was estimated by using either Bradford's reagent Folin-Phenol reagent [16], with bovine serum albumin as the standard. Amino acid content of seedlings was measured by the following method [17]. Sugars were estimated by the standard method of [18].

2.5. Replications and Statistical Analysis

All experiments were repeated 3 to 5 times on different days. The average values ±SE are presented. Statistical analysis of the data was done using the software Sigma plot (version 10.0).

3. Results

Agronomic performance of both legume plant species of faba bean (Vicia faba L.) and pea (Pisum sativum L.) were significantly influenced by the treatments of organic (cow dung) and inorganic fertilizer (urea and potassium chloride) over the control.

3.1. Percentage of Germination

The average cumulative germination percentage with organic and inorganic treatments was compared with control of bean and pea and is summarized (Figure 1 (a)). It was clearly observed that organic fertilizer (cow dung) showed high frequency of 92-96% germination when compared with inorganic fertilizer (urea and potassium chloride) of (80-86%) after 35 days of germination over the control of only 82% in faba bean and pea.

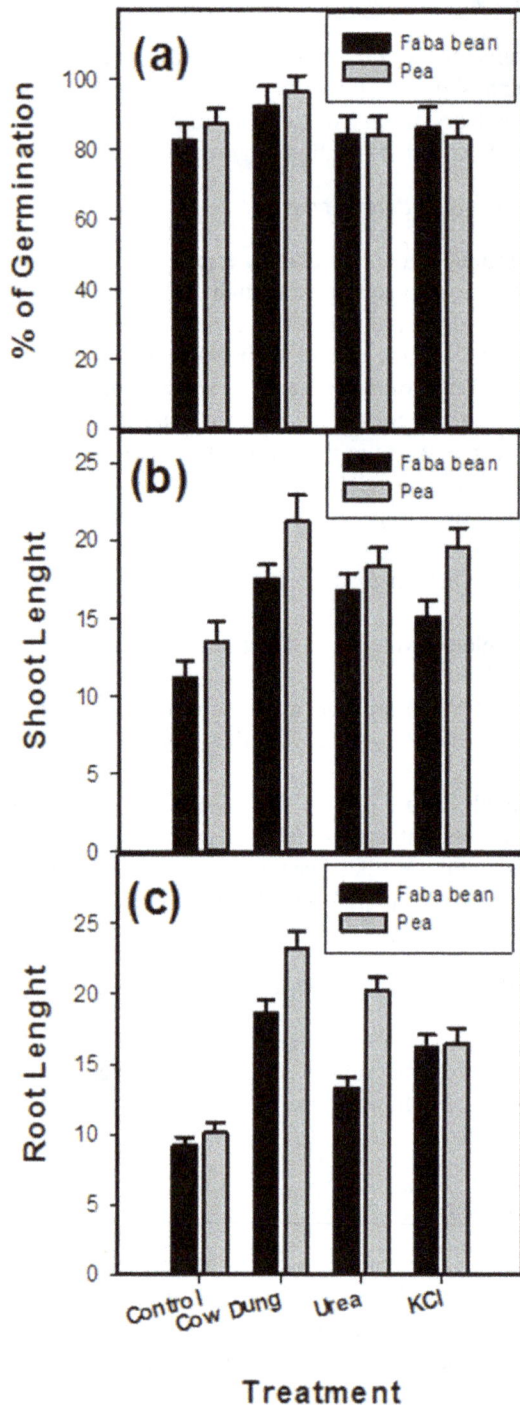

Figure 1. *Effect of organic and inorganic fertilizers on faba bean and pea plants (a) % of germination (b) shoot length (c) root length. The experiments were done on at least three different days and the average values are ± SE represented.*

3.2. Shoot Length and Root Length

Further experiments was designed to determine the changes particularly the mean shoot length and root length in the seedlings of 35 days old faba bean and pea with different treatments of organic, inorganic fertilizers over the control (Figure 1 (b and c)). The seedlings were pulled from the soil without the damage of root and measured with measuring scale. The length of the shoot increased in case of faba bean seedlings from organic fertilizers over inorganic fertilizers and control respectively. In the control the length of the shoot was observed to be the least. Similar observations are seen in the pea seedlings with increase in shoot length from organic fertilizer over inorganic fertilizers and control.

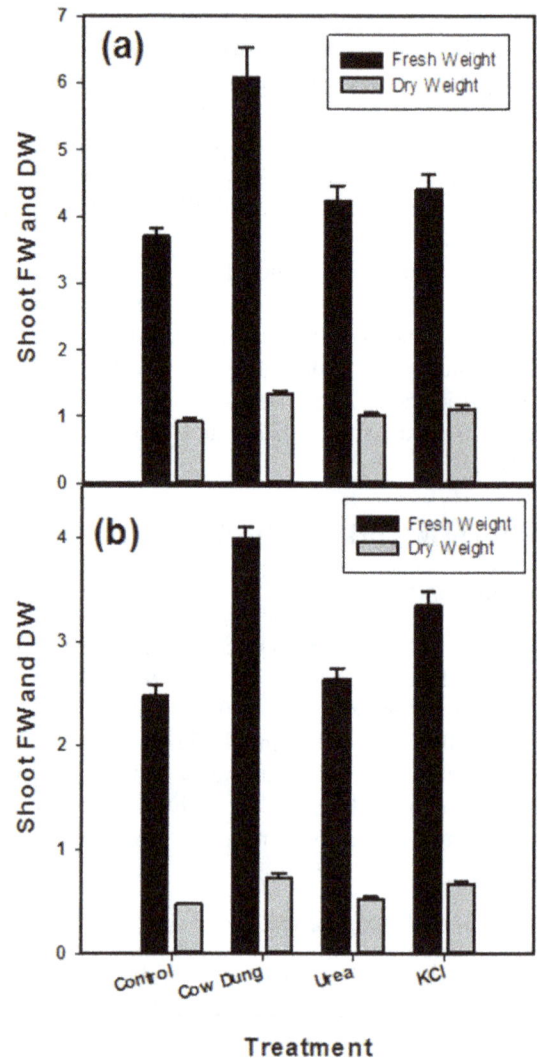

Figure 2. *Effect of organic and inorganic fertilizers of (a) shoot fresh weight and dry weight of faba bean (b) shoot fresh weight and dry weight of pea. The experiments were done on at least three different days and the average values are ± SE represented.*

3.3. Biomass

There was a marked difference of fresh weight and dry weight biomass between organic and inorganic fertilizers over control in both faba bean and pea plants. The average values of fresh weight and dry weight biomass of faba bean and pea plants treated with organic are higher than those with inorganic and control (Figure 2 (a and b) and Figure 3 (a and b). There is no much significant difference between bean and pea seedlings in fresh weight biomass compared to that of dry weight biomass, where as significant difference was observed treated with organic fertilizer.

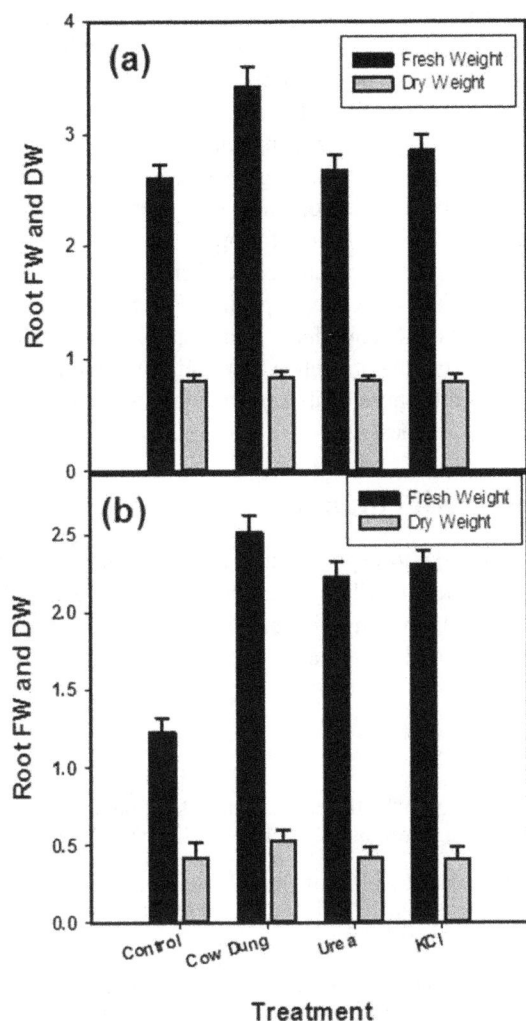

Figure 3. *Effect of organic and inorganic fertilizers of (A) root fresh weight and dry weight of faba bean (B) root fresh weight and dry weight of pea. The experiments were done on at least three different days and the average values are ± SE represented.*

3.4. Chlorophyll Estimation of Bean and Pea Seedlings

Significant difference was observed in both faba bean and pea plants grown in soils treated with organic and inorganic over the control (Table 1). The highest content of chlorophyll-a, chlorophyll-b, total chlorophyll were recorded in both faba bean and pea plants grown in soil treated with cow dung over the inorganic fertilizers like urea and potassium chloride. Similarly, the lowest chlorophyll-a, chlorophyll-b, total chlorophyll were recorded in both faba bean and pea grown without fertilizers (control). Indication of high levels of chlorophyll a, b and total chlorophyll content is a result of effective photosynthetic and metabolic activity. The disappearance of chlorophyll is the one of the most prominent phenomenon of an advanced age and rate of chlorophyll degradation.

3.5. Total Sugars, Soluble Protein and Amino Acids

Increased levels of total sugars were seen in the plants grown in soils treated with cow dung over control. Lowest levels of total sugars were observed in urea and potassium chloride. Similarly, total soluble proteins and amino acids were high in case of organic fertilizer cow dung over the inorganic fertilizers like urea and potassium chloride and control (Table 2).

4. Discussion

In present study, all the treatments with organic and inorganic fertilizers recorded higher germination percentage compared to control samples. The maximum, seed germination was observed in organic fertilizer (cow dung) treatment (Figure 1 (a)). These results are almost similarly agreed with the findings of [19] in *Albizia* and [20] in *Triticum*. Improved seed germination might be attributed to the role of some important microorganism which influence in enhancing the easy availability of nitrogen, phosphorus and potassium in the soil and making available to the germinating seed with consequent enhancement in the cell metabolic activity resulting in higher germination [21-22, 8]. According to [23] the nutrients present in organic fertilizers can be released only when its decomposition takes a long period of time.

Table 1. *Effect of organic and inorganic fertilizers on the chlorophyll a, chlorophyll b and total chlorophyll in different plant species of faba bean and pea.*

Treatment	Faba bean			Pea		
	Chlorophyll a (mg/g FW)	Chlorophyll b (mg/g FW)	Total chlorophyll (mg/g FW)	Chlorophyll a (mg/g FW)	Chlorophyll b (mg/g FW)	Total chlorophyll (mg/g FW)
Control	1.461 ± 0.041	0.661 ± 0.012	1.792 ± 0.045	0.871 ± 0.015	0.422 ± 0.009	1.476 ± 0.068
Cow dung	1.512 ± 0.061	0.837 ± 0.011	2.434 ± 0.051	1.423 ± 0.068	0.531 ± 0.010	1.784 ± 0.052
Urea	1.312 ± 0.075	0.804 ± 0.022	2.237 ± 0.052	1.272 ± 0.054	0.417 ± 0.008	1.433 ± 0.054
Potassium chloride	1.437 ± 0.052	0.786 ± 0.019	2.146 ± 0.041	1.326 ± 0.061	0.395 ± 0.005	1.571 ± 0.046

Table 2. *Effect of organic and inorganic fertilizers on the total sugars, soluble proteins and amino acids in different plant species of faba bean and pea.*

Treatment	Faba bean			Pea		
	Total Sugars (mg/g FW)	Soluble proteins (mg/g FW)	Amino acids (mg/g FW)	Total Sugars (mg/g FW)	Soluble proteins (mg/g FW)	Amino acids (mg/g FW)
Control	4.427 ± 0.097	2.621 ± 0.042	1.512 ± 0.041	4.321 ± 0.081	2.423 ± 0.035	1.325 ± 0.021
Cow dung	4.921 ± 0.073	3.101 ± 0.051	1.684 ± 0.032	4.634 ± 0.062	3.011 ± 0.043	1.412 ± 0.024
Urea	4.312 ± 0.062	2.912 ± 0.047	1.574 ± 0.037	4.199 ± 0.057	2.721 ± 0.032	1.401 ± 0.022
Potassium chloride	4.321 ± 0.075	3.011 ± 0.053	1.594 ± 0.023	4.152 ± 0.055	2.812 ± 0.025	1.392 ± 0.027

Effect of organic and inorganic fertilizers on shoot length and root length, of faba bean and pea were significant. The parameters showed best performance in organic fertilizer over the inorganic fertilizers, and the lowest were in control (Figure 1 (b and c)). Compared to shoot length, the root length had shown a drastic increase in both faba bean and pea seedlings treated with organic over inorganic fertilizers and control. Similar observations were also made according to [24] using biomeal as organic fertilizer in crop plant like carrot. According to [10], increase in N.P.K led to a significant increase in plant height and grain yield. Organic fertilizer cow dung is primarily responsible for the increase in biomass of both faba bean and pea plants. Among nitrogen, phosphorus and potassium, the essential nutrient required by crop plants, Nitrogen in the most commonly deficient in tropical soil [25], which is much available through the decomposition of organic matter.

It is also found (Figure 2 (a and b) and Figure 3 (a and b)) that application of cow dung and inorganic fertilizers like urea and potassium chloride significantly increased fresh and dry weight of both shoot and root in faba bean and pea. On the other hand, the lowest fresh and dry weight of shoot and root were seen from control, respectively. These results are consistent with the findings of [26], they reported that vermin compost, neem cake and FYM (farm yard manure) in carrot. A similar observation was also reported by [27] from a study on the effect of few eco-friendly manures on the growth attributes of carrot and [28] from agricultural waste composts. The results obtained from the present study indicate that the application of cow dung would be sufficient enough for proper growth shoot and root of faba bean and pea plant respectively.

The effect of cow dung and inorganic fertilizers like urea and potassium chloride on photosynthetic pigments of faba bean and pea was presented in Table 1. The highest content of chlorophyll a (1.512 ± 0.061 mg/g fr. wt.), chlorophyll b (0.837 ± 0.011mg/g fr. wt.) and total chlorophyll (2.434 ± 0.051 mg/g fr. wt.) were recorded in faba bean over pea grown in soil treated with cow dung and inorganic fertilizers. Similarly, the lower levels chlorophyll a, chlorophyll b, and total chlorophyll content (Table 1) were recorded without fertilizers (control). The indication of better photosynthetic and cell metabolic activity are mainly due to chlorophyll a, b and total chlorophyll content. The gradual ceasing of chlorophyll is the one of the most prominent phenomenon of an advanced age and rate of chlorophyll degradation [29-31]. The presence or absence of chlorophyll in plant greatly affects the production of secondary metabolites viz., proteins, glycosides, tannins, carotenoids etc. and other essential plant constituents [32]. It has been proved that chlorophyll play an important role in the ATP generation and prevention of essential plant constituents [33]. Similarly, the higher sugar, proteins and amino acid content of seedling was observed at application of organic fertilizer (cow dung) over the inorganic fertilizers and control (Table 2). However, similar results were observed that soluble sugar content was higher under organic fertilization compared to inorganic fertilization. This drift may be due to the lesser starch content when soils treated with without fertilizer and inorganic fertilizers which implies the starch metabolism and poor translocation of sugar to growing part [34].

According for our findings, it would be liable to recommend the use of organic fertilizers cow dung for farmers seeking a better yield for optimum growth of legume plant species. However, inorganic fertilizers, also help in growth and yield of legume species. The reason for this is noteworthiness the cow-dung treatment produced a better yield than no-treatment. Hence cow dung (organic manure) is recommended because of its easy availability, environmental and cost effectiveness.

Acknowledgements

We would like to express our gratitude to Dr. P.D. Sharma from Department of Plant Sciences, College of Agriculture, Arba Minch University for providing inorganic fertilizers which he procured from India.

References

[1]   A.T. Ayoub. "Fertilizers and the Environment". Nutrient Cycling in Agroecosystems. 1999. Vol. 55: pp. 117-121.

[2]   A. Mahajan, R.D. Gupta, and R. Sharma. "Bio-fertilizers-A way to sustainable agriculture". Agrobios Newsletter. 2008. Vol. 6: pp. 36-37.

[3]   N. K. Fageria, V.C. Baligar, and Y. C. Li, "The role of nutrient efficient plants in improving crop yields in the twenty first century". Journal of Plant Nutrition. 2008. Vol. 31: pp. 1121-1157.

[4]   E. A. E. Elsheikh. Environmental Soil Ecology, Khartoum University Press. 2011.

[5]   T. Abera and D. Feyisa. Faba bean and field pea seed proportion for intercropping system in horro highlands of western Ethiopia. African Crop Science Journal. 2009. Vol. 16: pp. 243–249.

[6]   W. G.O. Solomon, R.W. Ndana, and Y. Abdulrahim. The Comparative study of the effect of organic manure cow dung and inorganic fertilizer N.P.K on the growth rate of maize (Zea Mays L.). International Research Journal of Agricultural Science and Soil Science. 2012. Vol. 2: pp. 516-519.

[7]   M. Berova G. Karanatsidis, K. Sapundzhieva, and V. Nikolova. Effect of organic fertilization on growth and yield of pepper plants (Capsicum annuum L.). Folia Horticulturae Ann. 2010. Vol 22: pp. 3-7.

[8]   J. Tanimu, E. O. Uyovbisere, S. W. J. Lyocks, and Y. Tanimu. Effects of Cow Dung on the Growth and Development of Maize Crop. Greener Journal of Agricultural Sciences. 2013. Vol. 3: pp. 371-383.

[9]   C. Tu, J. B. Ristaino, S. Hu. Soil microbial bio-mass and activity in organic tomato farming systems: Effects of organic inputs and straw mulching. Soil Biology Biochemistry. 2006. Vol 38: pp. 247-255.

[10] J. Mani. Early events in environmental stresses in plants: Induction mechanisms of oxidative stress. In: D. Inzè and M.V. Montague (eds.) Oxidative stress in plants. 2002. Taylor and Francis, New York. p. 217-246.

[11] B. S. Ewulo, S. O. Ojeniyi, and D. A. Akanni. Effect of poultry manure on selected soil physical and chemical properties, growth, yield and nutrient status of tomato. African J. Agriculture Research. (2008). Vol 3(9): pp. 612-616.

[12] Agnote. Fertilizer applications, Carol Rose, Extension Agronomist. NSW Department of Primary Industries, Kempsey DPI 496, Edited by Michel Dignand, Information Delivery Program, Wagga, USA. 2004.

[13] I. A. S. Gudugi. Effect of cow dung and variety on the growth and yield of Okra (*Abelmoschus esculentus* L.) European Journal of Experimental Biology. 2013. Vol 3(2): pp. 495-498.

[14] D. I. Arnon. Copper enzymes in isolated chloroplasts. Polyphenol oxidase in *Beta vulgaris*. Plant Physiology. 1949. Vol 24: pp. 1-15.

[15] U. K. Bansal, R. G. Saini, and A. Kaur. Genetic variability in leaf area and chlorophyll content of aromatic rice. International Rice Research Notes. 1999. Vol 24: pp. 21-28.

[16] O. H. Lowry, N. J. Rosenbrough, A. L. Farr, and R. J. Randall. Protein measurement with the Folin phenol reagent. Journal of Biological Chemistry. 1951. Vol 193: pp. 265-275.

[17] S. Moore, and W. H. Stein. Photometric ninhydrin method for use in the chromatography of amino acids. Journal of Biological Chemistry. 1948. Vol 76: pp. 367-388.

[18] A. J. Willis, and E. W. Yemm. The micro-estimation of sugars in plant tissues. New Phytologist. 1955 Vol. 54: pp. 20-22.

[19] P. Kumudha, and M. Gomathinayagam. Studies on the effect of biofertilizers on germination of *Albizia lebbek* (L.) Benth. seeds. Advanced Plant Science. 2007. Vol 20: pp. 417-421.

[20] M. Ram, R. Dawari, and N. Sharma. Direct, residual and cumulative effects of organic manures andbiofertilizers on yields, NPK uptake, grain yield and economics of wheat (*Triticum aestivum* L) under organic forming of rice-wheat cropping system. Journal of Organic Systems. 2014. Vol 9: pp. 16-30.

[21] R. Copper. Bacterial fertilizers in the Soviet Unions. Soil Fertility. 1979. Vol 22: pp. 327-333.

[22] M. Ram, R. Dawari, and N. Sharma. Effect of organic manures on basmati rice (*Oryza sativa* L.) under organic forming of rice–wheat cropping system. International Journal of Agricultural and crop sciences. 2011. Vol 3: pp. 76-84.

[23] C. R. Frink, P. E. Waggoner, and J.H. Ausubel. Nitrogen fertilizer: Retrospect and prospect. Proceedings of National Academy of Science. USA. 1999. Vol 96(4): pp. 1175-80.

[24] H. M. Zakir, M. N. Sultana, and K. C. Saha. Influence of Commercially Available Organic vs Inorganic Fertilizers on Growth Yield and Quality of Carrot. J. Environ. Science and Natural Resources. 2012. Vol 5(1): pp. 39 – 45.

[25] K.A. Okeleye, and C.O. Alofe. Effect of nitrogen fertilizer source on dry matter accumulation and grain yields of open pollinatinated maize (*Zea mays* L.). Samaru Journal Agriculture Research. 1995. Vol 12: pp. 87-98.

[26] R. N. Sunanda, and K. Mallareddy. Effect of different organic manures and inorganic fertilizers on growth, yield and quality of carrot (*Daucus carota* L.). Karnataka Journal of Agricultural Sciences, 2007. Vol 20(3): pp. 686-688.

[27] B. Vijayakumari, R. Hiranmaiyadav, and M. Sowmya. A study on the effect of few eco-friendly manures on the growth attributes of carrot (*Daucus carota* L.). Journal of Environmental Science and Engineering. 2009. Vol 51(1): pp. 13-16.

[28] K. J. Rao, Ch. S. R. Lakshmi, and A.S. Raju. Evaluation of manurial value of urban and agricultural waste composts. Journal of Indian Society of Soil Science. 2008.254(3): pp. 295-299.

[29] I. Ahuja, and C.P. Mallik. Effect of Brassino steroida and Paleobutrazole on chlorophyll content in development of *Brassic tuornefortii*. Journal of Plant Science Research. 1977. Vol 13: pp. 31-34.

[30] A. R. Chopade, N. S. Naikwade, A. V. Nalawade, V. B. Shinde, and K.B. Burade. Effects of pesticides on chlorophyll content in leaves of medicinal plants. Pollination Research. 2007. Vol 26(3): pp. 491-494.

[31] S. Rajasekaran, P. Sundaramoorthy, and S. K. Ganesh. Effect of FYM, N, P fertilizers and biofertilizers on germination and growth of paddy (*Oryza sativa*. L). International Letters of Natural Sciences. 2015. Vol. 35: pp. 59-65.

[32] O.P. Singh, T. P. Singh, and A. L. Yadav. Variability and co-heritability and estimates for agronomical and quality traits in Opium poppy (*P. somniferum* L.). Sci. Cult. 1999. Vol 64(3-4): pp. 107-109.

[33] C.K. Kokate, S.B. Golbale, and Purohit. Textbook of pharmacognosy. Nirali Prakashan. Pune. 1998. pp. 17-18.

[34] M. H. Ibrahim, H. Z. E. Jaafar, E. Karimi, and A. Ghasemzadeh. Impact of Organic and Inorganic Fertilizers Application on the Phytochemical and Antioxidant Activity of Kacip Fatimah (*Labisia pumila Benth*). Molecules. 2013. Vol 18: pp. 10973-10988.

25

Evaluation of Plant Growth, Yield and Yield Attributes of Biofield Energy Treated Mustard (*Brassica juncea*) and Chick Pea (*Cicer arietinum*) Seeds

Mahendra Kumar Trivedi[1], Alice Branton[1], Dahryn Trivedi[1], Gopal Nayak[1], Sambhu Charan Mondal[2], Snehasis Jana[2,*]

[1]Trivedi Global Inc., Henderson, USA
[2]Trivedi Science Research Laboratory Pvt. Ltd., Bhopal, Madhya Pradesh, India

Email address:
publication@trivedisrl.com (S. Jana)

Abstract: The present study was carried out to evaluate the effect of Mr. Trivedi's biofield energy treatment on mustard (*Brassica juncea*) and chick pea (*Cicer arietinum*) for their growth, yield, and yield attributes. Both the samples were divided into two groups. One group was remained as untreated and coded as control, while the other group (both seed and plot) was subjected to Mr. Trivedi's biofield energy treatment and referred as the treated. The result showed the plant height of mustard and chick pea was increased by 13.2 and 97.41%, respectively in the treated samples as compared to the control. Additionally, primary branching of mustard and chick pea was improved by 7.4 and 19.84%, respectively in the treated sample as compared to the control. The control mustard and chick pea crops showed high rate of infection by pests and diseases, while treated crops were free from any infection of pests and disease. The yield attributing characters of mustard showed, lucidly higher numbers of siliquae on main shoot, siliquae/plant and siliquae length were observed in the treated seeds and plot as compared with the control. Moreover, similar results were observed in the yield attributing parameters of chick pea *viz.* pods/plant, grains/pod as well as test weight of 1000 grains. The seed and stover yield of mustard in treated plots were increased by 61.5% and 25.4%, respectively with respect to the control. However, grain/seed yield of mustard crop after biofield energy treatment was increased by 500% in terms of kg per meter square as compared to the control. Besides, grain/seed yield of chick pea crop after biofield energy treatment was increased by 500% in terms of kg per meter square. The harvest index of biofield treated mustard was increased by 21.83%, while it was slight increased in case of chick pea. In conclusion, the biofield energy treatment could be used on both the seeds and plots of mustard and chick pea as an alternative way to increase the production and yield.

Keywords: Mustard, Chick Pea, Biofield Energy Treatment, Growth, Yield, Yield Attribute

1. Introduction

Grain legumes being the major protein source in human and animal nutrition, play a major key role in crop rotations across the world. Among the various oilseed crops, mustard is one of the important because of its potential utilities in the growing biofuels industries [1]. It is widely used as a condiment and as edible oil. The pungency of mustard oil is due to the presence of allyl-isothiocyanate. The low pungency of mustard oil can be obtained after inactivating the myrosinase enzyme present in it and used as a filler component in various processed meat products [2]. Glucosinolates are the major class of bioactive phytocontituents mainly rich in mustard [3]. Mustard seed extract has the potential chemo-preventive and chemotherapeutic activities *in vitro* by scavenging the hydroxyl radicals; it also induces apoptosis of cancer cells [4]. It is also reported that the antioxidant activities of mustard seeds extract [1, 5]. Crop rotation along with other crops also improve the soil fertility, and reduces weeds, pest, and diseases [6]. Chick pea (*Cicer arietinum*) is the major legume in the vegetarian diet with high carbohydrate content. It is one of the drought resistant crops, and considered as an important legume in the newly cultivated land. Chick pea is the third most widely grown grain legume after bean and soybean in the world. Due to its very high protein concentration (approximate 19.3-25.4%), its agronomical importance is demanding for human and animal diet as an alternative protein source. Utilization of nitrogen was reported with enhanced yield, and yield attributes in legume [7]. The National Center for Complementary and Integrative Health (NCCIH), allows the use of Complementary and Alternative Medicine (CAM) therapies such as biofield energy as an alternative in the healthcare field. About 36% of US citizens regularly use some form of CAM [8], in their daily activities. CAM embraces numerous energy-healing therapies; biofield therapy is one of the energy medicine used worldwide to improve the overall human health. Mr. Trivedi's unique biofield treatment (The Trivedi effect®) has been extensively contributed in scientific communities in the field of agricultural science [9-12] and chemical science [13].

Due to the necessity of mustard and chick pea as the food resource, and the improvement in overall productivity of these two plants, an effective control measure need to be established. Under these circumstances, the present work was undertaken to evaluate the effect of biofield energy treatment on mustard and chick pea in relation to growth, yield, and yield attributes.

2. Materials and Methods

The seeds and plots of both mustard and chick pea were selected for the study. Field experiments on mustard and chick pea were conducted at the Agricultural Research Farm of the Institute of Agricultural Sciences, Banaras Hindu University, Varanasi, India during winter season. The experiments on both mustard and chick pea were performed un-replicated with gross plot size of 12.0 m x 6.0 m. One portion of both mustard and chick pea (seeds and plots) was considered as control; no biofield energy treatment was given. Besides, equally divided other portion (seeds and plots) was subjected to Mr. Trivedi's biofield energy treatment. Mustard crop was sprayed with insecticide (0.125% Rogor) against the aphid, while no plant protection was given to chick pea. Mustard and chick pea were received two and one irrigation, respectively.

2.1. Biofield Energy Treatment Strategy

The above assigned both seeds and plots of both mustard and chick pea were subjected to Mr. Trivedi's biofield energy treatment under ambient conditions. Mr. Trivedi provided the treatment to the seeds through his inherent unique energy transmission process without touching the seeds or lands. Afterward, both the control and the treated samples were assessed for growth, yield, and yield attributes of both mustard and chick pea plant.

2.2. Growth, Yield, and Yield Attributes of Mustard

Biofield treated mustard seeds were allowed to germinate until ready to be transplanted according to the season. As a control, untreated mustard seeds were allowed to germinate in the same manner and transplanted alongside the treated plots in a randomized fashion. Overall, the plant height, primary and secondary branches, seed/grain yield, and harvest index of the control and treated mustard crops were calculated [14].

2.3. Growth, Yield, and Yield Attributes of Chick Pea

Both the control and biofield treated chick pea seeds were permitted to germinate until they ready to be transplanted to the particular season. After germination both plantlets were transplanted in the pre-defined plots separated with an imaginary barrier. The plant height, primary and secondary branches, seed/grain yield, and harvest index of the both control and treated chick pea crops were noted [14].

3. Results and Discussion

3.1. Growth, Yield, and Yield Attributes of Mustard

Allelopathy is the process of plant communication system through the direct or indirect, detrimental or advantageous effects of one plant to another. They communicate through the release of allelochemicals i.e. the secondary metabolites or waste products of plants into the environment through leaching, root exudation, volatilization and decomposition of plant residues. The mustard plant belongs to *Brassicaceae* family cited as allelopathic crop [15]. The growth, yield, and yield attributes of mustard seedling data of control and treated samples are shown in Table 1.

Table 1. Growth, yield attributes and yield of control and biofield treated mustard.

Group	Plant height (cm)	Branches/plant		Siliquae/plant		Siliquae length (cm)	1000 seed wt. (g)	Seeds yield		Stover yield		Harvest index (%)
		Primary	Secondary	Main shoot	Total			kg/plot*	q/ha	kg/plot	q/ha	
Control	141.4	5.4	11.5	25.5	176.2	4.96	5.34	3.66	7.56	14.15	30.64	19.79
Treated	160.0	5.9	11.7	36.0	191.4	5.47	5.41	5.91	12.21	18.60	38.42	24.11

*Net plot size 11×4.4 = 48.4 m².

The effect of biofield energy treatment and its related data are presented in Table 1, which revealed marked difference in plant height of treated mustard at maturity as compared with the control. Plants obtained from the biofield treated seeds and plot grew taller and were recorded 13.2% higher plant height than the control plants. Primary branching in treated plots were improved by 7.4%, while slight increase was reported in secondary branches as compared with the control. Among the yield attributing characters, lucidly higher number of siliquae on main shoot, siliquae/plant and siliquae length were observed in treated seeds and plot as compared with the control. The seed and stover yield of mustard in treated plots were increased by 61.5% and 25.4%, respectively with respect to the control. However, grain/seed yield of mustard crop after biofield treatment was increased by 500% in terms of kg per meter square (Fig. 1). The harvest index of treated mustard was increased by 21.83% as compared to the control.

Use of fertilizers, pesticides, and nutrient management has been well reported as they play a key role in increasing and stabilizing the productivity of mustard [16]. The study results concluded, that the biofield energy treatment could be a new and safe approach in term of growth and yield of mustard crop.

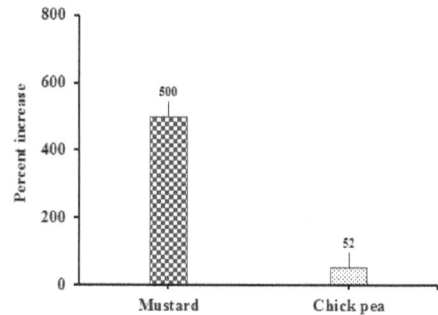

Figure 1. *Effect of biofield energy treatment on percent increase in grain/seed yield of mustard and chick pea crops.*

3.2. Effect on Growth, Yield, and Yield Attributes of Chick Pea

The results related with growth and yield attributes of chick pea before and after the biofield energy treatment are presented in Table 2.

Table 2. Growth, yield attributes and yield of control and biofield treated chick pea.

Group	Plant height (cm)	Branches /plant	Pods/ plant	Grains/ pod	1000 seed wt (g)	Grain yield		Straw yield		Harvest index (%)
						kg/plot	q/ha	kg/plot	q/ha	
Control	30.9	12.6	21.8	1.38	162.8	1.92	4.15	3.53	7.64	35.20
Treated	61.0	15.1	38.4	1.56	195.9	8.91	19.30	16.0	34.63	35.78

*Net plot size $11 \times 4.2 = 46.2$ m^2.

The results showed marked difference in growth characters of biofield treated chick pea as compared with the control. The linear growth as well as total number of branch/plant were recorded at harvest, and were found considerably higher in treated seeds and plot as compared with the control. The plants obtained from the treated seeds and plot were increase by 97.41% as compared the control plant. Branches per plants were also improved in biofield treated group by 19.84% as compared with the control. Similar results were noticed in yield attributing parameters *viz.* pods/plant, grains/pod as well as test weight of 1000 grains. Considerable infestation of wilt leading was observed which leads to plant mortality in untreated plots sown with normal seeds. Due to the better plant stand as well as growth and yield attributing characters, grain and straw yields of biofield energy treated plots increased by 365.1% and 353.3%, respectively. To improve the overall yield of chick pea, salinity mediated productivity have been reported with better growth [17]. The experimental results suggested biofield treated chick pea showed better yield as compared with the control. Moreover, grain/seed yield of chick pea crop after biofield energy treatment was also increased by 500% in terms of kg per meter square (Fig. 1). The harvest index was slightly increased in the case of treated chick pea as compared to the control.

The biofield treated crops had dark green colored leaves with a thick consistency being more in numbers, as compared with the control crops. The control mustard crop showed high rate of infection by pests and diseases, and leaves were reported with survival rate hardly by 40%, while biofield treated mustard was free from any kind of diseases or pests attack, and leaves were quite thick, large, dark green in color, and more secondary and tertiary branches. Similarly, biofield treated chick pea showed high survival rate after germination, free from any kind of infections, the canopy of plant was better as compared with the control. Overall, the treated crops showed high yield as compared with the control. However, crops from all the treated seeds were found with a very thick population and free from the diseases and pests attack as compared with the respective control. In biofield treated seeds, there was no airborne infection observed which defies the laws of aerobiology.

The canopy of the biofield energy treated trees was more than the double as compared to the control; and had more secondary and tertiary branches. Leaf area was significantly more in the treated crop, which was well indicated with more grain/seed yields. Leaf area is directly related with the final productivity of the crop [18]. The longevity of the all crops in the treated plot was found to be increased, hence fruiting period has also been extended resulting in higher yield. Weed or unwanted plant growth was not seen in the treated plot, whereas in the control plot even after spraying weedicides (three-time) the weeds were continuously required to be removed approximately four times manually. It was reported that climatic change can influence the flowering time, and overall productivity of crops [19], biofield treated crops resulted in better flowering, which was directly related to overall productivity.

However, biofield treatment has been reported an improved overall plant health of *Withania somnifera* and *Amaranthus dubius*. Leaf, stem, flower, seed setting, and immunity parameters were reported with enhanced effect after biofield treatment. Concentrations of chlorophyll a, chlorophyll b and total chlorophyll were consistently higher in treated plants along with genetic variability using RAPD DNA fingerprinting [12]. The impact of biofield treatment on the yield of ginseng, blueberry [10], and growth and yield of lettuce and tomato were reported [9]. Similar results were observed in our experiment with biofield treated mustard and chick pea. The results are well supported with the reported literature in terms of growth and yield of crops. Based on these results, it is expected that biofield energy treatment has the scope to be an alternative approach to improve the plant growth, yield, yield attributes, and development of crops.

4. Conclusions

Based on the study outcome, the biofield energy treated mustard and chick pea showed significant improvement of overall yield of the treated crops as compared to the control. The seed and stover yield of mustard in treated plots were increased by 61.5% and 25.4%, respectively as compared to the control. The percentage increase in yield was maximum in case of mustard (500%) in the biofield treated seed as compared to the control. Linear growth, plant height, branches, and grain/seed yield of mustard and chick pea were consistently increased in all the biofield treated crops, without any precautionary measures such as pesticides, fungicides, and organic additives. The harvest index of treated mustard was increased by 21.83%, whereas the harvest index was slightly increased in the treated chick pea as compared to the control. The chick pea plants obtained from the treated seeds and plot were increase by 97.41% as compared the control plant. Additionally, the growth and yield attributing characters, grain and straw yields of biofield energy treated chick pea were increased by 365.1% and 353.3%, respectively as compared to the control. Overall, Mr. Trivedi's biofield energy treatment resulted in improved yield in multiple kinds of crop, suggested the significant application of biofield treatment in agriculture sector instead of chemical measures to improve the overall productivity. In conclusion, the present investigation demonstrates that Mr. Trivedi's unique biofield treatment could be utilized as an alternate therapeutic approach concurrent with other existing approach to improve the productivity of mustard and chick pea in the field of agriculture in the near future.

Abbreviations

NCCIH: National Center for Complementary and Integrative Health; CAM: Complementary and Alternative Medicine.

Acknowledgements

Financial assistance from Trivedi science, Trivedi testimonials and Trivedi master wellness is gratefully acknowledged. Authors thank Agricultural Research Farm of the Institute of Agricultural Sciences, Banaras Hindu University, Varanasi, India for their support.

References

[1] Dubie J, Stancik A, Morra M, Nindo C (2013) Antioxidant extraction from mustard (Brassica juncea) seed meal using high-intensity ultrasound. J Food Sci 78: E542-E548.

[2] Tsuruo I, Yoshida M, Hata T (1967) Studies on the myrosinase in mustard seed part I. The chromatographic behaviors of the myrosinase and some of its characteristics. Agr Biol Chem 31: 18-26.

[3] Bones AM, Rossiter JT (1996) The myrosinase-glucosinolate system, its organization and biochemistry. Physiol Plant 97: 194-208.

[4] Bassan P, Sharma S, Arora S, Vig AP (2013) Antioxidant and in vitro anti-cancer activities of Brassica juncea (L.) Czern. seeds and sprouts Int J Pharma Sci 3: 343-349.

[5] Amarowicz R, Wanasundara UN, Karamac M, Shahidi F (1996) Antioxidant activity of ethanolic extract of mustard seed. Nahrung 40: 261-263.

[6] Chemining wa GN, Vessey JK (2006) The abundance and efficacy of Rhizobium leguminosarum bv. viciae in cultivated soils of the eastern Canadian prairie. Soil Biol Biochem 38: 294-302.

[7] Palta JA, Nandwal AS, Kumari S, Turner NC (2005) Foliar nitrogen applications increase the seed yield and protein content in chickpea (Cicer arietinum L.) subject to terminal drought. Australian J Agric Res 56: 105-112.

[8] Barnes PM, Powell-Griner E, McFann K, Nahin RL (2004) Complementary and alternative medicine use among adults: United States, 2002. Adv Data 343: 1-19.

[9] Shinde V, Sances F, Patil S, Spence A (2012) Impact of biofield treatment on growth and yield of lettuce and tomato. Aust J Basic Appl Sci 6: 100-105.

[10] Sances F, Flora E, Patil S, Spence A, Shinde V (2013) Impact of biofield treatment on ginseng and organic blueberry yield. Agrivita J Agric Sci 35.

[11] Lenssen AW (2013) Biofield and fungicide seed treatment influences on soybean productivity, seed quality and weed community. Agricultural Journal 8: 138-143.

[12] Nayak G, Altekar N (2015) Effect of biofield treatment on plant growth and adaptation. J Environ Health Sci 1: 1-9.

[13] Trivedi MK, Tallapragada RM, Branton A, Trivedi D, Nayak G, et al. (2015) Characterization of physical, spectral and thermal properties of biofield treated 1,2,4-Triazole. J Mol Pharm Org Process Res 3: 128.

[14] Patil SA, Nayak GB, Barve SS, Tembe RP, Khan RR (2012) Impact of biofield treatment on growth and anatomical characteristics of Pogostemon cablin (Benth.). Biotechnology 11: 154-162.

[15] Turk MA, Tawaha AM (2002) Inhibitory effects of aqueous extracts of barley on germination and growth of lentil. Pak J Agron 1: 28-30.

[16] Shekhawat K, Rathore SS, Premi OP, Kandpal BK, Chauhan JS (2012) Advances in agronomic management of Indian mustard (Brassica juncea (L.) Czernj. Cosson): An overview. Int J Agron 2012: 14. Article ID 408284.

[17] Asha Dhingra HR (2007) Salinity mediated changes in yield and nutritive value of chickpea seeds. Indian J Pl Physiol 12: 271-275.

[18] Devendra R, Urs YSV, Kumar MU, Sastry KSK (1983) Leaf area duration and its relationship to productivity in early cultivars of rice. Proc Natl Acad Sci USA 49: 692-696.

[19] Craufurd PQ, Wheeler TR (2009) Climate change and the flowering time of annual crops. J Exp Bot 60: 2529-2539.

26

Physicochemical Analysis of Pomegranate of Gilgit Baltistan, Pakistan

Faisal Abbas[1], Nawazish Ali[1], Yawar Abbas[2], Attarad Ali[3], Naveed Hussain[1], Tanveer Abbas[1], Abdul-Rehman Phull[4], Islamuddin[5]

[1]Department of Agriculture and Food Technology, Karakoram International University, Gilgit, Pakistan
[2]Department of Earth & Environmental Sciences, Bahria University, Islamabad, Pakistan
[3]Department of Biotechnology, Quid-i-Azam University, Islamabad, Pakistan
[4]Department of Biology, Kongju National University, Gongju, Republic of Korea
[5]Rescue 1122 Gilgit, Pakistan

Email address:
abbas.qasimi@gmail.com (F. Abbas), amjadmalik747@gmail.com (N. Ali), yawar_zaid@yahoo.com (Y. Abbas),
attarad.ali@kiu.edu.pk (A. Ali), naveed_kiu@yahoo.com (N. Hussain), abbasglt110@gmail.com (T. Abbas),
ab.rehman174@gmail.com (Abdul-Rehman P.), sanil_110@yahoo.com(Islamuddin)

Abstract: Juice can be considered as an important and functional ingredient in food products. The aim of current study was to screen and compare the physico-chemical properties of some indigenous species of pomegranate in Gilgit-Baltistan (GB) Pakistan. Fruits were collected from three tehsil regions of GB i.e. Bagrote, Jalalabad and Heramosh valleys. The fruits were washed, peeled off and arils were separated. Fresh juice was prepared from the arils and physico-chemical properties were evaluated. The pH of juice was found in the range of 2.4 (Sour, Jutial Gilgit) to 3.9 (sweet, Jalalabad). Comparative to other areas, pomegranate species of the Jutial exhibited higher total soluble solids (TSS) as 11.5 (sour) 14.5 (sweet) 14.2 (doom). The proximate reducing sugar analysis showed the higher content of reducing sugars in Sweet >Doom >Sour varieties. Lowest average ash and moisture content was observed in sour and higher was determined in sweet varieties.

Keywords: Physico-Chemical, Gilgit-Baltistan (GB), Pomegranate, Nutrients, Juice, Tehsil, TSS

1. Introduction

Pomegranate (Punicagranatum L.) belongs to the Punicacea family (Hardeet al., 1970). It is one of the important and commercial horticultural fruits which is generally very well adapted to the Mediterranean climate (Biale, 1981). It has been cultivated extensively in Iran, India and some parts in the U.S.A (California), China, Japan and Russia. Pomegranate fruits are consumed fresh or processed as juice, jellies and syrup for industrial production (Hodgson, 1917). Different parts of its tree (leaves, fruits and barks) have been used traditionally for their medicinal properties and for other purposes such as in tanning (Rania and Ne´jib, 2007).

Historical evidence reveals that its primary origin is in Iran where it has been grown in every area, both coastal and mountainous areas. It has now been spread to all other regions of the world (Kumar, 1990). Turkey was analyzed for the chemical components, some of which was found for nutritional and health. Four samples were found to contain glucose syrup one of which was found to have very high amount of hydroxymethylfurfural (152498 mg/kg). Twelve pomegranate (Punica granatum L.) cultivars obtained from different growing regions of Iran were analyzed for their physical and chemical properties. These properties included fruit fresh weight, volume and density, peel thickness, soluble solids (TSS), titratable acidity (TA), EC, pH, vitamin C, ellagic acid content of juice and peel, total antioxidant activity of peel and juice and etc. Fruit weight ranged from 103.38 to 505.00 g and fruit volume from 99.41 to 547.88 cm. Similarly, average fruit density ranged 0.91 g.cm G3to 1.04 g.cm G3 and peel thickness of the fruit was recorded from 1.60 to 6.01 mm. Reducing sugars (Vahid et al., 2009).

Different parts of its tree (leaves, fruits and bark skin) have been used traditionally for their medicinal properties and for other purposes such as in tanning (Rania *et al.,* 2007). The edible part of the fruit contains considerable amounts of acids, sugars, vitamins, polysaccharides, polyphenols and important minerals. (Gil, *et al.,*2000).The pomegranate fruit is widely considered as a "healthy" fruit due to its biological actions, most of them attributed to its phenolic content (Lansky and Newman, 2007). Studies about pomegranate polyphenols have shown prevention of cardiovascular, cancer diseases and neurological damage in humans (Aviram *et al.,* 2002).

2. Materials and Methods

2.1. Study Area

The present study was conducted in the Department of Agriculture and Food Technology, Karakorum International University Gilgit. The present study was carried out during 2010 growing season pomegranate trees grown in different areas of three districts of Gilgit Baltistan. All treatments were carried out in first week of September to last week of October.

2.2. Collection of Sample

Thirty normal size pomegranates were obtained from mature fruits growing in Bagrote, Jalal Abad and Heramoshvallies of district Gilgit. Commercially ripe fresh fruits were harvested during September and October from different mature trees randomly selected to represent the population of the plantation. The fruits was harvested at commercially maturity stage and transported to the Food Technology Laboratory of Karakorum International University Gilgit.

2.3. Separation of Seeds

Fruits were cleaned and washed to remove all foreign matters such as dust, then peeling was done manually by knife and grains or aerials were separated by hands.

2.4. Juice Extraction and Filtration

The pulp was extracted from the seeds of pomegranate varieties. Electric blender of good quality was used to crush seeds. Extracted juice was filtered by means of muslin cloths to separate juice from the pulp, fiber and seed particles.

2.5. Analysis of Physico Chemical Parameters of Pomegranate Juice

Following chemical parameters of juice of different pomegranate varieties was measured.
PH:
The pH values of samples were measured by using pH meter (Inolab) according to AOAC (1990) method No.981-12.
Principle:
The basic principle of electrometric pH meter is determination of the activity of hydrogen ions by potentiometer measurement using a standard hydrogen electrode and reference electrode. The glass electrode is

commonly used. The pH value of an aqueous solution is defined by the equation:

$$pH = -log10AH+$$

Where, AH+ = the activity of hydrogen ions in the solution in g-moles/l. The electromotive force (emf) produced in the glass electrode system varies linearly with pH. This linear relationship is described by plotting the measured emf against the pH of different buffers. Sample pH was determined by extrapolation.

2.6. Materials

PH meter, conical flask, Balance, Funnel, Shaker, Distilled water, soft tissue paper, Samples, Buffer Solutions (pH 4.0, 10.0and pH 7.0,)

2.7. Procedure

The pH meter is standardized with standard buffers of pH 4, and 7 respectively. Before taking each reading electrode is washed with distilled water and then dried with soft tissue paper. The electrode is simply dipped in sample till the pH meter gives final reading that is considered being the pH value of sample.

2.8. Total Soluble Solids

Total soluble solids content expressed as Brix was determined by using refracto meter, (Atago 3810-Japan) as described by AOAC (1984) at temperature (20 C). Refractometer, Distilled water, Spatula, Soft tissue paper and Samples were used. Refractometer gives rapid and accurate ratio of TSS present in sample. TSS is expressed in terms of Brix. The representative sample is placed on dry refractometer prism and refractometer is calibrated and readings were taken directly. Refractometer prism is washed with distilled water and dried with soft tissue paper after each reading.

2.9. Total Titratable Acidity

Total titratable acidity of samples was determined by following standard AOAC (1984) method by titrating against strong alkali solution. Burrette stand, 50ml Burrette, Conical flasks, volumetric flasks, Filter cloth, Distilled water, Funnel, Graduated cylinder, 10-ml sample were used. NaOH (0.1 N) 2-g pellets of NaOH are dissolved in 1000 ml distilled water, The resulting solution will be 0.1 N NaOH normal solution (Base) were used during analysis. 0.1g phenolphthalene is dissolved in 50ml distilled water and 50ml alcohol to prepare Phenolphthalene indicator. 5g sample is taken in volumetric flask and the volume is made with distilled water up to 100ml. Filter the sample if needed. The diluted sample (10ml) is taken in a conical flask and 2-3 drops of phenolphthalene indicator and titrated against 0.1N NaOH solution which is filled in a burrette. Continue titration till the solution persists pink color. Appearance of pink color is the end point and the color is persisted for 15-20 seconds. Finally the ml of 0.1N NaOH used is recorded for all the samples and acidity is calculated by using the following formula,

Acidity = F x T x 0.1N NaOH x 100 x 100 / L x M

Where, F is factor of acid (citric acid) which is equal to 0.0064.

2.10. Reducing Sugars

AOAC (1984) recommends following procedure for the determination of reducing sugars present in samples. 50ml Burette, Stand, conical flask, volumetric flask, distilled water, filter cloth, 100ml-graduated cylinder, test tube holder, funnel, and 5g sample was used. Fehling Solution A: 34g $CuSO_4.5H_2O$ is dissolved in 500ml of distilled water. Fehling solution B: 173g Sodium Potassium Tarter ate and 50g of NaOH is dissolved in 500ml distilled water and as Indicator: Methylene Blue. Fruit grounded sample (5g) is taken and diluted up to 100ml with distilled water. The burrette is filled with this solution. 5ml fehling A and 5ml fehling B with 10ml of distilled water is taken in conical flask and boiled, On boiling it was titrated against the sample solution from the burrette till color changes to dark brown or red. 2-3 drops of methylene blue is added as indicator till dark brown or red color persisted. The ml of sample is then recorded and reducing sugar percentage is calculated by using the following formula.

Xml of sample solution =0.05g of reducing sugar

100ml of sample solution contains = 100 x 0.05 / Xml = Yg of reducing sugar

This 100ml of sample solution was prepared from 5g sample
So,

5g sample contains = Yg of reducing sugar

% Reducing sugar in sample = Yg X 100 / 5

2.11. Ash Content

Ash is the inorganic residue remaining after the complete oxidation of organic matter in foodstuff. AOAC (1990) recommends following procedure for determination of ash content. Balance precisian, china dishes, electrical muffle furnace, oven, spatula, desiccator, test tube holder, and 10-g sample was used during chemical analysis of samples. China dishes were taken and weighted properly and weight is noted down. 10-gm of sample is added in china dish. Then charred the sample or dry in oven at 70^0C for 2-4 hours. In temperature up to 100^0C and left it for 24 hours. After 24 hours dry matter is obtained. The sample must be dried in oven till it gives constant weight of sample. After drying the sample 1-gm dry sample is taken in china dish and incarnated in muffle furnace at 550^0Cfor 12-18 hours. Turn off the muffle furnace after the required time and wait till the temperature inside furnace dropped to at least 250^0C. Door of muffle furnace is opened carefully to avoid losing ash that may be fluffy. Ash obtained is transferred in desicator for cooling. After cooling ash is

weighted and calculated on the basis of following formula :

% Ash = Initial Weight - Final weight / Initial weight x 100

2.12. Moisture Content

Sample was dried in the oven provided with opening for ventilation and maintained at 130c for 60 minutes. The loss in sample weight is expressed as percentage moisture. China dish, Desiccators, Silica granules, Rubber gloves were used. Perten laboratory mill 3100, Analytical Balance, Accuracy +/- 0.0001g, Hot air oven was used. Clean moisture dishes were taken and dried in oven at 130c for 30 minutes. Moisture dishes were taken out and placed in desiccators and weighed soon after they reach at room temperature.10g of well-mixed sample was added to the each moisture dish and recorded the weight. Removed from scale and covered with lid.

Placed the moisture dish in oven uncovered for 60 minutes (60 minutes dry period Begin when oven temperature is usually 130c. Afterward removed the sample from oven and covered with lid. Place the sample in desiccators for cooling. Gloves were used for sample transfer room oven to the desiccators.

Weighed the sample till after reached at room temperature.

$$\% \text{ moisture} = \frac{\text{wt of original sample} - \text{wt of dried sample}}{\text{Wt of original sample}} \times 100$$

3. Result

3.1. Physico Chemical Analysis of Pomegranate

Table 1. Total Soluble Solid Of Pomegranate.

Varieties	Jalalabad	Bagrote	Jutal	Mean
Sour	11	11	11.5	11.1
Sweet	14	14	14.5	14.1
Doom	14.1	14.1	14.2	14.1

3.2. Above Readings Is Mean of Three Replication

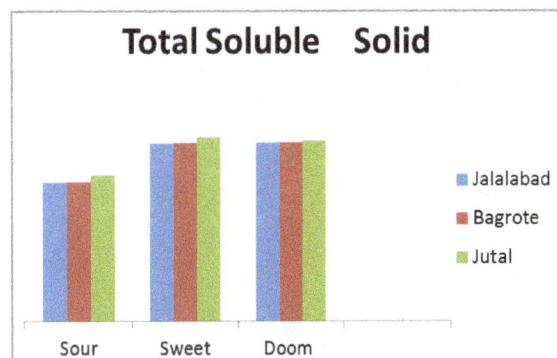

Figure 1. Total Soluble Solid of Pomegranate.

Table 2. pH of pomegranate.

Varieties	Jalalabad	Bagrote	Jutal	Mean
Sour	2.5	2.6	2.4	2.5
Sweet	3.9	3.6	3.86	3.78
Doom	3.2	3.2	3.1	3.1

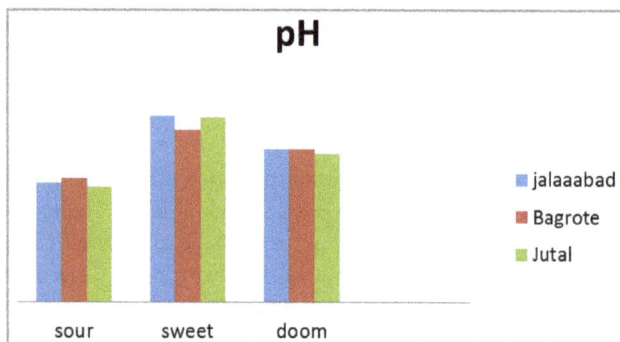

Figure 2. *pH of pomegranate.*

Table 3. *Ttitratable Acidity in pomegranate.*

Varieties	Jalalabad	Bagrote	Jutal	Mean
Sour	0.867	0.879	0.885	0.876
Sweet	0.362	0.382	0.399	0.381
Doom	0.645	0.654	0.665	0.654

3.3. Above Readings Is Mean of Three Replication

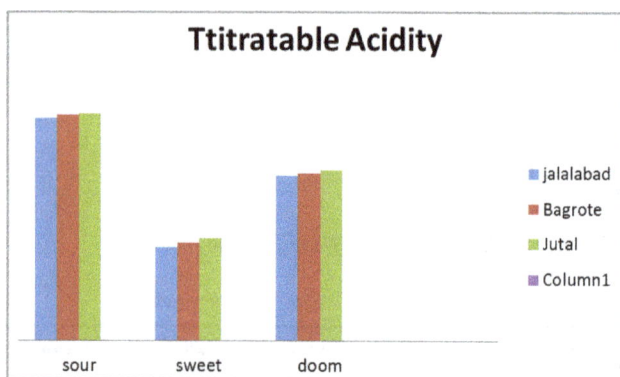

Figure 3. *Ttitratable Acidity in pomegranate.*

Table 4. *Reducing sugar in Pomegranate Fruit.*

Varieties	Jalalabad	Bagrote	Jutal	Mean
Sour	07.00	7.18	7.27	7.15
Sweet	11.99	12.25	12.65	12.29
Doom	10.05	10.08	10.2	10.11

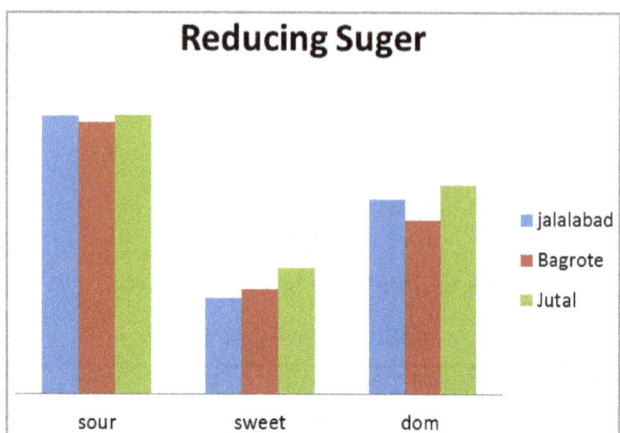

Figure 4. *Reducing sugar in pomegranate fruit.*

Table 5. *Moisture in Pomegranate Fruit.*

Varieties	Jalalabad	Bagrote	Jutal	Mean
Sour	73.82	72.24	73.1	73.05
Sweet	77.5	76.3	77.2	77
Doom	75.3	74.6	74.2	74.83

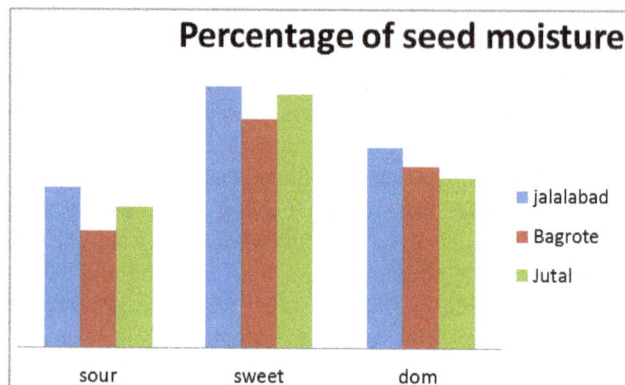

Figure 5. *Percentage of seed moisture in pomegranate.*

Table 6. *Pecentage of seed Ash in pomegranate.*

varieties	Jalalabad	Bagrote	Jutal	Mean
sour	0.59	0.56	0.52	0.556
sweet	0.64	0.6	0.69	0.64
Doom	0.55	0.59	0.55	0.56

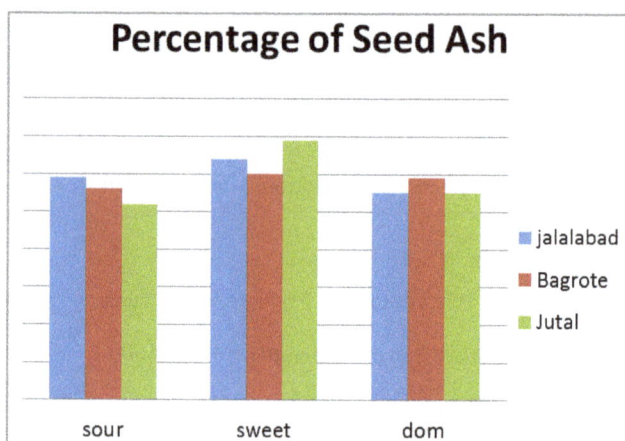

Figure 6. *Percentage of seed ash in pomegranate.*

4. Discussion

4.1. Physico Chemical Analysis of Pomegranate

The chemical composition of fruit differs depending on the cultivar, growing region, climate, maturity, cultural practice and storage conditions (Melgarejo *et al.*, 2000; Nanda *et al.*, 2001; Barzegar*et al.*, 2004; Miguel *et al.*, 2004b; Fadavi *et al.*, 2005).

The results for total soluble solids, pH, titrable acidity, and maturity index of the pomegranate from the different Localities.

Total soluble solids%: In Jalalabad valley the highest total soluble solids content in 14 ∘Brix (Sweet), 14.1∘Brix (Doom), 11.0∘Brix (Sour). In Bagrote valley the highest total soluble solids content in 14.00∘Brix (Sweet), 14.1∘Brix (Doom), 11.0

∘Brix (Sour). In Jutal valley the highest total soluble solids content in 14.5 ∘Brix (Sweet), 14.2 ∘Brix (Doom), and 11.5∘Brix (Sour). Over all the Total soluble solid is high in Jutal because due to the clay soil as shown in Table 1. While our results were in agreement with values (10–16.5 ∘Brix) reported by (Fadavi et al,. 2005).

The total soluble solids observed in the present study were found to be in typical range as reported by other researchers (Radunić, Mira, et al. 2015).

Acidity%:

Data in Table 3 show that, pomegranate cultivars Titratable Acidity % in Jalalabad valley the highest are 0.36% (sweet), 0.64 % (Doom), 0.86 % (Sour). Acidity % in Bagrote valley is highest 0.382%in (sweet), 0.654% in (Doom), and 0.87% in (Sour). Acidity % in Jutal valley is highest 0.399% in (sweet), 0.665% in (Doom), and 0.885% in (Sour).

The minimum acidity of pomegranate juice is 0.35% which was reported by (Mustafa Ozgen *et al.*, 2008).The titratable acidity content varied from 0.33, similar results were also reported by Fadavi *et al.* (2005).

According to (Melgarejo, 1993), the predominant acid of this fruit is malic acid. In studies carried out with the Mollar variety quantities of malic and citric acid. The content of malic acid ranged between 0.143% and 0.249% and the citric acid between 0.147% and 0.4 % (Sharms *et al.,* 1997).

Reducing sugar: Data in Table 4: show that, in Jalalabad pomegranate juice reducing sugar content are 11.99 %(sweet), 10.05 %(Doom), 7.7 % (Sour) and, in Bagrote pomegranate cultivars reducing sugar content are 12.25 %(sweet), 10.08 %(Doom), 7.18 % (Sour).In Jutal pomegranate cultivars reducing sugar content are 12.65 %(sweet), 10.20 %(Doom), 7.27 % (Sour).

Reducing sugar ranged 13.89 % this finding is (Al-Kahtani and Saxena *et al,.* 1987)

Our results are in agreement with findings of a study carried out in Egypt (I.E. Abd El-Rhman, 2010).

pH: Data in Table.2 show that, in Jalalabad pomegranate cultivars pH are 3.92 in (sweet), 3.2 in (Doom), and 2.5 in (Sour). In Bagrote pomegranate cultivars pH are 3.68 in (sweet), 3.22 in (Doom), 2.6 in (Sour). InJutal pomegranate cultivars pH are 3.86 in (sweet), 3.11 in (Doom), and 2.45 in (Sour).

The pH ranged from 2.75 to 4.14, the findings of this research are more than average (3.34) reported by Mustafa *et al.*, 2008.

The pH values ranged between 3.16 and 4.09 the pH values obtained in the current study are greater than those reported by Cam et al. (2009) on pomegranate cultivars grown in Turkey.

The pH range (2.75–4.14) is appeared as representative range as reported in different studies and the pH values recorded in in the current study were also observed in the same range (Radunić, Mira, et al. 2015).

4.2. Physical Analysis of Pomegranate Fruit

Seed Ash%:

Data in Table 9: show that, in Jalalabad pomegranate cultivars seed Ash % are 0.64% in (sweet), 0.55 % in (Doom),

and 0.59% in (Sour). In Bagrote pomegranate cultivars seed Ash % are 0.60 % in (sweet), 0.59% in (Doom), and 0.56% in (Sour). In Jutal pomegranate cultivars seed Ash % are 0.69% in (sweet), 0.55% in (Doom), and 0.52% in (Sour).

Seed moisture %:

Table 5: the seed moisture percentage of the studied in Jalalabad pomegranate cultivars 77.5% (Sweet), 75.3% (Doom), 73.82% (Sour).The seed moisture percentage of the studied in Bagrote pomegranate cultivars 76.3% (Sweet), 74.6% (Doom), 72.24% (Sour). The seed moisture percentage in Jutal pomegranate verities 77.2% (Sweet), 74.3% (Doom) and 73.11% (Sour).

The moisture content of the seeds up to 81.53% was reported by Amit Parasahar (2010). Evaluations of various important components are also reported previously to highlight the important pomegranate characteristics (Zaouay Faten et al., 2012).

5. Conclusion

In the current study physico-chemical properties of pomegranate indigenous to Gilgit-baltistan were determined. The evaluated parameter includes the total soluble solids, acidity, reducing sugurs, pH, seed ash and seed moisture. Significant difference in the physicochemical properties such as sugurs, soluble solids, seed moisture and pH were observed. This study is important in exploring the knowledge of indigenous plants and also increases the information of pomegranate associated properties in specified study region. Whereas, difference in physico-chemical characteristics of the indigenous pomegranate varieties showing the genetically diversity among the species of this Gilgit-baltistan (Rajasekar, Dhivyalakshmi, et al. 2102). Furthermore, such studies may help to the producers or companies for selecting the better indigenous pomegranate specie having better composition.

References

[1] AOAC. 1990, 1984. Official methods of analysis, association of analytical chemist. (15th Ed) Virginia, 22201, Arlington, USA.

[2] Aviram M, Volkova N, Coleman R, Dreher M, Reddy MK, Ferreira D, Rosenblat M. 2008.Pomegranate phenolics from the peels, arils, and flowers are antiatherogenic: studies in vivo in atherosclerotic apolipoprotein e-deficient (E 0) mice and in vitro in cultured macrophages and lipoproteins. J. Agric Food Chem 56 (3).

[3] Biale, J. B., 1981. Respiration and ripening in fruitsretrospect and prospect. In J. Friend and M. J. Rhodes (Eds.), Recent advances in the biochemistry of fruits.

[4] Barone, E., T. Caruso, F. P. Marra and F. Sottile, 2001. Preliminary observations on some Sicilian pomegranate (Punicagranatum L.). Journal of American Pomological Society, 55(1):4-7.

[5] Gil, M. I., C. Garcia-Viguera, F. Artes and F. A. Tomas-Barberan, 1995. Changes in pomegranate juice pigmentation during ripening. Journal of the Science of Food and Agriculture, 5(68): 77-81.

[6]   Gil, M. I., F. A. Tomas-Barberan, B. Hess-Pierce, D. M. Holcroft and A.A. Kader, 2000. Antioxidant activity of pomegranate juice and its relationship with phenolic composition and processing. Journal of Agriculture and Food Chemistry, 48: 4581-4589.

[7]   Hodgson, R. W., 1917. The pomegranate. Calif. Agric. Expt. Sta. Bul., 276: 163-192. Nagy, P., P. E. Shaw and W.F. Wordowski, 1990. Fruit of Tropical and Subtropical Origin. Florida Science Source, Florida, USA., pp: 328-347.

[8]   Harde, H., W. Schumacher, F. Firbas and D. Deffer, 1970. Strasburg's Textbook of Botany. Chaucer, London. vegetables. London: Academic Press, pp: 1-39.

[9]   Kumar, G. N. M., 1990. Pomegranate. In S. Nagy, P. E. Shaw, and W. F. Wardowski (Eds.), Fruits of tropical and subtropical origin Auburndale, FL: Ag Sciences, Inc., pp: 328-347.

[10] Radunić, Mira, Maja Jukić Špika, Smiljana Goreta Ban, Jelena Gadže, Juan Carlos Díaz-Pérez, and Dan MacLean. "Physical and chemical properties of pomegranate fruit accessions from Croatia." Food chemistry 177 (2015): 53-60.

[11] Rania, J., H. Ne´jib, M. Messaoud, M. Mohamed and T. Mokhtar, 2007. Characterization of Tunisian pomegranate (Punicagranatum). cultivars using amplified fragment length polymorphism analysis Scientia Horticulturae.

[12] Rajasekar, Dhivyalakshmi, Casimir C. Akoh, Karina G. Martino, and Daniel D. MacLean. "Physico-chemical characteristics of juice extracted by blender and mechanical press from pomegranate cultivars grown in Georgia." Food Chemistry 133, no. 4 (2012): 1383-1393.

[13] Vahis Akbarpour, Khodayar Hemmati and Mehdi Sharifani. j. Agric. environ. Sci., 6(4): 411-416, 2009.

[14] Zaouay, Faten, Pedro Mena, Cristina Garcia-Viguera, and Messaoud Mars. "Antioxidant activity and physico-chemical properties of Tunisian grown pomegranate (Punica granatum L.) cultivars." Industrial Crops and Products 40 (2012): 81-89.

27

Communicating Food Quality and Safety Standards in the Informal Market Outlets of Pastoral Camel Suusa and Nyirinyiri Products in Kenya

Madete S. K. Pauline[1, *], Bebe O. Bockline[2], Matofari W. Joseph[1], Muliro S. Patrick[1], Mangeni B. Edwin[3]

[1]Department of Dairy and Food Science and Technology, Egerton University, Nakuru, Kenya
[2]Department of Animal Sciences, Egerton University, Nakuru, Kenya
[3]Livestock Sector, FAO Somalia, Nairobi, Kenya

Email address:
mdtpauline@yahoo.com (M. S. K. Pauline)

Abstract: The foods pastoral women process using indigenous knowledge have potential to enhance food security to households and health benefits to consumers but safety and quality concerns of consumers presents market barriers. This could be addressed through communicating food quality and safety standards. However, there are challenges in reaching the actors producing, processing and trading camel *Suusa* (spontaneously fermented milk) and *Nyirinyiri* (deep fried meat) because they are predominantly in the informal markets. This study identified communication strategies used to promote uptake of food quality and safety standards and level of awareness of actors along the value chains using data from survey, Focus Group Discussion (FGD) and Participatory appraisal. Results indicated low level of awareness among actors in the informal markets of Camel *Suusa* and *Nyirinyiri*. This can be attributed to underutilization of communication strategies to promote uptake of food quality and safety standards in the informal markets.

Keywords: Communication, Food Standards, Pastoral Women, Indigenous Technologies

1. Introduction

Food safety and quality standards are requirements in food systems for public health concerns at country of global level. Informal market outlets of camel *Suusa* and *Nyirinyiri* are some of the markets accessible only to the camel keeping communities. This is as a result of safety and quality concerns arising from growing application of agri-food value chain concept (Delia *et al.*, 2010; Akweya *et al.*, 2012) beyond the camel eating communities and niche markets. In rural households, most foods consumed are processed using indigenous technologies. This is true in the arid and semi-arid lands (ASAL) of Kenya where livestock and livestock products, mainly meat and milk are central in the diets and livelihoods.

In the arid and semi-arid lands (ASAL) of Kenya, pastoral households consume a substantial amount of food that is processed using indigenous technologies, especially meat and milk products. Of importance in the diets and livelihoods of pastoral households are camel meat (*Nyirinyiri*) and camel milk (*Suusa*) processed by women using indigenous knowledge food processing (IKFP). Processed camel meat (*Nyirinyiri*) and camel milk (*Suusa*) are also considered important for household food security and income from sales in the rural and urban market outlets. However, the pastoral women who play key role in indigenous knowledge food processing (IKFP) often have their families' food and income insecure because consumer acceptability of their IKFP products is low beyond traditional camel keeping communities and the quality is not ascertained.

Women's needs and knowledge should be taken into account when planning livestock projects and services. These will enhance their delivery of livestock services in pastoral areas. Identifying and supporting the roles, decision-making and capabilities of women as livestock owners, processors and users of livestock products are key aspects to promote women's economic and social empowerment and consequently a rural women's ability to break the cycle of poverty (IFAD, 2009, ILRI, 2008). Policy decisions should take women's roles and needs into account, and encourage a gendered perspective in livestock service delivery planning (Watson, 2010).

The existing policies on food safety discriminate against indigenous and informal food processors in favour of modern and formal processors. There is therefore the need to bridge the gap in uptake of food safety and quality standards in both informal and formal markets in order to break market barriers to niche markets for IKFP. This can be achieved through enhanced effective communication of the food safety and quality standards to actors in the value to increase awareness and uptake of the standards.

This study determined communication strategies that regulators use to promote awareness and uptake of food safety and quality standards and requirements in the informal market outlets of indigenous produced camel *Suusa* and *Nyirinyiri*.

2. Methodology

This study used survey, Focus Group Discussions (FGD) and participatory rapid appraisal approaches to sample actors along the camel Suusa and Nyirinyiri value chains in Isiolo, Marsabit and Nairobi (Eastleigh) where consumption of camel products is predominant in Kenya. Sampling targeted the value chain regulators, supporters and operators to identify

communication strategies which they use in creating awareness on food safety and quality standards. Data collected was subjected to cross tabulation, Chi-square test statistics and post hoc Anova.

3. Results and Discussion

3.1. The Age Bracket and Level of Education of Actors of Suusa and Nyirinyiri Value Chain

Indigenous Knowledge Food Products are obtained using traditional knowledge unique to a given society or culture. From figure 1 below majority of the actors of *Suusa* and *Nyirinyiri* value chains are above 50 years of age. These individuals are highly skilled and knowledgeable in milk and meat production, *Suusa* and *Nyirinyiri* processing as well as the benefits of consuming such products. The least age bracket with 4 individuals is composed mainly of the transporters of fresh unprocessed milk and meat from the grazing sites and also transportation of Suusa and Nyirinyiri to town. Transportation of milk is mostly done by young men at different times of the day; early morning, midday and at night.

Figure 1 below illustrates the age bracket under which the actors of *Suusa* and *Nyirinyiri* value chain fell (n=70). Of the 70, majority of the actors 28 fell in the age bracket of 51-60 years followed by 41-50 years with 21 actors. 3 of the actors were between the ages of 61-70 with at least four of them being above 70 years of age.

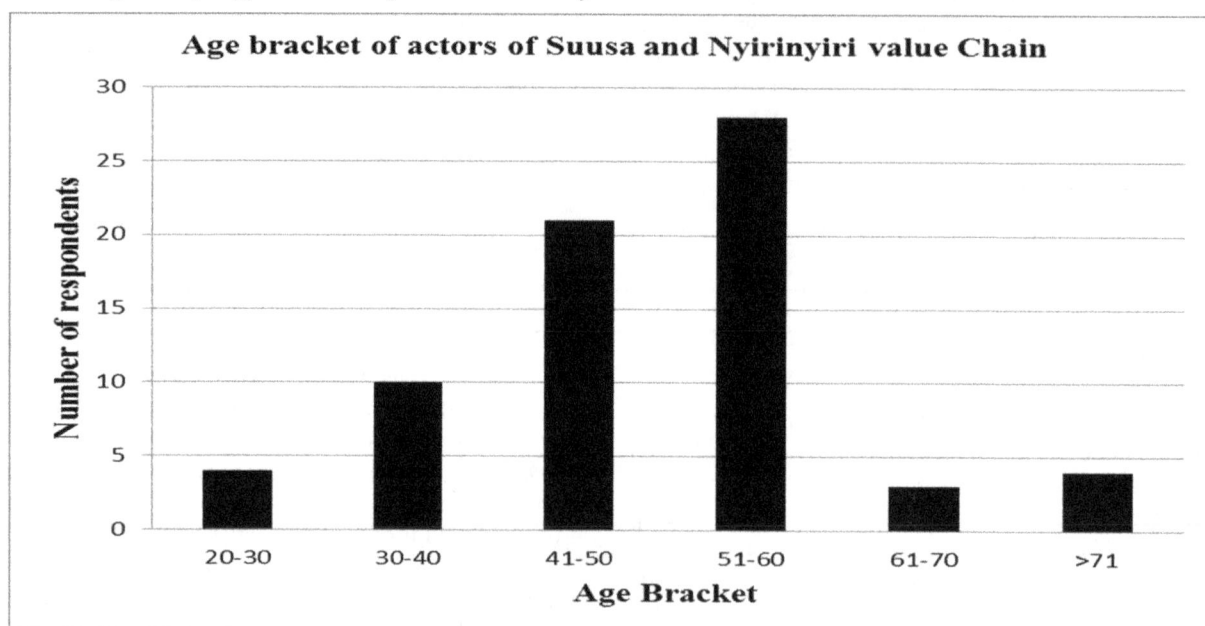

Figure 1. The age bracket of actors of Suusa and Nyirinyiri value chain.

The age of the value chain actors was complimented by the level of education and exposure of actors of *Suusa* and *Nyirinyiri* value chain to food quality and safety standards. Figure 2 below highlight the level of education of actors of *Suusa* and *Nyirinyiri* value chain. Of the actors of *Suusa* and *Nyirinyiri* value chain (n=70) only 37% had access to primary education, 7% secondary education, 1% tertiary

education, with majority of them at 52% having no access to education. 3% of the actors mentioned that they had a different form of education. Training done by Vital Camel Milk and VSF were termed as other forms of education. Through these trainings some food quality and safety aspects were passed onto actors of *Suusa* and *Nyirinyiri* value Chains.

Level of Education of actors of *Suusa* and *Nyirinyiri* Value Chain

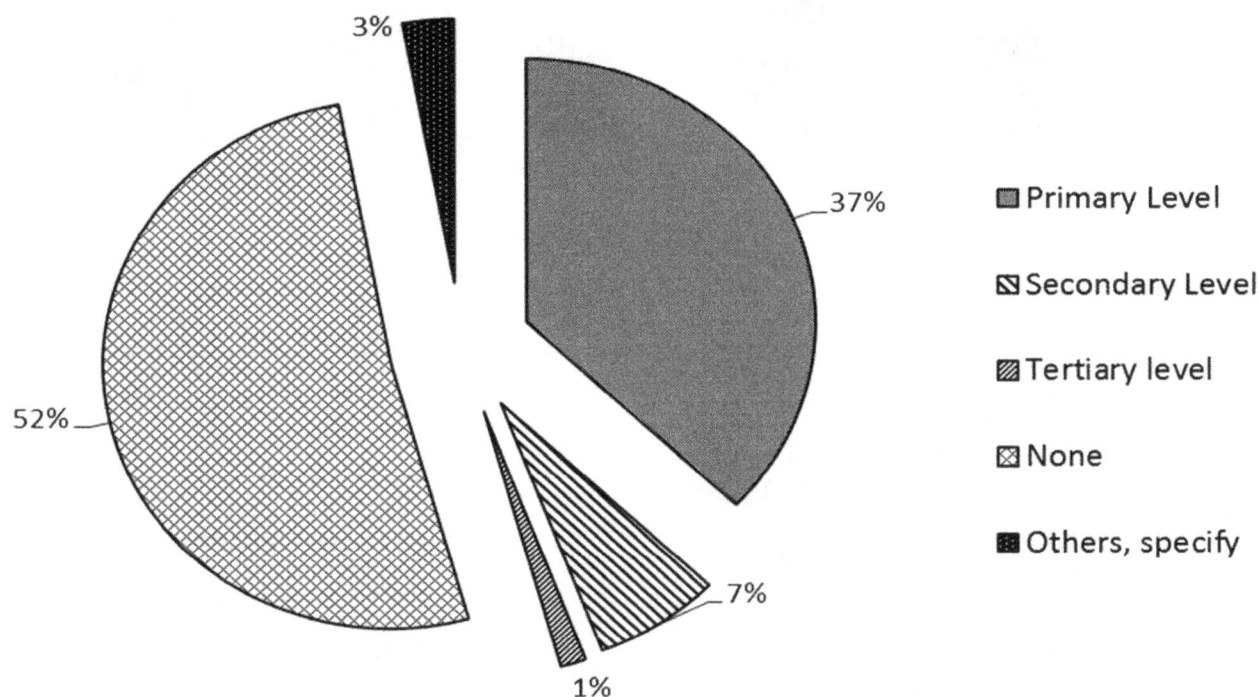

Figure 2. *The Level of Education of Actors of Suusa and Nyirinyiri value chain.*

Socio-economic characteristics of a farmer such as level of poverty, farming experience, age and education influence the adoption food quality and safety standards (Hudson and Hite, 2003). Looking at figure 2 above low levels of education or no education at all is hampering the understanding and uptake of food quality and safety standards in the informal market outlets of indigenous knowledge food products. Hence, for Camel *Suusa* and *Nyirinyiri* processors to adopt food safety and quality standards depends on factors such as access to institutional services, in-put supply markets and effective communication strategies (Khan 2005).

Age and level of education is related to production experience and also directly influences information access for formal and inform market outlets camel processors. In general, older farmers are less willing to try out new innovations or take risks compared to younger farmers, older farmers are less likely to engage in simultaneous receiving and providing of information, perhaps due to their low ability to communicate (Katungi, 2006).

Individuals can perceive the risk from the same hazard very differently. Some of the public may disagree with risk assessors and managers regarding important hazard characteristics, the relative magnitude or severity of the risks associated with those hazards, the priority of risks, and other issues. Other segments of the public also may not pay attention to risk information if the message does not address their actual concerns, but instead addresses only technical risk assessments provided by the experts (FAO/WHO, 2002).

3.2. Level of Awareness About Food Safety and Quality Standards

Figure 3 illustrates level of awareness on food safety and quality standards among the actors in *Suusa* and *Nyirinyiri* value chains. Of all the sampled actors (n=70), the level of awareness was low (23%) and peaked at 30% among processors and retailers (market traders) with transporters being the least aware (13%).

For those actors aware of the food safety and quality standards, specific aspects known to them were further examined and results are illustrated in Figure 4. Actors' awareness was highest on packaging conditions (25%) and on processing hygiene and purchasing conditions (19%) and lowest on environmental and transportation hygiene (6%).

In Kenya, the standard requirement bar transportation of milk with other goods and use non-food grade containers (Lusato, 2006). However, these requirements are often violated as products are traditionally transported in plastic jericans with other goods and animals to the destined markets in vans or buses from Isiolo and Marsabit to Nairobi urban markets.

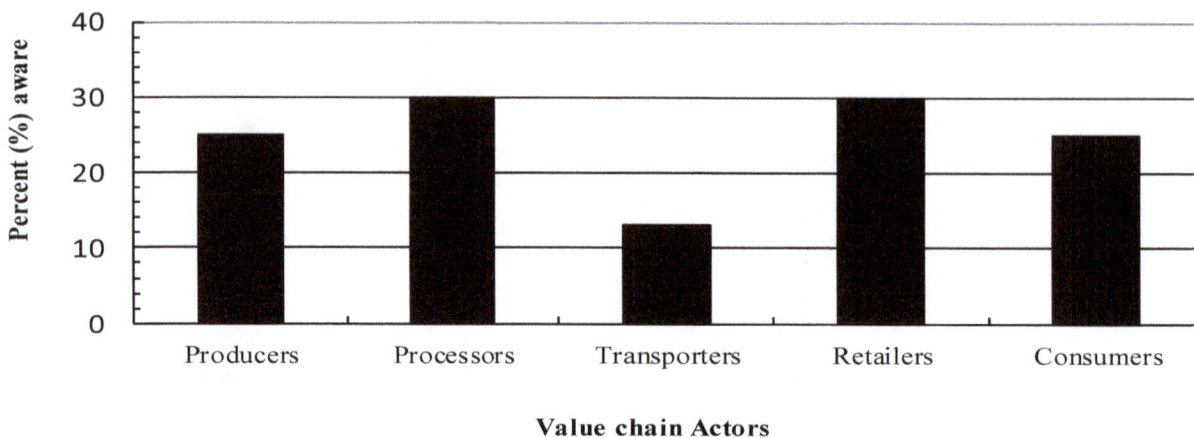

Figure 3. *Awareness about food safety and quality standards among actors in value chains of camel Suusa and Nyirinyiri sampled in the e informal markets outlets.*

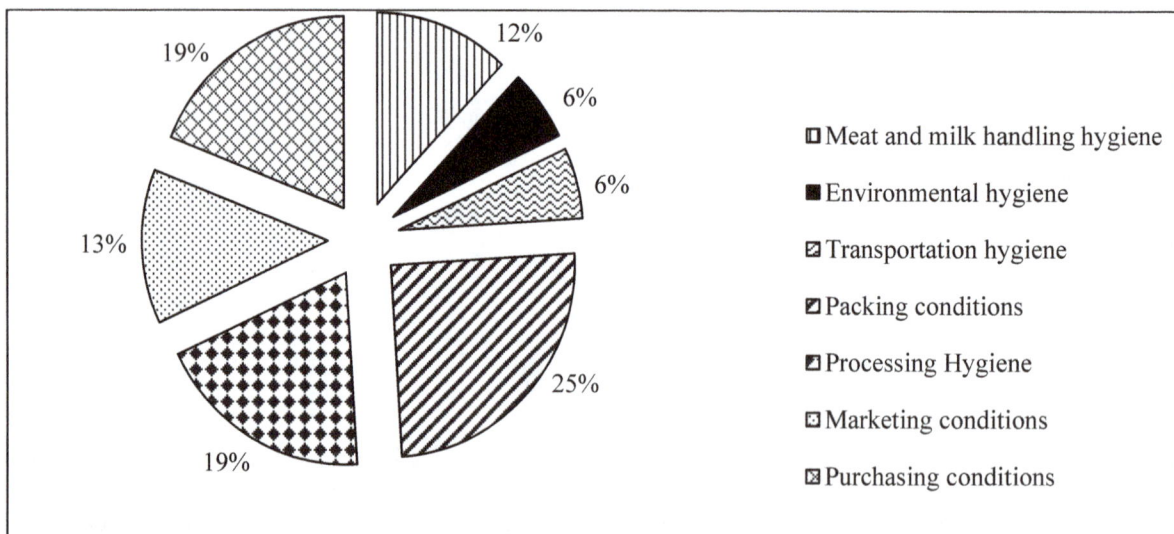

Figure 4. *Specific aspects know to the actors of Suusa and Nyirinyiri value chain.*

Plate 1. *Handling practices of the milk and meat products in pastoral areas.*

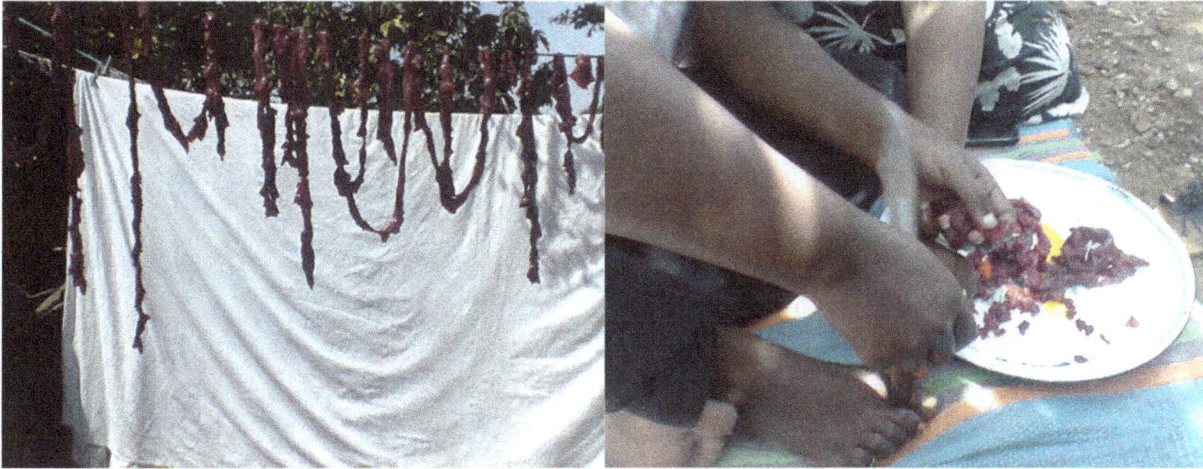

Plate 2. Highlight the handling hygiene of fresh camel meat just before processing using indigenous knowledge.

Plate 3. Captures Eastleigh market in Nairobi for Suusa and Nyirinyiri.

Plate 1 illustrates the poor milk and meat handling hygiene practices at production among the sample actors using unclean plastic containers which is attributable to lack of water.

Herders do milk camels on transit; use discolored dirty plastic containers without adequate washing of hands, udder and containers before milking. The milking environment is generally unhygienic with a lot of animal waste and polythene wastes. Although milk was commonly boiled before consumption, handling practices and wide spread lack of refrigeration across the value chain potentially pose increased health risks. In meat handling the slaughter houses had partial covered roofing, insufficient and unclean water supply, unwashed and cleaned slabs and workers with dirty protective clothing. These observations reflect unhygienic handling of carcasses related to the environment (Plate 1-3).

3.3. Communication Strategies for Creating Awareness on Food Safety and Quality Standards

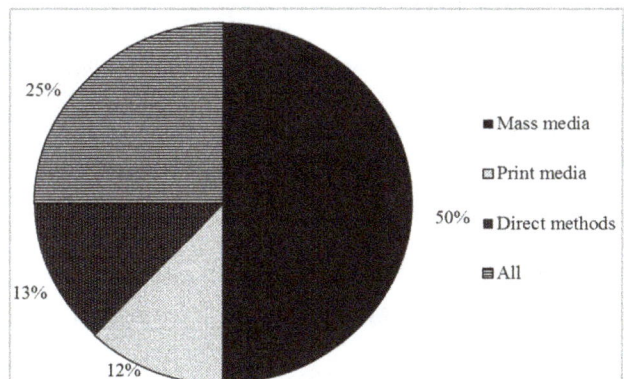

Figure 5. Communication strategies used by regulators and information service providers in the formal market outlet

Figure5 shows that of the various communication strategies used by regulators and information service providers in promoting uptake of food quality and safety standards in the informal markets, mass media dominated (43%) complemented with print media and direct contacts (29%). Maina (2006) noted that using radio has achieved impressive results in the delivery of useful information to poor people. FM stations have become handy tools in small scale agriculture in rural areas and in facilitating dissemination of market information but have been less exploited in communicating food quality and safety standards in the informal market outlets of *Suusa* and *Nyirinyiri* to raise awareness and upscale uptake.

Survey results in figure 5 could not identify any communication strategy specifically targeted to promoting awareness and uptake of food safety and quality standards in the informal market outlets of camel *Suusa* and *Nyirinyiri*. The communication strategies mentioned in Figure 5 are best suited to communicating formal food quality and safety standards as part of risk communication in formal market outlets. Moreover, the standards being communicated are generated are in line with international standards mainly because they deal with large industries and firms. Indigenous knowledge has not been in cooperated into their knowledge source hence it has been given less attention. This therefore calls for an in-depth knowledge on how to integrate indigenous and scientific knowledge with an aim of developing safety and quality standards suited for the informal market outlets of camel *Suusa* and *Nyirinyiri*.

4. Conclusion and Recommendation

Adoption of food quality and safety standards remains elusive in the informal market outlets. Communication strategies used by regulators do not adequately cater for the informational needs in the informal market outlets of camel *Suusa* and *Nyirinyiri*. This study aimed at identifying communication strategies used in communicating food quality and safety standards in line with the awareness on food quality and safety standards in the informal market outlets. There is a general low level of awareness about food safety and quality among actors in the value chains of camel *Suusa* and *Nyirinyiri* in the informal market outlets. This can be attributed to underutilization of formal communication strategies to promote uptake of food quality and safety standards in the informal markets. There is the need therefore for reviewing formal communication strategies with potential to reaching more actors in the informal markets of Indigenous Knowledge Food Products (IKFP). With increased outreach there is a perceived benefit for both pastoral women processors and consumers informal market outlets.

Acknowledgements

This work was supported by a grant from RUFORUM which the authors are grateful to. The authors would also like to thank Egerton University and pastoral women processors in ASAL areas of Isiolo and Marsabit Kenya and consumers of indigenous camel products of *Suusa* and *Nyirinyiri*.

References

[1] Akweya B. A., Gitao C. G., Okoth M. W. (2012). The acceptability of camel milk and milk products from North Eastern province in some urban areas Kenya. African Journal of Food Science Vol 6(19) pg 465-473.

[2] Delia G, Makita K., Kang'ethe E. K., Bonfoh B. (2010). Safe Food, Fair Food: Participatory Risk Analysis for improving the safety of informally produced and marketed food in sub Saharan Africa http://www.ilri.org/ilrinews/index.php/archives/tag/delia-grace

[3] FAO/WHO. (2002). Post- harvest technology and food quality. Part 6.Science and technology for sustainable development FAO/WHO. Joint FAO/WHO Food Standards Programme Report of the Sixteenth Session of FAO/WHO Coordinating Committee for Africa, Codex Alimentarius Commission FAO. Rome. 25 – 28 January 2005.

[4] Hudson, D. and Hite, D. (2003). Willingness to Pay for Water Quality Improvements: The Case of Precision Application Technology. Journal of Agricultural Resource Economics, 27: 433–449.

[5] IFAD. (2009). Rota, A. and Sperandini, S. "Value chains, linking producers to the markets", in Livestock Thematic Papers: Tools for project design. Rome: International Fund for Agricultural Development (IFAD).

[6] International Livestock Research Institute (ILRI). (2008). Policy change in dairy marketing in Kenya: Economic impact and pathway to influence research. www.ilri.org/ilribriefs/index.php/pdf. Retrieved on 16/03/2013.

[7] Katungi, E. (2006). Gender, Social Capital and Information Exchange in Rural Uganda IFPRI and Melinda Smale, IFPRI (International Food Policy Research Institute) CAPRi Working Paper (59), University of Pretoria. Uganda. Retrieved on March 01, 2012 from: <http://www.capri.cgiar.org/pdf/capriwp59.pdf>

[8] Khan, S. A. (2005). Introduction to Extension Education in Memon, R. A. and E. Bashir (eds.). Extension Methods (3rd ed.). National Book Found, Islamabad, Pakistan

[9] Lusato K. R. (2006). Hygienic milk handling, processing and marketing: reference guide for training and certification of small-scale milk traders in Eastern Africa. ILRI. Nairobi, Kenya.http://www.ilri.org/Link/Publications/Theme3/Trainer GuideVol-1_Cmprss.pdf.

[10] Maina w. Lucy. (2006). African media initiative Kenya. Research findings and conclusion 2006.BBC world service. http://africanmediainitiative.org/content/2013/07/22/AMDI-BBC-summary-report.pdf.

28

Study on Present Status of Fish Biodiversity in Wetlands of Sylhet District, Bangladesh

Mohammed Ariful Islam[1], Md. Zahidul Islam[1], Sanzib Kumar Barman[1], Farjana Morshed[1], Sabiha Sultana Marine[2]

[1]Department of Aquatic Resource Management, Faculty of Fisheries, Sylhet Agricultural University, Sylhet, Bangladesh
[2]Department of Fisheries Technology and Quality Control, Faculty of Fisheries, Sylhet Agricultural University, Sylhet, Bangladesh

Email address:
arif.fisheries@gmail.com (M. A. Islam)

Abstract: The study was conducted to identify the present status of fish biodiversity in the wetlands of Sylhet district for a period of 12 months from January 2014 to December 2014. It was done by questionnaire interviews (QI) of fishers, focus group discussions (FGD), and key informant interviews (KII) and secondary data collection. During the study period, a total of 58 fish species under 21 families were recorded. The species availability status was remarked in four categories and obtained as 24 commonly available, 16 moderately available, 18 rarely available species. Highest number of commonly available species was found in October to December and lowest number of commonly available species was observed in March to April. Among 54 threatened fish species listed by IUCN Bangladesh, about 30 species were found 10-15 years ago in those wetlands but only 23 were found during the study period. It is revealed that there has been gradual reduction of fish diversity in the wetlands of the area of Sylhet district and average fish catch per fisherman per day was also reduced. Community based fisheries management, fishing gears maintenance, sanctuary establishment and management, implementation of fish acts and regulations, stocking of fish fingerling in the open waters, dredging of beels and raising public awareness can play a great role in conserving fish biodiversity.

Keywords: Biodiversity Reduction, Species Availability, Threatened Species, Conservation, Questionnaire Interview

1. Introduction

Aquatic biodiversity has enormous economic and aesthetic value and is largely responsible for maintaining and supporting overall environmental health *(Hossain, 2012)*. Wetlands are one of the world's key natural resources *(Khan et al., 1994)*. It is the transition between land and water and is the most productive ecosystems in the world. The biodiversity of the wetland ecosystem is variable in the world; it encompasses the range of living things, the degree of genetic variation, and the wealth of different habitats within a particular ecosystem.

The Ramsar Convention (1971) has defined wetlands as – "areas of marsh, fen, peat land, or water, whether natural or artificial, permanent or temporary, with water that is static or flowing, fresh, brackish or salt, including areas of marine water the depth of which at low tide does not exceed six meters." Bangladesh is a home to at least 265 freshwater fish species *(Rahman, 2005)*. Huge number of wetlands in various forms viz. rivers, haors, boars, beels, pond, ditch etc. support these large number of fish species. Among them, wetlands of Sylhet district are so important that contributes a huge amount of fish to the people of the country. The major wetlands in Sylhet district are Hailkar, Jilkar, Patharchauli, Jainkar, Chauldhani, Balai, Muria, Erali and Damrir haors.

At present time, reduction in the abundance and kinds of fish species from the inland waters or wetlands of Bangladesh is a burning issue in the country *(Galib et al. 2009, Imteazzaman and Galib 2013)*. However, a total of 54 fish species of Bangladesh have been declared threatened by IUCN *(IUCN Bangladesh 2000)*. All these findings clearly indicate the need for water body specific detailed biodiversity studies which is essential to assess the present status of fish biodiversity and sustainable management of a body of water *(Galib et al. 2013a; Imteazzaman and Galib 2013)*. Though such type of research efforts are much common in neighbor countries like India *(Dahanukar et al. 2012, Kharat et al. 2012, Baby et al. 2010, Jadhav et al. 2011, Patra 2011, Johnson and Arunachalam 2009, Heda 2009, Saha and Bordoloi 2009)* but very few in Bangladesh.

To the best knowledge of the authors no previous research work has been conducted on fish biodiversity of Wetlands of Sylhet Districts, Bangladesh. So, this study will be very significant for the assessment of present status of fish biodiversity of Wetlands of Sylhet district, Bangladesh.

2. Methodology

2.1. Description of Study Area

Wetland fish species diversity was recorded in Sylhet district in Bangladesh from January to December in 2014. The selected wetlands were Hailkar, Jilkar, Patharchauli, Jainkar, Chauldhani, Balai, Muria, Erali and Damrir haors. The geographical location of the Sylhet district is an area of 3452.07 km^2 (1332.85 sq. miters); it is bounded by Sunamgonj district on the west, Moulvibazar district on the south and Habogonj district on the south-west. The sites of the study are illustrated in Figure 1.

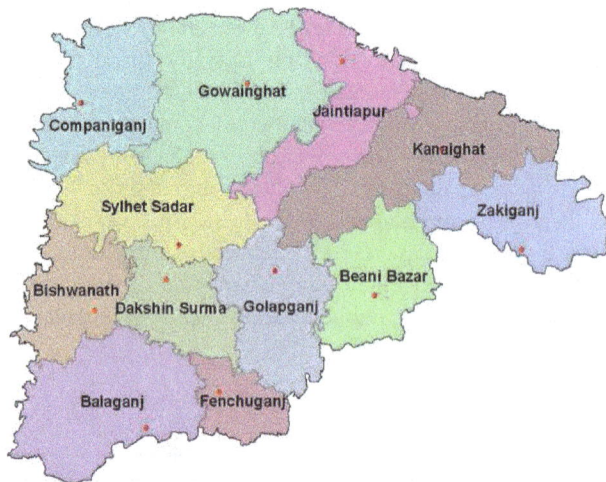

Figure 1. Location of wetlands at different upazilas of Sylhet district.

2.2. Data Collection

The study was based on field survey method where an appropriate questionnaire was prepared and used for collecting data from villagers of the surrounding study area under Sylhet district. During collection of data, both primary and secondary sources were considered to interpret the results.

Primary data were collected from 40 randomly selected fishermen through questionnaire interviews (QI) and focus group discussions (FGD) where Upazila Fisheries Officer (UFO), union parishad chairman & members, leaders of the fisher community, fish market leaders, fish traders, fry traders and community people of the selected wetlands area were also present. The secondary information was collected from upazila fisheries office under Sylhet district, district fisheries office of Sylhet, books, journals and others. After collecting data, it was cross-checked through key informant interviews (KII) with Upazila Fisheries Officer (UFO), District Fisheries Officer (DFO), school teachers, local leaders and NGO workers in the study area. Finally data were analyzed by using Microsoft office excels 2010.

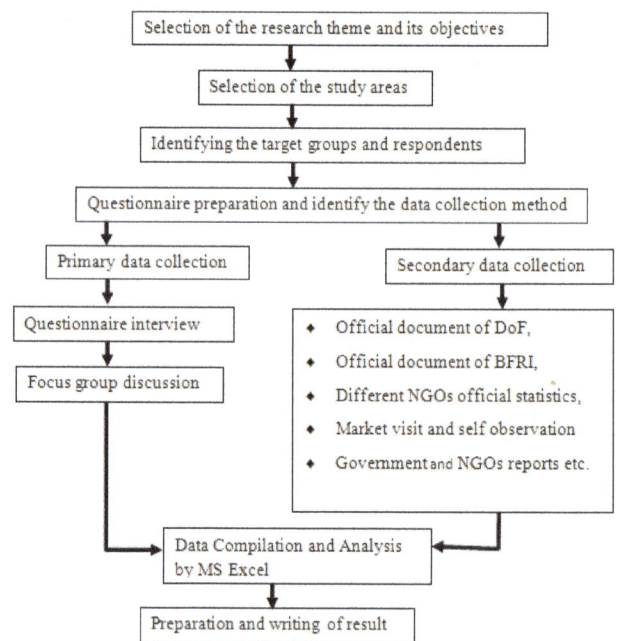

Figure 2. The design of the present study involved the steps.

3. Results and Discussion

According to the statement of local fishermen, a total of 58 fish species under 21 families were recorded. The recorded fish species with accurate taxonomy were identification by the cross-matching of definite fish characters from the text book, named "Freshwater Fishes of Bangladesh" (A. K. Ataur Rahman, Department of Fisheries, Matshya Bhaban, Dhaka). The recorded fish species are represented in the following table with their biodiversity status and IUCN status.

Table 1. List of recorded fish species with their status.

Sl. No.	Family	Local Name	Common Name	Scientific Name	Biodiversity status	IUCN Status
1	Cyprinidae	Rohu	Indian Major Carp	*Labeo rohita*	CA	NO
2	Cyprinidae	Carpio	Common Carp	*Cyprinus carpio*	CA	EX
3	Cyprinidae	Kalibaush	Black Rohu	*Labeo calbasu*	CA	EN
4	Bagridae	Bujuri	Long Bled Catfish	*Mystus tengra*	CA	NO
5	Bagridae	Tengra	Striped Dwarf Catfish	*Mystus vittatus*	CA	NO
6	Siluridae	Boal	Freshwater Shark	*Wallago attu*	CA	NO

Sl. No.	Family	Local Name	Common Name	Scientific Name	Biodiversity status	IUCN Status
7	Siluridae	Modhu Pabda	Butter Catfish	Ompok pabda	CA	EN
8	Siluridae	Pabda	Pabo Catfish	Ompok pabo	CA	EN
9	Clariidae	Magur	Walking Catfish	Clarius batrachus	CA	NO
10	Heteropneustidae	Shing	Stinging Catfish	Heteropneustes fossilis	CA	NO
11	Cyprinidae	Tit Punti	Ticto Barb	Puntius ticto	CA	VU
12	Cyprinidae	Teri Punti	One Spot Barb	Puntius terio	CA	NO
13	Cyprinidae	Jat Punti	Spot Fin Swamp Barb	Puntius sophore	CA	NO
14	Channidae	Taki	Spotted Snakehead	Channa punctatus	CA	NO
15	Channidae	Shol	Snakehead Murrel	Channa striatus	CA	NO
16	Mastacembelidae	Guchi Baim	Striped Spiny Eel	Macrognathus pancalus	CA	NO
17	Mastacembelidae	Tara Baim	One Striped Spiny Eel	Macrognathus aculeatus	CA	VU
18	Anabantidae	Baro Khalisha	Striped Gourami	Colisa fasciatus	CA	NO
19	Anabantidae	Chota Khalisha	Honey Gourami	Colisa chuno	CA	NO
20	Anabantidae	Koi	Climbing Perch	Anabas testudineus	CA	NO
21	Ambassidae	Gol Chanda	Indian Glass Fish	Parambassis ranga	CA	VU
22	Nandidae	Meni	Mud Perch	Nandus nandus	CA	VU
23	Cobitidae	Gutum	Guntea Loach	Lepidocephalichthys guntea	CA	NO
24	Palaemonidae	Sada Icha	Prawn	Macrobrachium sp.	CA	NO
25	Cyprinidae	Catla	Indian Major Carp	Catla catla	MA	NO
26	Cyprinidae	Mrigal	Indian Major Carp	Cirrhinus cirrhosus	MA	NO
27	Cyprinidae	Grass Carp	Grass Carp	Ctenopharyngodon idella	MA	EX
28	Cyprinidae	Silver Carp	Silver Carp	Hypophthalmicthys molitrix	MA	EX
29	Cyprinidae	Mola	Carplet	Amblypharyngodon mola	MA	NO
30	Cyprinidae	Darkina	Flaying Barb	Esomus danricus	MA	DD
31	Channidae	Cheng	Asiatic Snakehead	Channa orientalis	MA	VU
32	Bagridae	Golsha	Long Whiskered Catfish	Mystus cavasius	MA	VU
33	Channidae	Gozar	Giant Snakehead	Channa marulius	MA	EN
34	Mastacembelidae	Baro Baim	Two-track Spiny Eel	Mastacembelus armatus	MA	EN
35	Sybranchidae	Cuchia	Gangetic Mud Eel	Monopterus cuchia	MA	VU
36	Ambassidae	Lamba Chanda	Elongated Glass Perchlet	Chanda nama	MA	VU
37	Gobiidae	Bele	Bar Eyed Goby	Glossogobius giuris	MA	NO
38	Notopteriidae	Foli	Bronze Featherback	Notopterus notopterus	MA	VU
39	Cobitidae	Bou Rani	Bengal Loach	Botio dario	MA	EN
40	Tetraodontidae	Potka	Ocellated Puffer fish	Tetraodon cutcutia	MA	NO
41	Cyprinidae	Goniya	Kuria Labeo	Labeo gonius	RA	EN
42	Cyprinidae	Bata	Minor Carp	Labeo bata	RA	EN
43	Cyprinidae	Dhela	Cotio	Osteobrama cotio	RA	EN
44	Cyprinidae	Chela	Finescale Razorbelly Minnow	Chela phulo	RA	NO
45	Cyprinidae	Shar Punti	Olive Barb	Puntius sarana	RA	CR
46	Bagridae	Rita	Rita	Rita rita	RA	CR
47	Bagridae	Ayre	Long Whiskered Catfish	Mystus aor	RA	VU
48	Bagridae	Gagla	Menoda Catfish	Hemibagrus menoda	RA	NO
49	Schilbeidae	Gharua	Gharua Bacha	Clupisoma garua	RA	CR
50	Schilbeidae	Batashi	Indian Potasi	Pseudeutropius atherinoides	RA	NO
51	Schilbeidae	Kajoli	Gangetic Ailia	Ailia coilia	RA	NO
52	Pangasidae	Thai Pangus	Sutchi Catfish	Pangasius hypophthalmus	RA	EX
53	Clupeidae	Chapila	Indian River Shad	Gaduasia chapra	RA	NO
54	Anabantidae	Lal Khalisha	Dwarf Gourami	Colisa lalia	RA	NO
55	Ambassidae	Lal Chanda	Indian Glass Perchlet	Parambasis lala	RA	EN
56	Cichlidae	Tilapia	Mozambique Tilapia	Oreochromis mossambicus	RA	EX
57	Hemiramphidae	Ekthute	Congaturi Halhbeak	Hyporamphus limbatus	RA	NO
58	Palaemonidae	Golda	Prawn	Macrobrachium rosenbergii	RA	NO

CA=Commonly available, MA=Moderately available, RA=Rarely available.
CR=Critically endangered, EN=Endangered, VU=Vulnerable, NO=Not threatened, DD=Data deficient and EX=Exotic species.

4. Recommendation

- Fish sanctuary should be established in the selected wetlands area before breeding season.
- No fishing in the fish sanctuary in a defined time (several months) should be ensured.
- Overfishing should be prohibited in the wetlands area.
- Banded fishing gears (specially banded fishing nets) should be prohibited in the selected areas.
- Community based fisheries management (CBFM) should be established in the haor areas.
- Alternative earning source of the people of haor areas should be provided during banded season of fishing.
- Increasing awareness among the much people of surrounding wetland areas.

5. Conclusion

Wetlands of Sylhet district are generally considered as a highly diversified zone of Bangladesh mainly for its rich aquatic biodiversity. According to the statement of the respondent fishermen, the total fish biodiversity is reducing drastically in the wetlands of Sylhet district due to lake of proper management. As a consequence, wetland ecosystem protection is important for species conservation and the protection of a sustainable environment.

References

[1] Baby F, Tharian J, Ali A and Raghavan R (2010). A checklist of freshwater fishes of the New Amarambalam Reserve Forest (NARF), Kerala, India. Journal of Threatened Taxa 2(12): 1330-1333.

[2] Dahanukar N, Paingankar M, Raut RN and Kharat SS (2012). Fish fauna of Indrayani River, northern Western Ghats, India. Journal of Threatened Taxa 4(1): 2310-2317.

[3] Galib SM, Naser SMA, Mohsin ABM, Chaki N and Fahad MFH (2013a). Fish diversity of the River Choto jamuna, Bangladesh Present status and conservation needs. International Journal Biodiversity and Conservation 5(6): 389-395. DOI: 10.5897/IJBC2013.0552.

[4] Hossain, M., 2012. Biodiversity of Threatened Fish Species of Choto Jamuna River in Badalgachhi Area under Naogaon District. MS Thesis, Department of Fisheries Management, Bangladesh Agricultural University, Mymensingh, pp: 30-53.

[5] IUCN Bangladesh (2000). Red book of threatened fishes of Bangladesh, IUCN- The world conservation union. xii+116 pp.

[6] Jadhav BV, Kharat SS, Raut RN, Paingankar M and Dahanukar N (2011). Freshwater fish fauna of Koyna River, northern Western Ghats, India. Journal of Threatened Taxa 3(1): 1449-1455.

[7] Johnson JA and Arunachalam M (2009). Diversity, distribution and assemblage structure of fishes in streams of southern Western Ghats, India. Journal of Threatened taxa 1(10): 507-513.

[8] Patra AK (2011). Catfish (Teleostei: Siluriformes) diversity in Karala River of Jalpaiguri District, West Bengal, India. Journal of Threatened Taxa 3(3): 1610-1614.

[9] Rahman, AKA (2005). Freshwater Fishes of Bangladesh, second edition. Zoological Society of Bangladesh, University of Dhaka, Dhaka, Bangladesh, 263 pp.

[10] Saha S and Bordoloi S (2009). Ichthyofaunal diversity of two beels of Goalpara District, Assam, India. Journal of Threatened Taxa 1(4): 240-242.

[11] Rahman, A. K. A. (1989). Freshwater Fishes of Bangladesh. Zoological Society of Bangladesh. Department of Zoology, University of Dhaka. Dhaka-1000.

[12] Khan, M. S., Haq, S., Rahman, A. A., Rashid S. M. A. and Ahmed. H., 1994. Wetlands of Bangladesh. Bangladesh Centre for Advanced Studies & Nature Conservation Movement.

29

Variations in Climate Change Indicators and Implications on Forest Resources Management in Taraba State, Nigeria

Agbidye Francis Sarwuan[1,*], **Emmanuel Zando Angyu**[1], **Egbuche Christian. Toochi**[2]

[1]Department of Forest Production and Products, University of Agriculture, Makurdi, Nigeria
[2]Department of Forest and Wildlife Technology, Federal University of Technology Owerri, Imo State, Nigeria

Email address:

fagbidye@yahoo.com (A. F. Sarwuan)

Abstract: This study was carried out to investigate the variation in climate change indicators of specific interest of rainfall and temperature) in Taraba State from 1991 to 2011. Secondary data were obtained on these climate change indicators from the Nigerian Meteorological Agency (NIMET), Jalingo, Taraba State, Nigeria. Data obtained on amount of rainfall, maximum and minimum temperature and temperature range were subjected to Analysis of Variance (ANOVA). The results showed no significant differences in all the variables considered ($p \leq 0.05$). This shows that the much talked about climate change in Taraba State is speculative and not based on empirical data. However, to forestall the possible menace of climate change in the study area, it was recommended that policies and programmes supporting afforestation and proper environmental management should be put in place. Also, there is need for a widespread enlightenment of the citizens on climate change to enable them to change some of the human behaviours contributing to climate change.

Keywords: Climate Change, Indicators, Temperature, Rainfall, Implications, Forest Resources

1. Introduction

Climate is the composite of the day-to-day weather condition and of atmospheric elements of a place or region, over a long period of time, at least about 30-35 years [1]. It is further accepted that climate change has been defined as a change in the state of the climate that can be identified (. by using statistical tests) by changes in the mean and/or the variability of its properties, and that persists for an extended period typically decades or longer [2]. Global climate change and warming however, is supported as the rising earth's surface temperature caused by the increase in the concentration and accumulation of green house gases in the atmosphere. Today, climate change is among the major global environmental problems threatening the survival of the entire race. It is seen as a key phenomenon of our times, as a set of events which affects forestry and may alter the lives of human kind in general [2].

Climate change is controlled by factors such as temperature, precipitation (rainfall), solar energy, humidity, wind, air pressure, altitude, latitude, distribution of land and water, mountain barriers and ocean currents [3]. Temperature

and rainfall are among the major factors that influence weather conditions of a place and has been documented that major indicators of climate change are the increase in the mean surface temperature [4] and increase in mean annual rainfall [5]. Global records of earth's surface temperature have indicated a rise of about 0.3-7°C in temperature in the last century [6]; [7]. Precipitation has increased by about 1% across the world in the last century, while high latitudes tend to see more rainfall. Temperature changes affect plants in the following ways: Increase in temperature may prevent some plants from growing in their natural areas; abnormal increases in temperature have adverse effects on productivity of forest crops and plants; thereby alters flowering pattern of crops and forest plants species, resultant starvation leads to death and migration of wild species; low temperature could lead to formation of ice which hinders the germination of plants and operation of some important soil microbes that act as decomposers among others. Apart from this, some authors have suggested that climate change could lead to dieback in existing or future forests due to water stress, insect infestations, or fires [8]; [9]; [10].

The association between climate and forestry has widely been discussed in the literature [11]; [12]. There is great concern over the implications of a changing climate to the forest industry, particularly since the species and provenances planted at present and in the recent past reflect the current climate. It is now widely recognized that climate change is likely to have strong influences on the structure and function of the forests [13]; [2]. These impacts can be categorized into three general areas that include forest productivity changes, ecosystem disturbances and changes in forest species distributions. Productivity changes are adjustments in the productivity of the forest which alter the growth rates of forest species (in either a positive or negative way), while changes in disturbance influence the standing stock of timber and non-timber species through pest infestation, forest fires, wind-throw, and ice damage. Changes in species distribution results from shift in climate, which ultimately alters the optimal geographic location of different species [14].

There have been widespread speculations about climate change all over the world which in most cases not backed by relevant data that has resulted in such questions as Is climate change real or speculative? This research sought to provide the answer to this question particularly as it relates to Taraba State. This study was therefore carried out to determine whether there is a significant change or variation in the amount and pattern of rainfall and temperature in Taraba State from the inception of the State in 1991 to 2011. The analyses of climate change indicators (particularly rainfall and temperature) in the study area are is of paramount importance to the people of the State. As an agrarian State, climate change will have negative impacts on the overall economy of the State. Therefore, this study was to throw more light on whether there is significant variation in the climate of the State judging from the analyses of rainfall and temperature figures. This will enable the people to take precautions against the possible effects of climate change in the area.

2. Methodology

2.1. The Study Area

Taraba State is located in the North-East geographical zone of the country, with its head-quarters in Jalingo. It has sixteen (16) Local Government Areas (LGAs). The State has a total land mass of 51,000 kilometer square. It lies roughly between latitude 6^030^1 and $9^{\circ}36^1$N and longitude 9^010^1 and 11^05^1E. It is bounded on the North-East by Adamawa State and the West and South-East by Plateau and Benue States respectively. On its east border is the Republic of Cameroun. According to the 2006 Census figures released by the National Population Commission (NPC), Taraba State has a population of 2, 294, 800 people [16].

The temperature of Taraba State ranges between 33°C and 37°C; however, in the driest month (March), it could rise to 40°C. The amount of rainfall in the State ranges between 1350mm in the North and 1650mm in the South. The rainy season starts in April and ends in October, while the dry season begins in November and terminates in March. The dry season reaches its peak in January and February when the

dusty North-East trade wind blows across the State. The climate, soil and hydrology of the State provide a conducive atmosphere for the cultivation of most staple food crops, grazing lands for animals and fresh water for fishing and forestry. As a result of its agrarian nature, the predominant population of the State engages in farming as an occupation. About three quarters (75%) of the people are farmers while an estimated one quarter (25%) are engaged in other economic activities [15].

The vegetation of Taraba state comprises of three types of vegetation zones namely: the guinea savannah, which is marked mainly by forest and tall grass is found in the Southern part of the state like Wukari, Takum and Dunga, the sub-sudan type characterized by short grasses is found in Jalingo, Lau, and Ardo-kola, interspersed with short trees, while the semi-temperate zone is found in the central part of the State [15].

2.2. Data Collection and Analyses

The data used for this research was secondary data of monthly rainfall and temperature from 1991 to 2011 which were obtained from the Nigerian Meteorological Agency, Jalingo, Taraba State. The data obtained were analyzed using both one way and two way analysis of variance (ANOVA) with mean separation by Fischer's Least Significance Difference (LSD). All probability levels were at 5% level of significance. The statistical package used was Genstat Discovery Edition 4.

3. Results

3.1. Mean Annual Maximum Temperature (°C)

The results on maximum temperature are presented in Table 1 and Figure 1. There were no significant differences noticed and the temperature trend remained somewhat stable since 1991 to 2011. However, a gradual increase was noticed in the years 1998 (Tmax(°C) = 34.02±3.14), 2003 (Tmax (°C) = 34.60±3.21), 2006 (Tmax(°C) = 34.23±3.33), 2009 (Tmax(°C) =34.04 ±3.10) and 2010 (Tmax(°C) = 34.42±3.44). A slight drop was also experienced in the maximum temperature of 1992 (Tmax(°C) = 32.99±2.11) and 1994 (Tmax(°C) = 32.85±2.66).

3.2. Mean Annual Minimum Temperature (°C)

As indicated in Table 1, the results on minimum temperature showed no significant differences. There was a slight increase in the minimum temperature in 1998 (Tmin (°C) = 23.08±2.39) and2005 (Tmin (°C) = 23.38±2.59).

3.3. Temperature Range (°C)

Results on temperature range within the period of study (1991 - 2011) also recorded no significant differences at 5% level of probability even though the results were fluctuating (Table 1, Figure 1). There was a slight rise in the temperature range in 2003 (Trange (°C) = 12.01± 3.84).

3.4. Rainfall (mm)

The results on rainfall as indicated on Table 1 and Figure 2 revealed that there were no significant differences in the rainfall distribution for the periods under study (1991-2011). Although a slight increase in rainfall distribution was recorded in 1993 (113.30 ± 109.70mm) and 2009 (129.00 ± 128.60mm), statistically, the differences were not significant.

Table 1. *Mean Annual Temperature and Rainfall Values in Taraba State from 1991-2011.*

Year	Tmin (°C)	Tmax(°C)	Trange (°C)	Rainfall (mm)
1991	22.58±2.34	33.27±2.70	10.69±3.59	88.47±88.46
1992	22.06±2.33	32.99±2.11	11.00±3.37	78.78±75.56
1993	22.66±2.08	33.03±2.02	10.37±2.49	113.30±109.70
1994	22.48±2.41	32.85±2.66	10.82±2.94	84.02±93.91
1995	22.46±2.59	33.70±2.70	11.24±3.40	102.70±108.50
1996	22.04±3.18	33.38±3.12	11.34±4.52	112.20±115.50
1997	22.34±2.04	33.15±2.01	10.81±2.88	72.58±78.15
1998	23.08±2.39	34.02±3.14	10.20±4.36	75.10±83.66
1999	22.83±2.40	33.73±2.56	10.91±3.04	90.92±94.83
2000	22.47±2.04	33.53±2.65	11.07±3.12	76.93±83.23
2001	22.27±2.81	33.89±2.94	11.62±3.88	85.15±110.40
2002	22.66±2.68	33.52±2.83	10.86±3.65	103.70±109.10
2003	22.51±2.32	34.60±3.21	12.01±3.84	59.88±66.83
2004	22.30±2.13	33.97±2.95	11.68±3.98	69.50±84.54
2005	23.38±2.59	33.87±3.16	10.49±3.42	71.03±78.54
2006	22.89±2.90	34.23±3.33	11.34±3.86	76.97±91.23
2007	22.54±2.33	33.82±3.04	11.28±3.82	79.32±71.87
2008	22.74±1.92	33.73±2.66	10.99±3.02	102.80±109.00
2009	22.49±2.36	34.04±3.10	11.55±3.85	129.00±128.60
2010	22.34±2.87	34.42±3.44	10.71±4.27	82.60±111.60
2011	22.63±2.33	33.85±3.08	11.37±3.28	76.85±87.44

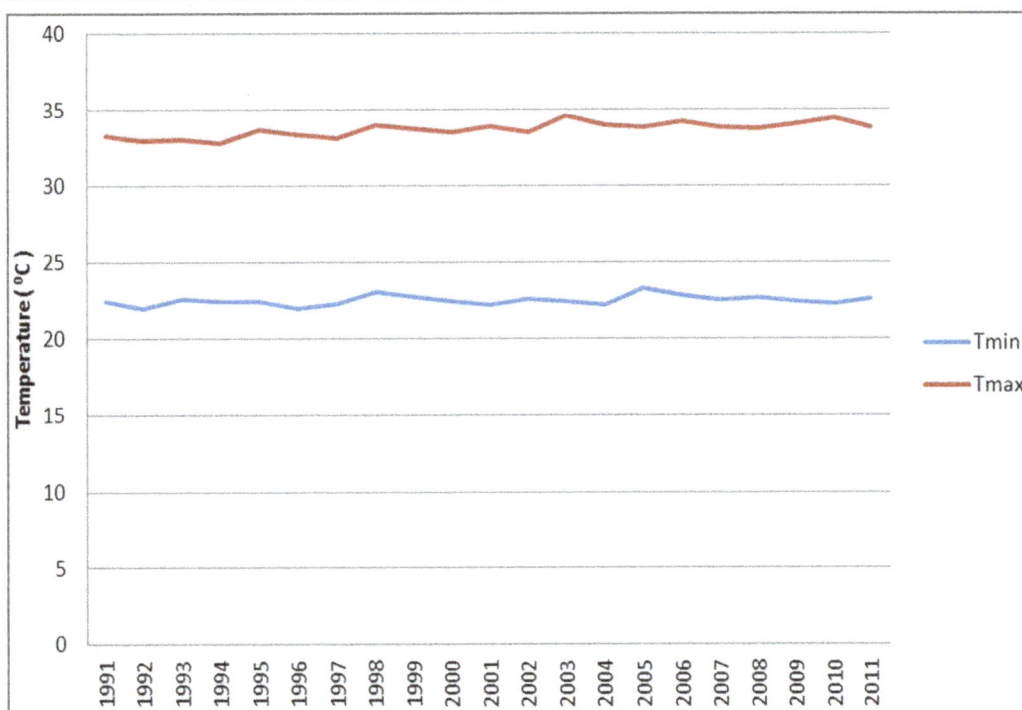

Figure 1. *Mean Annual Temperature (°C).*

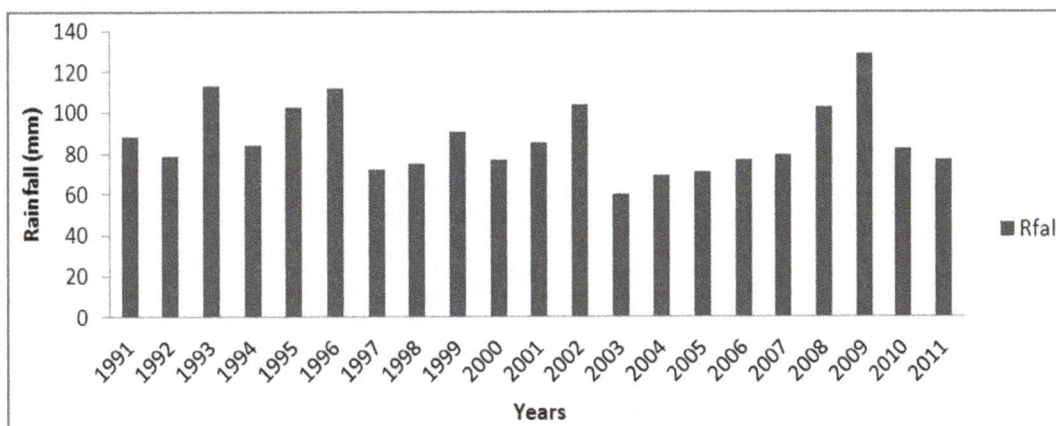

Figure 2. *Mean Annual Rainfall (mm).*

4. Discussion

This study has revealed a steady rise in temperature in Taraba State during the period under review even though significant differences were not observed. This is observed is in agreement with the global trend [2]. Similarly, [17] also recorded a slight difference in temperature within the late 1940s and early 1960s in Nigeria. The rise in the minimum temperature also agreed with [18] that the earth has warmed by 0.6±0.2°C on the average since 1900. The findings also conform to those of [19] who reported that the climatic records and the statistical manipulation of daily records of temperature, rainfall and other parameters have shown considerable variations over the years.

The rise in temperature range recorded in the study was in conformity with [20], that in Africa, it has been predicted that by 2100, temperatures could rise between 2°C to 6°C relative to what is obtained at present. Despite the fact that that there were no significant differences in the trends, the area under study (Taraba State) still experienced some level of warmth within the years of study (1991-2011). It is worth mentioning that if the temperature trends continue to increase, there is the possibility of Taraba State experiencing a warmer climate in the near future?

The increase in the amount of rainfall recorded in Taraba State within the period under study agrees with [20] who predicted that there would be corresponding increases in mean global precipitation (from 2.5 – 5.1%) with regional differences that include significant localized reductions in annual precipitation. Also, the distribution of rainfall across the months showed higher concentrations mostly around July, August and September in most of the years. Some of the years may not have much rainfall, but because of such concentration of rainfall intensity, the soil may not be able to accommodate moisture any longer, and this could lead to high runoff that makes the volume of water in the rivers rise and consequently overflow as flood, causing all sorts of destruction; as has been experienced severally in the study area (Taraba State). This indicates that the effects of climate change could be enormous all over the world. For example, [21] wrote that United States will suffer a wide range of increasingly extreme weather events in the coming decades as a result of global warming from drought and excessive heat to wildly destructive hurricanes and record floods triggered by intense rains. Furthermore, [22] noted that impacts of climate change include increasing water scarcity and flood risk, along with decline in water quality.

5. Conclusion

Slight variations were noticed in the temperature trends (maximum, minimum and temperature range) and also in the rainfall distribution within the period under investigation (1991 and 2011). The variations in temperatures (maximum, minimum and range) and the rainfall distribution in the study area (Taraba State) showed that climate change is unequivocal and climate change indicators are becoming stronger; and therefore, the effects of climate change on forest resources and the environment as a whole are generally becoming threatening in the study area (Taraba State).

Recommendations

To forestall the possible menace of climate change in the study area, the following are hereby recommended:

- Policies and subsidy programmes supporting private forestry management and reforestation should be put in place.
- Ministry of environment should enforce various laws on conservation and proper management of the environment.
- There should be the integration of climate change impacts information into the general planning process; and
- There is need for a widespread enlightenment of the citizens on climate change to enable them to change some of the human behaviours contributing to climate change.

References

[1] I. J. Ekpoh. Environmental Change and Management. St. Paul's Printing Company, Calabar, Nigeria. 2002. pp. 63-94.

[2] Intergovernmental Panel on Climate Change (IPCC). Impacts, Adaptation and Vulnerability. Working Group II Report. 2007. (available at www.ipcc.ch).

[3] World Resources Institute (WRI). World Development Reports. Development and the Environment. The World Bank, OUP, Oxford, U.K. 1992. pp. 160-177.

[4] S. I. Udofia. Economic Approaches in Valuing the Effects of Climate Change on Environment .The Nigerian Journal of Forestry. 2001. 31(1&2): pp105-110.

[5] W. P. Cunningham and M. A. Cunningham. Principles of Environmental Science Inquiry and Applications, 2nd edition. McGraw-Hill Companies Inc. New York, U.S.A. 2004. pp. 201-210.

[6] M. L. Wilki. Climate Change, Forests and SIDs. The International Forestry Review. 2002. 4(4): 313-316.

[7] D. K. Asthana and M. Asthana. Environment: Problems and Solutions. S. Chand and Company LTD; Ram Nagar, New Delhi, India. 2006. pp. 208-218.

[8] A.M. Solomon and A.P. Kirilenko. Climate Change and Terrestrial Biomass: what if trees do not migrate? Global Ecology and Biogeography Letters. 1997. Vol.16, 139-148.

[9] D. Bachelet, R.P. Nelson, J.M. lenihan, R.J. Drapek. Regional Differences in the Carbon source-Sink: Potential of Natural Vegetation in the U.S.A. Environmental Management, 2004. 33: 23 - 43.

[10] M. W. Scholze, N. W. Knorr, Arnell, and I.C. Prentice. A Climate risk Analysis for world Ecosystems. Proceedings of the National Academy of SEED-UNDP (2000): Sustainable Energy Strategies: Materials for Decision – Makers. 2006. 208pp.

[11] L. C. Nwoboshi.. Tropical Silviculture, Principles and Techniques. Ibadan University Press, Nigeria. 1982. pp. 333.

[12] T. O. Amusa. The Role of the Forests in the Amelioration of the Nigerian Environment. The Nigerian Field. Bachelet, d., R.P. Nelson, T. Hickler, R.J. 2002. Vol.67 part 1. Pp 31-43.

[13] R. T. Watson, I. R. Noble, B. Bolin, N.H Ravindranath, D.J. Verardo, and D.J. Dokken (eds). Land use, Land-use Change, and Forestry, 2000. Special Report of the Inter-Governmental Panel on Climate Change. Cambrige University Press, Cambrige.

[14] A. L. Westerling, H.G. Hidalgo, D.R. Cayan, and T.W. Swetnam. Warming and Earlier Spring Increase, Western U.S. Forest Wild life Activity. Science. 2006. 313: 940-943.

[15] Ministry of Information and Social Development, Jalingo, Taraba State. 2007.

[16] National Population Commission, NPC. Federal Republic of Nigeria Official Gazette 2007. No. 24 Volume 94 Lagos, Nigeria.

[17] J. O. Ayoade. Climate Change, Vantage Publishers, Ibadan. 2003. pp. 45-46.

[18] Commonwealth Scientific and Industrial Research Organization (CSIRO). Climate Change Projections for Australia. Summary Document CSIRO, Cauberra. Ekanade, O., Ayanlade, A. and Orimoogunje, I.O.O. (n. d). Geospatial Analysis of Potential Climate Impacts on Coastal Urban Settlements of Nigeria for the 21st Century. GSDI 10 Conference Proceedings. 2001. pp. 12.

[19] E. O. Aina and S.A. Adejuwon. Regional Climate Change: Implication on Energy Production in the Tropical Environment in: Jerome, C. and Umolu, P.E. (eds) The Proceeding of the International Workshop on Global Climate Change: impact on Energy Development, March 28-30, 1994, Lagos, published by Dunatech Nigeria limited.

[20] M. Hulme. Climate perspective in Sahelian Desiccation: 1973-1998. Global Environmental Change. 2001. 11: 19-29.

[21] C. Miller. Climate Change. A Publication of Newser U.S. 2008. pp. 2.

[22] J. Anderson, K. Arblaster and J. Bartely. Climate Change - Induced Water Stress and its Impacts on Natural and Managed Ecosystems. A publication of Institute for International and European Environmental Policy. 2008. pp.1-3.

30

Land Suitability Evaluation for Plantation Forest Development Based on Multi-criteria Approach

Aminuddin Mane Kandari[1], Safril Kasim[2], Muh. Aswar Limi[3], Jufri Karim[4]

[1]Spesifications Agroclimatology, Department of Environmental Science, Faculty of Forestry and Environmental Science, Halu Oleo University, Southeast Sulawesi, Indonesia
[2]Spesifications Agroforestry, Department of Environmental Science, Faculty of Forestry and Environmental Science, Halu Oleo University, Southeast Sulawesi, Indonesia
[3]Spesifications Management, Department of Agribusiness, Faculty of Agriculture, Halu Oleo University, Southeast Sulawesi, Indonesia
[4]Spesifications GIS, Department of Geography, Faculty of Science and Mining Technology, Halu Oleo University, Southeast Sulawesi, Indonesia

Email address:

manekandaria@yahoo.com (A. M. Kandari), safrilkasim1970@gmail.com (S. Kasim), aswar_agribusiness@yahoo.com (Muh. A. Limi), jufrikarim@yahoo.com (J. Karim)

Abstract: Information of land suitability is needed to prioritize suitable forest plantation for land use development. This is important to increase land productivity and eficiency on forest management decision making process. This research aimed to: (1) to evaluate land suitability based on pedo-agroclimate characteristics for plantation forest development; (2) to identify the farming-based socio-cultural and economic characteristics; (3) to determine the priority level of trees that will be cultivated through plantation forest development. This research was conducted from January to May 2015 in four districts namely: Kapontori, Lasalimu, South Lasalimu, and Siotapina (KALALASSI region). These are located in Buton Island, Southeast Sulawesi Province, Indonesia. The research method used was a spatial research method using GIS [1];[3]. The research has three main activities: data collection, evaluation, and mapping. Data collection included data on land biophysics, climate, and forest management development using survey method. Land evaluation was carried out on FAO method [14] and [15]. The major of trees were determined using LQ method [16] and the priority level of trees determine using MCDM method through the application of AHP [3];[4]. The last but not the least, spatial data development was used to map recommended forest land uses. The results showed that: (1) there were three major of trees in the research area, namely: Teak, Mahogany and Silk Tree; (2) based on land suitability classification, there are two classes found in the KALALASSI region, namely: moderately suitable (S2) which were located 3,836.05 ha for Teak and Mahogony, and Marginally Suitable (S3), which are located 3,343.45 ha for Teak, 3,467.20 ha for Mahogony, and 10,106.22 ha for Silk Tree; (3) the sequence of trees priority in KALALASSI region is Teak, Silk Tree, and Mahogony, then based on these recommendation, forest land uses and management plan were developed.

Keywords: Pedo-Agroclimate, Land Suitability, GIS, MCDM Method, Trees Priority, Forest Plantation, Development

1. Introduction

Forest is one of the natural resource that is important for human life. Trees provide many benefits in human life, especially for producing oxygen and absorbing carbon, thus making human beings can breathe well and remain comfortable living environment. Besides that, if properly managed, forest will provide economic benefits through job creation while maintaining environmental advantages. In the other words, the depletion of forest resource can lead to unavailability of oxygen and an increased in carbon concentration in the atmosphere, thus causing global warming, which in turn is definitely an uncomfortable environment even human life can be destroyed.

On the other hand, forest development sector encounters many obstacles and threats, including highly sensitive of forestry land use which can be rapidly change to other land uses, such as housing, infrastructure, and agriculture land, thus threatening a sustainability of forest community development [17]. Futhermore, according to [12] the conversion of forested land to other land uses will decrease the ecological benefits, such as hydrological roles, erosion control, reducing surface runoff and soil fertility control. Thus the need to continuously maintaining existing forest and the development of forest plantation efforts should be sustainably conducted in order to obtain socioeconomic and environmental benefits.

Buton regency is one of the potential areas in Southeast Sulawesi Province for the development of plantation forest because there is still many uncultivated dryland areas, especially in the KALALASSI region which includes the districts of Kapontori, Lasalimu, South Lasalimu and Siotapina. This region is dominated by dry land with an average yearly rainfall of < 2000 mm. The development of plantation forest area in this region is potential. This is due to land availability and well adapted of trees to relatively low rainfall. To deal with a relative low rainfall in this area, the soil structure needed to properly managed so that infiltration capacity of soil can increase as well as soil aeration. [30] states that the trees growth are affected by soil structure. This is because the rough structure of soil has a better aeration and infiltration capacity compare to the clay soil.

According [25] the development of sustainable forest plantation area can be managed through corporate agribusiness system approach in overall management process ranging from commodities selection to production techniques (land preparation, planting, maintenance, harvesting and post-harvesting). [28] stated that there are at least five main aspects that must be met in order to achieve sustainable plantation of forest management, namely: (1) the rigidity and safety of forest resources, (2) the sustainability of production, (3) flora and fauna conservation aspect; and biodiversity as well as various line functions of the forest for the environment, (4) the economic benefits for the nation's development and community participation, (5) institutional aspect. Forestry plants can only be well grown and achieve optimal productivity when supported by land characteristics particularly suitable climatic and soil conditions. According to the [13], there are three aspects of climate that affect plant growth and productivity of the forest, namely changes of air temperature, changes of precipitation and atmospheric CO_2 concentration. [29] states that the effective use of the land does not closely related to the climate change, however unappropriate land use can cause climate change.

According to [30] crop management will be more efficient and sustainable when in accordance with the biophysical and preferences of society toward plant species developed. [36] states that due to the differences in land capability, thus land use planning should be taken into account both biophysical suitability of the land, society preference and economic impact through land cultivation.

Land use planning that involves multifactors and multicriterias, especially biophysical aspect based on pedo-agroclimate, and socio-economic and cultural aspects will be appropriate solution for land use development in the study area [2];[19]. However, a proper land use planning has

not been developed due to lack of spatial and attribute data.

Based on aboved explanations, the need for conducting land suitability analysis for plantation forest management employing multi-criteria approach is urgent. This research aimed to: (1) to evaluate land suitability based on pedo-agroclimate characteristics for plantation forest development; (2) to identify the farming-based socio-cultural and economic characteristics; (3) to determine the priority level of trees that will be cultivated through plantation forest development.

2. Study Area

Buton regency is situated in the southeastern part of Sulawesi Island, Indonesia, and geographically located between 05°03'32" and 05°39'55" South Latitude and 122°40'44" and 123°13'26" East Longitude [6]. This research was conducted on four districts out of seven districts in Buton regency, including Kapontori (38,450.00 ha) [7], Lasalimu (33,130.00 ha) [8], South Lasalimu (17,090.00 ha) [9] and Siotapina (24,760.00 ha) [10], which is abbreviated by the term KALALASSI region (113,430.00 ha).

3. Methodology

This research using multiple criteria that include secondary data and primary data. Secondary data were collected from various sources that support the analysis and presentation of data / information area, both of which have shaped the basic data directly related to the biophysical aspects of land based pedo-agroclimate, as well as from the cultural and social aspects of farm-based economy, including demographics.

The research employs a spatial analysis method, using the ArcView GIS software. The research was conducted in three main: (1) survey, (2) evaluation, and (3) mapping.

These three main steps were performed in five stages: (a) preparation, (b) collection of total area and land biophysic data (physio-topography, climate, soil) and land use, (c) input and data analysis, (d) interpretation for land evaluation, and (e) development of land suitability map for forest plantation (f) determine of priority level of trees and mapping (Figure 1).

Figure 1 shows the schematic diagram of research activities. At the early stage, preparation was conducted to collect data and information on secondary and primary data. Secondary data were directly collected from the sources, while primary data were collected using survey method and interview, either directly or through questioners. Land biophysic data were directly collected in the field on each research site, based on technical guidelines provided by the Research Center for Soil and Agroclimate [32]; [31]; [27].

Method for soil sampling and analysis was based on Soil Survey Staff [33]; [34]; [35].

Similarly, social and economic data which has been collected and analyzed by appropriate method. Analysis of data to determine land suitability classes using FAO method [14] and [15]. The major of trees were determined using Location Quotient (LQ) method [16] and the priority level of trees determine using Multiple Criteria Decision Making (MCDM) method through the application of Analysis Hierarchy Process (AHP) [3];[4]. The last but not the least, spatial data development was used to map recommended forest plantation land uses.

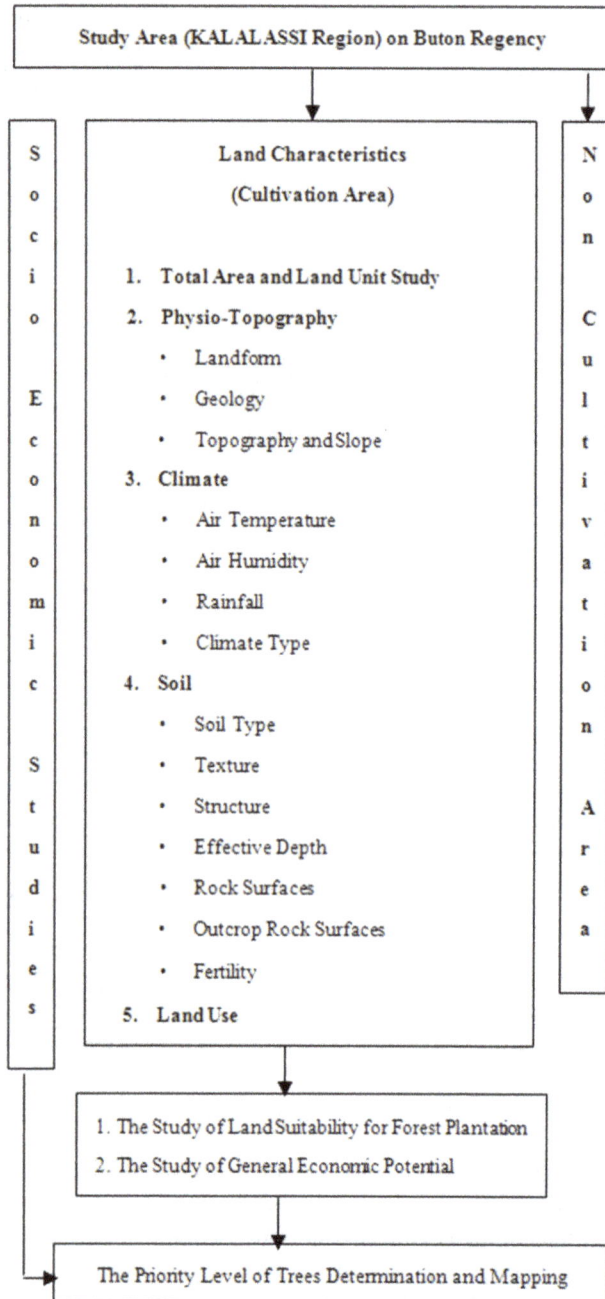

Figure 1. Diagram of Research Activity.

4. Result and Discussion

4.1. Land Characteristics

4.1.1. Total Area and Land Unit Study

Map interpretation, field observation and GIS analysis identify the total area of the study ((KALALASSI region) is 113,430.00 ha. There is still 38.88% of the total area

(35,027.43 ha) can be developed as plantation with 214 land units, covering: Kapontori 73 units (12,458.86 ha or 35.57%), Lasalimu 65 units (9,436.40 ha or 26.94%), South Lasalimu 52 units (8,709.84 ha or 24.87%), and Siotapina 24 units (4,422.33 ha or 12.63%).

4.1.2. Physio-Topography

(i). Landform

GIS analysis interpretation and field observation show that there were five landform categories on the recerch region with a total area coverage is different, namely: Hill system (H) 15,255.72 ha; Piedmont Land (P) 11,213.94 ha; Uplifted limestones (C) 4,965.02 ha; Marine Territory (B) 2,040.85 ha; and Aluvium Land (A) 1,551.89 ha. All categories are included in the four districts with different coverage areas, for example category hills system (H) which has the highest coverage area was in Kapontori (6,016.53 ha), and the lowest was in Siotapina (677.08 ha).

(ii). Geology

Interpretation of satellite images and regional geological map was relevant to the field observation of geological formation on research locations, in which the geological formation in the research region was relatively complex and they were grouped into nine formations: wapulaka (Wp) 13,452.31 ha; sampolakosa (Sm) 13,187.90 ha; tondo (Td) 4,461.36 ha; aluvium (Al) 2,358.81 ha; limestones (Bg) 727.49 ha; ultramafic kanpa (UbKp) 435.32 ha; winto (Wt) 225.43 ha; ultramafik (Uf) 166.26 ha; tobelo (Tb) 12.55 ha. Not all of these formations are in all research location, such as the formation Wt and Bg only in Lasalimu and Kapontori, Tb only in Siotapina, UbKp and Uf only in Kapontori. Geology formation that is located in four research location namely Wp, Td, and Al, with an area of coverage vary from one location to others. Geological formation with the highest coverage area was Wp (13,452.31 ha or 38.41%), found in all research location with the highest area is in Kapontori (6.629.22 ha) and the lowest in Siotapina (695.85 ha).

(iii). Topography and Slope

The results showed that the topography in the research region are relatively varied, ranging from flat areas (flats), wavy (undulating), corrugated (rolling), rolling (hilly) to mountained (mountainous). The Slope ranges on the research region are varied that are grouped in to seven classed: 41-60% (12,974.82 ha); 9-15% (6,755.97 ha); 16-25% (5,573.40 ha), > 60% (5,069.29 ha), 2-8% (2,972.42 ha); <2% (1,671.13 ha), and 26-40% (10.41 ha). Slope range with the highest coverage area is 41-60%, covering 12,974.82 ha (37.04%), and existed in all research locations, in which the highest coveraged area was in Kapontori (7,373.64 ha, and the lowest was in Siotapina (224.94 ha).

4.1.3. Climate

(i). Temperature

Climatic data analysis showed that the overall air temperature in the study area ranges from 26 - 28oC to

28.1-30oC, but the most widespread is the range of 28.1-30oC (33,503.83 ha or 95.65%), which cover broad areas of study, while the temperature range 26-28oC is spread only in Kapontori with coverage 1,523.60 ha (4.35%).

(ii). Humidity

Climatic data analysis showed that the humidity in the study area consists of three categories, namely 83%; 87%, and 94%, and which is humidity rate 87% dominated the research region 20,915.98 (59.71%) and spread over to almost all of the research region. Humidity rate 83% 12,943.49 (36.95%) is only spread in Kapontori (11,290.90 ha) and in Lasalimu (1,652.59 ha).

(iii). Rainfall

There are four yearly rainfall categories over the research region, namely: 1,500 - 1,750 mm year-1. Rainfall 1,645.5 mm year-1 is dominated the research region (21,188.19 ha or 60.49%). This is spread over in almost of the study area, except in Kapontori. Whereas rainfall with 1,552.4 mm year-1 is spread only in Kapuntori 12,315.64 ha (35.16%) and in Lasalimu (1,380.38 ha). Besides, there is also 1,747.2 mm year-1 only in Kapontori (1,523.60 ha).

(iv). Climate Type

Accumulation of all climate conditions, particularly monthly rainfall, creates a certain climate type. Based on Schmidth-Fergusson classification method in the research region, there are two types: C with a value of Q = 55.22 (WMR = 6.7 and DMR = 3.7), which means a rather wet area with jungle vegetation, includes districts Kapontori. In addition, there is a climate type D with a value of Q = 75.22 (WMR = 5.7 and DMR = 4.3), which means the area is rather dry with forest vegetation season, covering districts Lasalimu, South Lasalimu, and Siotapina.

4.1.4. Soil

(i). Soil Type

The soil type ranges on the research region area varied, grouped into six classed: eutropepts 9,591.94 ha (27.38%); tropudalfs 9,202.67 ha (26,27%); rendolls 8,713,00 ha (24.87%); tropudults 3,434.83 ha (9.81%); dystropepts 3,033.28 ha (8.66%); and hydraquents 1,051.72 ha (3.00%). The six types of soil are located in all research location, except in Kapontori. Tropudults and Dystropepts types are located in Siotapina. Eutropepts is the widest area and is located in Kapontori (5,617.69 ha), Lasalimu (1,642.60 ha), South Lasalimu (1,587.72 ha), and Siotapina (743.93 ha).

(ii). Soil Texture

The results showed that the soil texture in the research location consists of eight broad categories with different extent, namely: clay (L) 10,137.66 ha (28.94%); sandy loam (SL) 6,951.01 ha (19.84%); Sandy Clay Loam (SCL) 6,546.81 ha (18.69%); Silty Clay (SiC) 3,955.07 ha (11.29%); clay loam (CL) 3,917.04 ha (11.18%); silty clay loam (SiCL) 1,777.03 ha (5.07%); loam sandy (LS) 1,642.12 ha (4.69%); and sandy clay (SC) 100.69 ha (0.29%). The soil texture with

the most extensive coverage is clay which spread over in the research location, which is 3,258.08 ha (Kapontori), 2,513.38 ha (Lasalimu), 2,307.87 ha (South Lasalimu), and 2,058.33 ha (Siotapina).

(iii). Soil Structure

The results showed that the soil structure in the area consists of three categories with a total area of coverage varies: blocky 21,341.79 ha (60.93%); granular 8,982.36 ha (25.64%), and crumb 4,703.28 ha (13.43%). The structure range with the highest coverage area is blocky and existed in all research locations, in which the highest coveraged area was in Kapontori (8,031.00 ha), South Lasalimu (6,090.64 ha), Lasalimu (5,329.41 ha), and the lowest in Siotapina (1,890.74 ha).

(iv). Soil Depth

Soil observation and measurement showed that the thickness of the solum and effective rooting depth in the research location consists of five categories with different scope: shallow (20 to 50 cm) 13,863.25 ha (39.58%); moderately deep (> 50 to 75 cm) 7,541.04 ha (21.53%); deep (> 75 to 100 cm) 6,005.93 ha (17.15%), very deep (> 100 cm) 4,965.34 ha (14.15%), very shallow (< 20 cm) 2,661.88 ha (7.60%). Soil depth of the dominant category in the research region is Shallow (20 to 50 cm), the widest available in Kapontori (5,536.94 ha), Lasalimu (4,394.33 ha), South Lasalimu (3,155.59 ha) and Siotapina (776.39 ha). While very deep soils are most widely available in South Lasalimu, Lasalimu, Kapontori and Siotapina which cover 2,123.00 ha, 1,327.38 ha, 808.02 ha and 696.94 ha respectively.

(v). Rock Surface

The results showed that the condition of the soil surface rocks in the study area consists of seven broad categories with different scope: very rocky (> 15-60%) 9,313.07 ha (26.59%); pretty rocky (> 3-15%) 7,385.79 ha (21.09%); a little rocky (0.01-0.1%) 6,938.76 ha (19.81%), a bit rocky (> 0.1 to 3%) 4,774.70 ha (13.63%); very little rocky (> 50-90%) 3,352.13 ha (9.57%); rocky (> 90%) 1,942.64 ha (5.55%), and not rocky 1,320.33 ha (3.77%). The broadest category of surface rocks is very rocky, of which is located in Kapontori (4,563.03 ha), Lasalimu (2,069.39 ha), South Lasalimu (2,020.47 ha), and Siotapina (660.18 ha). While the not rocky is the smallest rocky surface which is relatively evenly distributed in all areas of study and covers 265.55 ha, 641.75 ha, 261.55 ha, 151.48 ha in Kapontori, Lasalimu, South Lasalimu and Siotapina respectively.

(vi). Surface Rock Outcrops

Land surface analysis showed that the condition of the ground surface rock outcrops in the study area consists of seven broad categories with different scope: pretty rocky (> 2 to 25%) 10,921.19 ha (31.18%); a little rocky (0.01 to 0.1%) 10,381.88 ha (29.64%); a bit rocky (> 0.1 to 2%) 4,773.72 ha (13.63%), very little rocky (> 50 to 90%) 3,389.44 ha (9.68%); very rocky (> 25 to 50%) 3,159.00 ha (9.02%), not rocky (0.00%) 1,320.33 ha (3.77%); and outcrop rock (> 90%)

1,081.88 ha (3.09%). The broadest category of surface rocks outcrops is pretty rocky, of which is located in Kapontori (4,883.87 ha), Lasalimu (2,816.99 ha), South Lasalimu (2,259.80 ha), and Siotapina (960.53 ha). While the outcrop rock is the smallest rocky outcrops surface which is relatively evenly distributed only in Siotapina (941.52 ha) and South Lasalimu (140.36 ha).

(vii). Soil Fertility

The results showed that the fertility of the soil in the research region consists of three broad categories with different scope: medium 26,205.85 ha (74.82%); high 8407.74 ha (24.00%); and low 413.83 ha (1.18%). From category soil fertility that is most widespread medium category, are the largest in Kapontori (9,018.85 ha), Lasalimu (6,651.56 ha), South Lasalimu (6,401.67 ha), and the smallest in Siotapina (4,133.77 ha). While soil fertility with a low category, only in South Lasalimu (348.09 ha) and in Kapontori (65.74 ha).

4.2. Land Use

The interpretation result satellite data and field observation revealed that there are extremely diversed land primary forest (Htpr) 1,590.90 ha (4.54%); mangrove (Mv) 826.66 ha

(2.36%); agricultural wetlands (Ptlb) 608.66 ha bushes (ptlkcs) 14,660.90 ha (41.86%); secondary dry forest (htlksk) 6,453.86 ha (18,43%); bushes (SmB) 5,165.44 ha uses over the research region, and they were classified in to ten categories, namely: dry land agriculture mixed with (14.75%); dry land agriculture (Ptlk) 4,732.75 ha (13.51%); (1.74%); open land area (Lt) 449.86 ha (1.28%); savana (Sv) 348.98 ha (1.00%), and swamp bushes (Sbr) 189.22 ha (0.54%). The widest land use is a mixture of dry land agriculture bushes, mostly located in Lasalimu (4,983.33 ha), while in South Lasalimu (4,925.59 ha), in Kapontori (2,425.89 ha), and in Siotapina (2,326.09 ha). Primary forest is only found in Kapontori (1,590.90 ha), shrub swamp only in Siotapina (189.22 ha), savana, open land area and field only in Kapontori and South Lasalimu.

4.3. Land Suitability Potential for Plantation Forest

There are three types of trees that has been evaluated using Land Suitability Classification for the plantation forest and management, namely Teak (Tectona grandis), Mahogany (Swietinia sp.) and Silk tree (Paraserianthes falcataria). Results of analysis of potential land suitability classification of the trees can be seen in Table 1.

Table 1. *Result of Potential Land Suitability for Three Forest Trees Type in the KALALASSI Region on Buton Regency.*

Forest Trees Type	Land Suitability Potential	Size Coverage on Study Area (Ha)				
		KA (Kapontori)	LA (Lasalimu)	LAS (South Lasalimu)	SI (Siotapina)	KALALASSI Region
Teak (Tectona grandis)	S2	688.46	1,089.71	1,363.80	694.08	3,836.05
	S3	165.58	1,377.80	1,228.81	571.26	3,343.45
	N	11,937.89	7,176.91	6,083.60	3,442.88	28,641.28
Mahogany (Swietinia sp.)	S2	688.46	1,089.71	1,363.80	694.08	3,836.05
	S3	530.16	1,136.97	1,228.81	571.26	3,467.20
	N	11,573.31	7,417.74	6,083.60	3,442.88	28,517.53
Silk Tree (Paraserianthes falcataria)	S3	1,334.87	2,693.66	3,654.62	2,423.07	10,106.22
	N	11,457.06	6,950.76	5,021.59	2,285.15	25,714.56

Source: GIS Analysis, 2015 Table 1 shows that the forest plantation of Teak, Mahogany and Silk Tree are principly feasible to be developed in the KALALASSI region of Buton Regency. Land suitability classes of those tree species are vary both between different trees at the same districts and similar trees in different districts.

4.3.1. Teak

Table 1 showed that the Teak has three classes of land Suitability, namely S2 class (Moderately Suitable), S3 class (Marginally Suitable), and N class (Permanently Not Suitable) with area of coverages vary both at the same districts and between one district to others. Potential land suitability of S2 class has a total coverage of 3,836.05 ha, which is the largest located in the District of South Lasalimu (1,363.80 ha) then folowed by Lasalimu District (1,089.71 ha), while in two other districts have small different of area coverages (694.08 ha in Siotapina District and 688.46 ha in Kapontori). Permanently Not Suitable class (N) has a total area of coverage 2,8641.28 ha, which is the largest located in the district of Kapontori (11,937.89 ha) and the smallest one is in Siotapina (3,442.88 ha). Map of the distribution of potential land suitability classes

for Teak Plantation Forest in the KALALASSI region is presented in Figure 2.

4.3.2. Mahogany

Mahogany species has three potential land suitability, namely S2 (Moderately Suitable), S3 (Marginally Suitable), and N (Permanently Not Suitable) with an area of coverage varies, both the same trees at different districts and different trees at the same districts in the KALALASSI region (Table 1). Similarly with Teak species, the potential S2 class covers 3,836.05 ha. The largest S2 class is located in the district of South Lasalimu (1,363.80 ha) and the smallest one is in the district of Kapontori (688.46 ha). However, the Permanently Not Suitable class covers a broad area of land (28,517.53 ha). The largest N class is located in the district of Kapontori (11,573.33 ha) and the smallest one is in the district of

Siotapina (3,442.88 ha). The potential land suitability map for the Mahogany Plantation Forest in the KALALASSI region

presented in Figure 3.

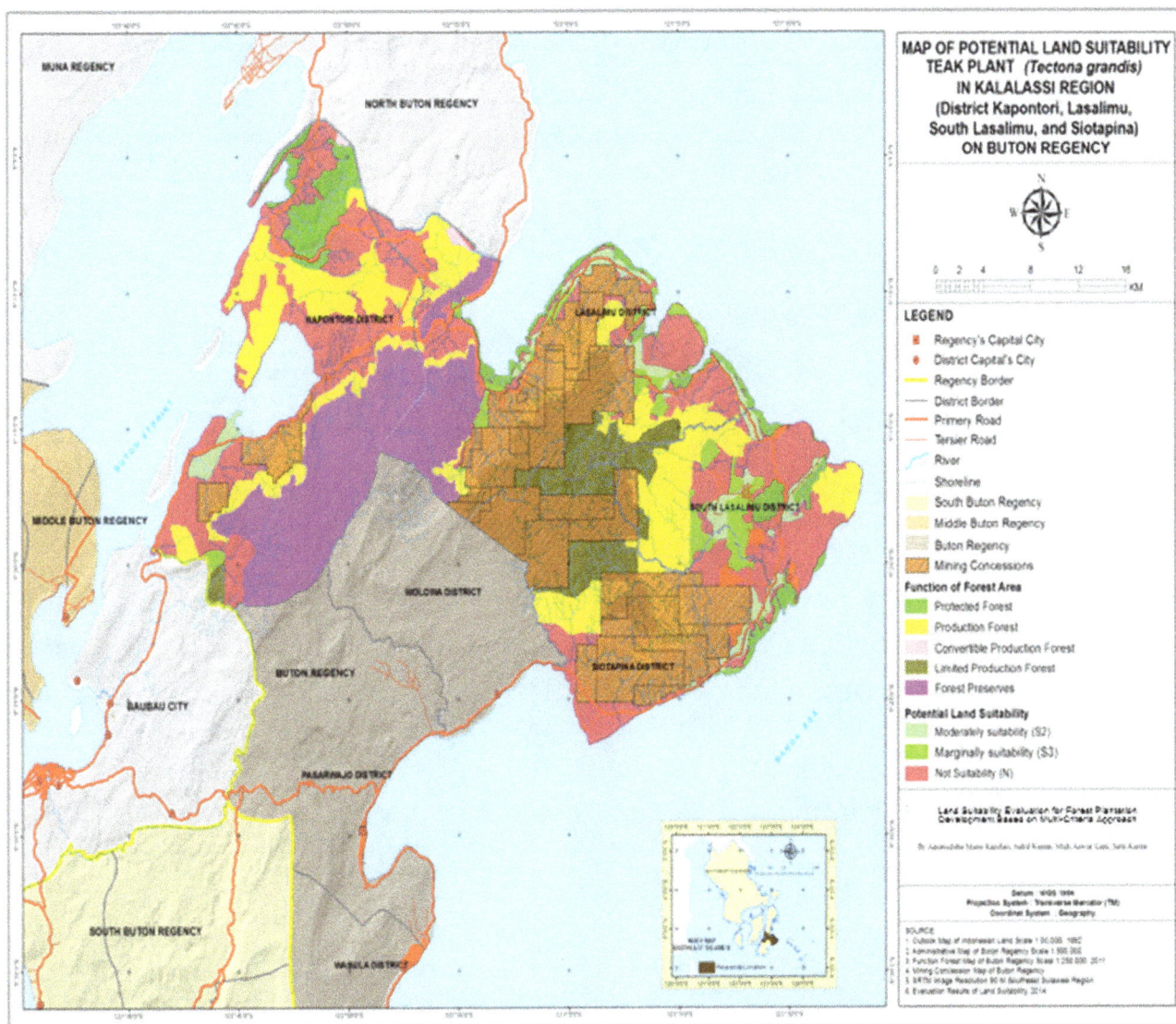

Figure 2. *Map of Land Units and Land Suitability Potential for Teak Plantation in the KALASSI Region on Buton Regency.*

4.3.3. Silk Tree

The potential land suitability for Silk Tree in KALALASSI region consists of only two classes, namely S3 (*Marginally Suitable), and N (Permanently Not Suitable)*. An area of coverages of potential land suitability of these two classes different, both at the same district and between one district to others in the KALALASSI region (Table 1). The total area coverage of potential land suitability for the S3 class Silk Tree is 10,106.22 ha which is the largest located in South Lasalimu District (3,654.62 ha) and the smallest one is in the Kapontori (1,334.87 ha). While the total area coverage of potential for N class is 25,714.56 ha, which is the largest is located in Kapontori District (11,457.06 ha) and the smallest one is in the Siotapina (2,285.15 ha). The potential land suitability map of the development Silk Tree plantation in KALALASSI region presented in Figure 4.

4.4. Determination of Trees Priority

Trees priority in this study determined used AHP (Analytic Hierarchy Process) method based on multi-criteria approach, namely: (1) policy direction, (2) land suitability classes, (3) infrastructure, (4) location quotient (LQ), (5) farmer preferency, (6) market opportunities, and (7) (R/C ratio) of the three trees mentioned above (Teak, Mahogany, and Silk tree) in the KALALASSI region of Buton Regency. Results of the AHP analysis can be seen Figure 5.

Figure 5 shows that Teak Plantation has a first priority to be developed in the KALALASSI Region on Buton Regency. Based on AHP analysis, Teak has the highest weight of number of criteria in four districts, and obtain with weight of 2,76 in the district of Lasalimu and South Lasalimu, while the lowest weight of 2.48 achieved in the Siotapina District. The second priority in the KALALASSI region is Silk Tree which has second weight of number of criteria in four districts. Silk Tree achieves 2,42 as the highest weight in South Lasalimu District, and 2,21 as the lowest in Siotapina District. Whereas the third priority is Mahogany which has the weight number of

criteria 2,23 as the highest in South Lasalimu District, and 2,04 as the lowest in Siotapina District. The mapping of trees priority level in KALALASSI region on Buton Regency can be seen Figure 6.

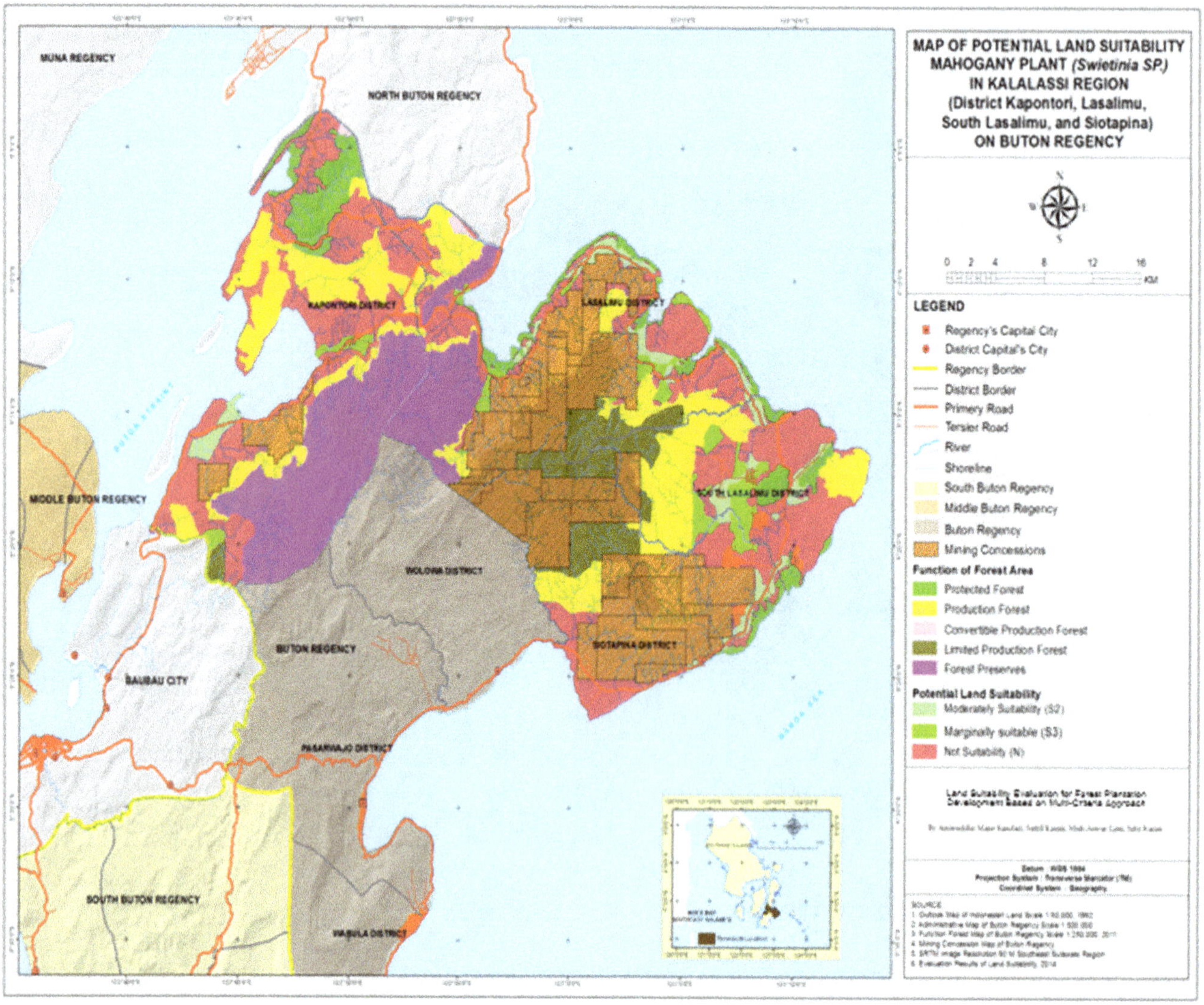

Figure 3. Map of Land Units and Land Suitability Potential for Mahogany Plantation in the KALASSI Region on Buton Regency.

4.5. Discussion

The results of land suitability classification for plantation forest depict that there is a wide opportunity to develop plantation forest in term of land availability and suitability in the KALALASSI region of Buton Regency, especially for the Teak, Mahogany and Silk Tree species. This is in line with the report [20] that local people of this region have cultivated these kinds of trees for generations. As mentioned aboved that Teak and Mahogany trees are moderately suitable (S2 class) to be cultivated in this region, which cover 3,836.05 ha. While the best suite for Silk Tree cultivation is S3 class (Marginally Suitable), which has total area of 10,106.22 ha coverage. However, this research also found that there is a broad area of land that is permanently Not Suitable (N) for Forest Plantation, especially for Teak, Mahogany and Silk Tree species, which cover 25,000-29,000 ha in each district. This can be related to

the facts the land characteristics in the study area, especially dealing with effective depth of soil. It has been explained aboved that soil effective depthness in the region is dominated by the shallow category. Moreover, climatic conditions of the area belongs to N class become barrier factor of this piece of land. The N class are dominated by dry climate with average yearly rainfall of <2000 mm and are included in climatic type C (rather wet) covering Kapontori district and climatic type D (slightly dry) covers Lasalimu, South Lasalimu and Siotapina districts. According [11] that climatic conditions can impact directly and indirectly on the plantation forest development.

The results also show that the productivity of these three species in four districts of the KALALASSI region is different. This fact is closely related to the suitability of the local climatic conditions with growing requirements needed by all these trees. This point of view is relevant to the statement [26] that components of natural climate factors become the main indicators to in plantation forest development. These are solar radiation, air temperature, evapotranspiration and precipitation. Moreover, the research also found that three species of trees in each district has different responds to the interaction between soil characteristics and climatic factors in the region. This fact is relevant to the explanation of [24] that the soil and climate factors, especially precipitation have significant roles to support plant growth process so as to achieve optimum growth and productivity must meet the requirements of soil and climate.

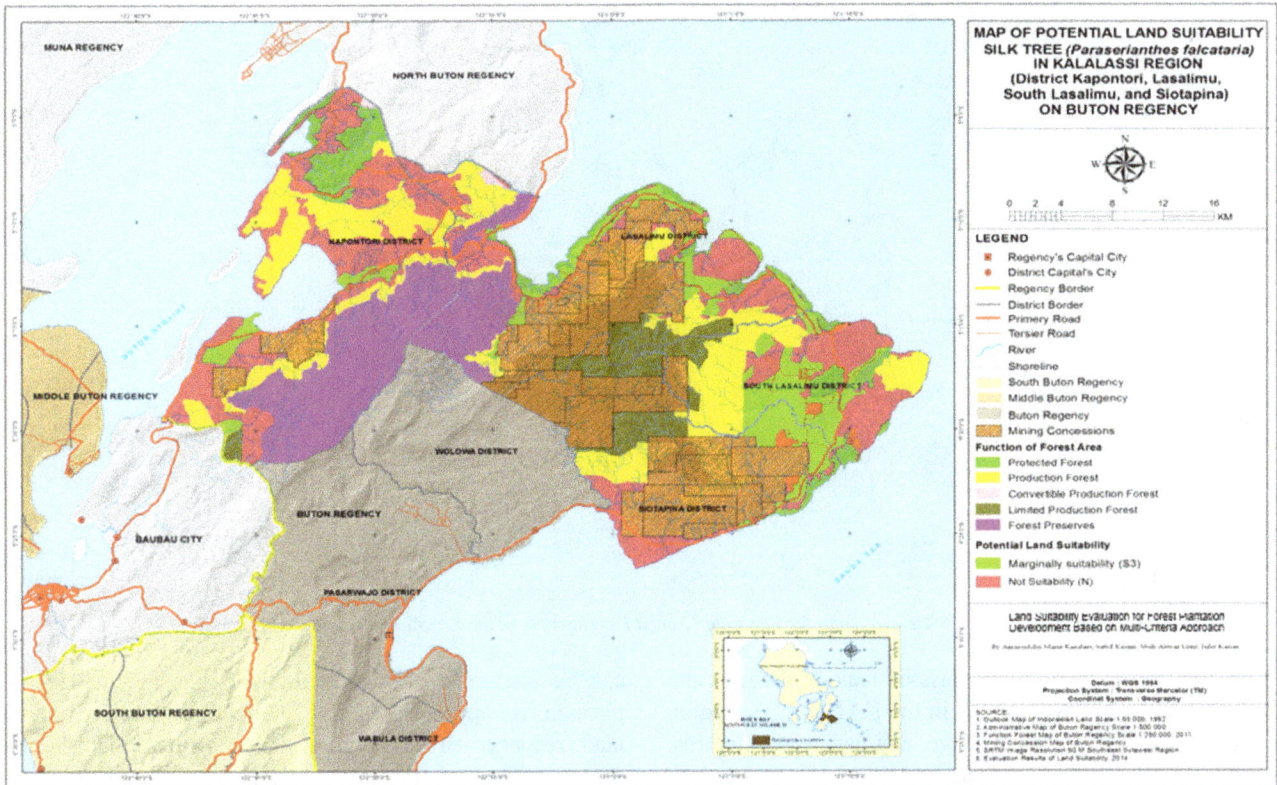

Figure 4. *Map of Land Units for Land Suitability Potential for Silk Tree Plantation in the KALASSI Region on Buton Regency.*

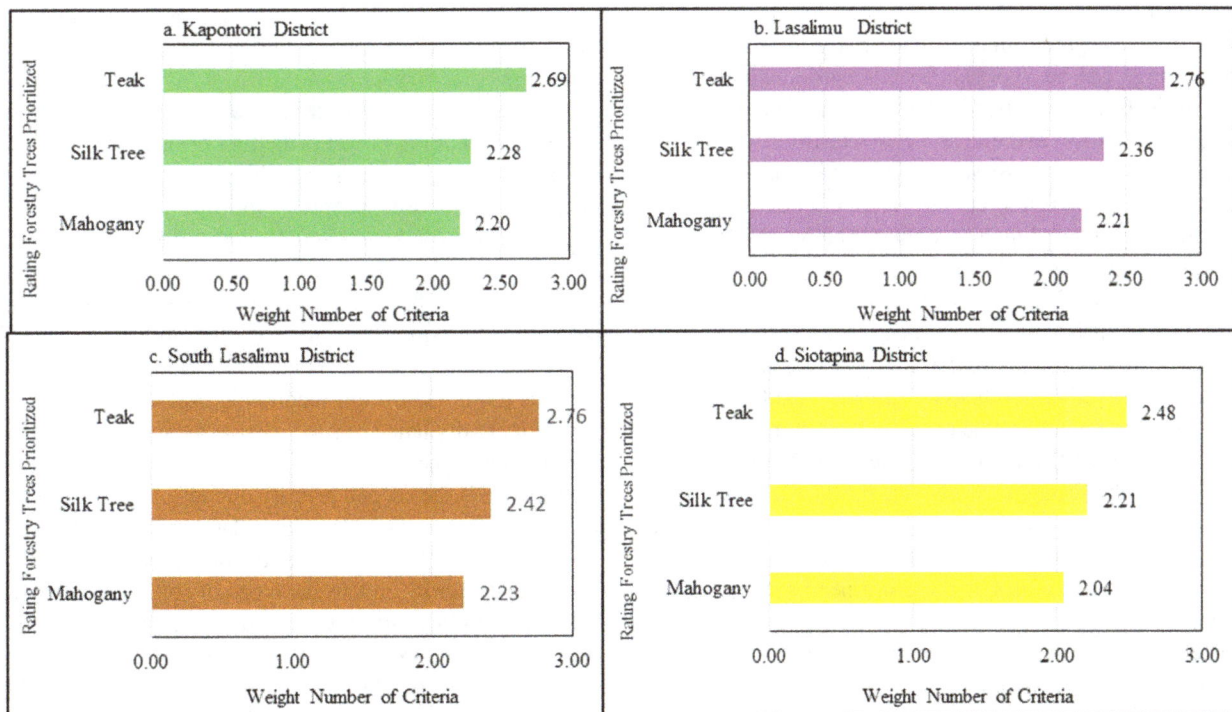

Figure 5. *Diagram Result of Priority Trees Based on Multi-Criteria at: (a). Kapontori District; (b). Lasalimu District; (c). South Lasalimu District; (d). Siotapina District; in KALALASSI Region on Buton Regency.*

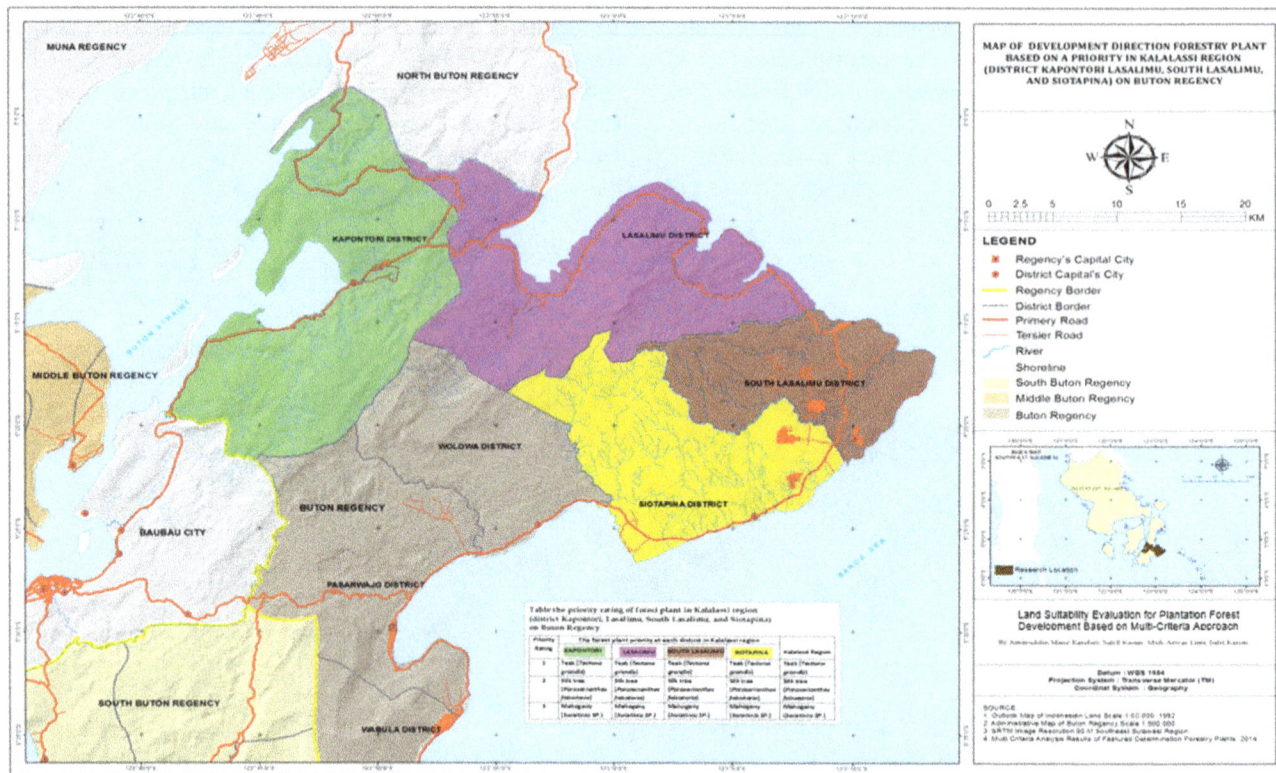

Figure 6. Map of Priority Level of Trees for Plantation Forest Development in the KALALASSI Region on Buton Regency.

A broad area of Not Suitable class of Teak, Mahogany and Silk Tree plantation development in the KALALASSI region can be caused by climate change and land use patterns. According [29] ineffective land use is closely related to climate change and can lead to a decrease in land productivity of the region. Therefore, land use management and development has primarily pay attention to the land use patterns so that negative impacts, especially in relation with climate change can be avoided. To this end, land suitability classification has a significant role in determining suitable plants match with soil and climatic characteristics.

Furthermore, the term of suitability is not only related to soil and climatic condition, but also it should be ecologically suitable for the surrounding environment and be economically viable and socially acceptable.

Effective depth of surface rock become one of the main limiting factors. It is therefore the need to solve this obstacle should be prioritized. This research found that rock surface with shallow effective depthness dominated land surface of the region, which is under this condition can minimize water holding capacity of the plan roots, then soil becomes more susceptible to the drought. Moreover, the fact that has been reported by [19], soil characteristics in the area of Buton Regency relatively have many limiting factors related to soil and climatic aspects, and it is therefore need solution-based approach.

This research also underlined that land resource analysis using spatial analysis method can efectively determine land use potency, then significantly support land use planning and development. In the case of conducting research, [5] states that using land use on research activities as a source of data

can be integrated efficiently for planning and management through GIS approach. A GIS can effectively explore potential land resouces of the region which will further facilitate determination of the appropriate type of plants to be developed and planning and decision making process. This was also in line with the statement [18] and [3] that the spatial method by utilizing GIS can perfectly evaluate land characteristics so we can easily develop appropriate land use planning.

The results of multi-criteria analysis using AHP showed that Teak Plantation Forest is the first priority to be developed in the KALALASSI region. This is due to Teak is more suitable and well adapted to the climate and soil characteristics of the region. It is therefore Teak suitability class (S2 and S3 class) covers the largest area of the region, comparing to Silk Trees and Mahogany. Besides, Teak is economically more viable and socially more acceptable. This is relevant to statement [23] that Teak is a kind of trees that has a high grade and high economic value, and also high suitability to dry land so as to increase the income of farmers if proprely managed.

Silk Tree is the second priority due to total coverages of its land suitability class (S3). This relates to the ability of Silk Tree to adapt with various types of soil. According to [21] that Silk Tree can grow on various types of soil, does not requires fertile soil and can grow well in dry soils, alkaline soils and even in salty soil during drainage sufficient. Finally, Mahogany is the third priority in KALALASSI region. Based on land suitability classification, this kind of tree has S2 class but the total area of coverage smaller than Silk Tree and also has lower economic value than Silk Tree. According to [22],

Mahogany can be well adapted with marginal condition of soil, but it has lower economic value compare with two Teak and Silk Tree. It is therefore, there are only a few local people have cultivated this species.

5. Conclusions

Based on the results of the research and discussion that has been described, it can be concluded as follows:

a. Land characteristics of KALALASSI Region on Buton Regency is dominated by dry land with marginal physiographic conditions, and various soil physical and chemical properties, and dry climate with an average yearly rainfall < 2000 mm and a climatic type belongs to C and D categories.

b. Plantation Forest of Teak, Silk Tree and Mahogany has a wide oppotunity to be developed in the KALALASSI region due to the suitability of soil, the availability of economic value and the acceptability of local farmers.

c. Forest Plantation Planning of Teak, Silk Tree and Mahogany if properly managed can bring about change of local farmers' prosperity through job creation and an increase of long term incomes.

Acknowledgment

First, the grateful thanks are delivering to the Head of Board Planning Agency of Buton Regency that has provided fund for the planning and implementation of the research activities. We also thank to the head of Forestry Department of Buton Regency who has allowed Team Research to access various data source from his office. A special thanks delivers to the head of Kapontori, Lasalimu, South Lasalimu, and Siotapina Districts as well as the government of all villages of KALALASI region both for their administerative and data supports. In particular thanks goes to The Rector of Halu Oleo University for his legal and moral supports in implementation and compliance of this research. Many thanks also go to the head of Soil Laboratory of Faculty of Agriculture, Hasanuddin University and the Head of GIS Laboratory of Faculty of Forestry and Environmental Science for their supports in soil analysis phase and map making activities process. Last but not the least, a very grateful thank goes to the research team for their dedication and time management so that this research can be accomplished on schedule.

References

[1] Baja, S. 2002. Aplikasi Sistem Informasi Geografi dan Analytic Hierarchy Process. Dalam Studi Alokasi dan Optimasi Penggunaan Lahan Pertanian. Warta Informatika Pertanian, Volume 11: 619-635.

[2] Baja, S. 2012a. Perencanaan Tata Guna Lahan dalam Pengembangan Wilayah. Pendekatan Spasial dan Aplikasinya. Penerbit ANDI, Yogyakarta. 378p.

[3] Baja, S. 2012b. Metode Analitik Evaluasi Sumber Daya Lahan: Aplikasi GIS, Fuzzy Set, dan MCDM. Penerbit: IDENTITAS. Universitas Hasanuddin, Makassar. 242p.

[4] Baja, S. 2013. Modul Pelatihan Analytic Hierarchy Process (AHP)-Expert Choice untuk Pengembangan Komoditas Unggulan Wilayah (Procedure Guidelines).Disampaikan pada Pelatihan Analytic Hierarchy Proces (AHP) dalam Pengambilan Keputusan Spasial, Diselenggarakan oleh Pusat Penelitian dan Pengembangan Wilayah, Tata Ruang dan Informasi Spasial (WITARIS) Universitas Hasanuddin, Makassar, 11 Maret 2013.

[5] Bobade S. V., B. P. Bhaskar, M. S. Gaikwad, P. Raja, S. S. Gaikwad, S. G. Anantwar, S. V. Patil, S. R. Singh, dan A. K. Maji. 2010. A GIS-based land use suitability assessment in Seoni district, Madhya Pradesh, India. National Bureau of Soil Survey and Land Use Planning, Tropical Ecology 51(1): 41-54, 2010 ISSN 0564-3295, © International Society for Tropical Ecology, www.tropecol.com, Nagpur 440 010, India.

[6] [BPS Buton] Badan Pusat Statistik Kabupaten Buton. 2014. Kabupaten Buton Dalam Angka. Pasarwajo: BPS Kab. Buton, CV. Kainawa Molagina Bau-Bau.

[7] [BPS Buton] Badan Pusat Statistik Kabupaten Buton. 2014. Kecamatan Kapontori Dalam Angka. Pasarwajo: BPS Kabupaten Buton.

[8] [BPS Buton] Badan Pusat Statistik Kabupaten Buton. 2014. Kecamatan Lasalimu Dalam Angka. Pasarwajo: BPS Kabupaten Buton.

[9] [BPS Buton] Badan Pusat Statistik Kabupaten Buton. 2014. Kecamatan Lasalimu Selatan Dalam Angka. Pasarwajo: BPS Kabupaten Buton.

[10] [BPS Buton] Badan Pusat Statistik Kabupaten Buton. 2014. Kecamatan Siotapina Dalam Angka. Pasarwajo: BPS Kabupaten Buton.

[11] Broadmeadow, M., Ray, D., 2005. Climate change and British woodlands. Forestry Commission Information Note, 16 pp.

[12] Diniyati, D., E. Fauziyah, T.Sulistyawati, B. Achmad., A. Badrunasar, Suyarno, dan E. Mulyati. 2006. Kajian Sosial Ekonomi Budaya dan Jasa Hutan Lindung Di Suaka Margasatwa Gunung Sawal. Laporan Penelitian. Loka Litbang Hutan Moonson Ciamis. Tidak Dipublikasikan.

[13] EPA Home, 2013. Climate Impacts on Forest. United States Environmental Protection Agency. EPA Climate Science Research, Last update on 9 September 2013.

[14] [FAO] Food Agricultural Organization. 1976. A Framework for land evaluation. FAO Soils Bulletin No.32, Food and Agriculture Organisation of the United Nations, Rome.

[15] Hardjowigeno S dan Widiatmaka. 2007. Kesesuaian Lahan dan Perencanaan Tataguna Lahan. Gadjah Mada University Press. Yogyakarta. 352 p.

[16] Hendayana R. 2003. Aplikasi Metode Location Quotient (LQ) dalam Penentuan Komoditas Unggulan Nasional. Informatika Pertanian 12 (1): 658-675.

[17] Herawati, T. 2001. Pengembangan Sistem Pengambilan Keputusan dengan Kriteria Ganda dalam Penentuan Jenis Tanaman Hutan Rakyat. Contoh Kasus di Kabupaten Ciamis Jawa Barat. Thesis. Program Pascasarjana. IPB. Bogor. Tidak dipublikasikan.

[18] Hezam M. Al., J. B. M. Akhir, S. A. Rahim, 2011. GIS-Based Sensitivity Analysis of Multi-Criteria Weights for Land Suitability Evaluation of Sorghum Crop in the Ibb Governorate, Republic of Yemen. J. Basic. Appl. Sci. Res., 1(9)1102-1111, 2011, TextRoad Publication, ISSN 2090-424X

[19] Kandari A. M. 2014. Optimasi Spatio-Temporal Lahan Pertanian di Wilayah Berpotensi Rawan Pangan. Disertasi. Tidak dipublikasi.

[20] Kasim S dan A.M. Kandari. 2015. Analisis Pengembangan Agroforestri di Kecamatan Lasalimu Kabupaten Buton. Majalah ECO GREEN. FHIL UHO ISSN: 2407 - 9049, Vol. 1, No: 1, April 2015. Hal. 55 – 64.

[21] Krisnawati, H., Varis, E., Kallio, M. and Kanninen, M. 2011a. Swietenia macrophylla King. ecology, silviculture and productivity. CIFOR, Bogor, Indonesia. © 2011 Center for International Forestry Research ISBN 978-602-8693-39-4.

[22] Krisnawati, H., Varis, E., Kallio, M. and Kanninen, M. 2011b. *Paraserianthes falcataria* (L.) Nielsen: ecology, silviculture and productivity. CIFOR, Bogor, Indonesia. © 2011 Center for International Forestry Research, ISBN 978-602-8693-41-7.

[23] Langenberger, G. and J. Liu, 2013. Performance of Smallholder Teak Plantations (Tectona Grandis) In Xishuangbanna, South-West China. Journal of Tropical Forest Science 25(3): 289–298 (2013).

[24] Lavalle, C., Micale F., Houston T. D., Camia A., Hiederer R., Lazar C., Conte C., Amatulli G., Genovese G. 2009. Climate change in Europe. Impact on agriculture and forestry. A. Review, Agron. Sustain. Dev. 29, 433-446.

[25] Mile, Y. M. 2007. Prinsip-Prinsip Dasar dalam Pemilihan Jenis, Pola Tanam dan Teknik Produksi Agribisnis Hutan Rakyat. Basic Principles on Species Choice and Production Techniques of Community Forestry Agribisnis INFO TEKNIS Vol. 5 no. 2, September 2007. Balai Besar Penelitian Bioteknologi dan Pemuliaan Tanaman Hutan. Balai Penelitian Hutan Ciamis.

[26] Mueller, L. U. Schindler, W. Mirschel., T. G. Shepherd, B. C. Ball., K. Helming., J. Rogasik, F. Eulenstein, H. Wiggering. 2010. Assesing the productivity functions of soils. A Review. Agron. Sustain. Dev. 30. 601-614. INRA.EDP. Sciences.

[27] Mulyani, A. 2001. Petunjuk teknis penyusunan peta pewilayahan komoditas pertanian berdasarkan zona agroekologi skala 1:50.000. Pusat Penelitian dan Pengembangan Tanah dan Agroklimat. Bogor.

[28] Nurtjahjawilasa, K. Duryat, I. Yasman, Y. Septiani, Lasmini, 2013. Konsep dan Kebijakan Pengelolaan Hutan Produksi Lestari dan Implementasinya (Sustainable Forest Management/Sfm). Modul. Natural Resources Development Center. Program Terestrial The Nature Conservancy Indonesia, November 2013, Jakarta.

[29] Rounselvell M. D. A., and D. S. Reay. 2009. Land use and climate change in UK. Land Use Policy 26S, S160-S169. Elsevier.

[30] Robert F. P. 2002. Effects of Soil Disturbance on the Fundamental, Sustainable Productivity of Managed Forests. USDA Forest Service Gen. Tech. Rep. PSW-GTR-183.

[31] Soekardi, M. 1993. Mengenal peta tanah. Pusat Penelitian Tanah dan Agroklimat. Bogor.

[32] Soekarman, 1993. Pengamatan tanah di lapang. Pusat Penelitian Tanah dan Agroklimat Bogor. Bogor.

[33] Soil Survey Division Staff, 1993. Soil survey manual. USDA Handbook No.18. United States Department of Agriculture, Washington DC.

[34] Soil Survey Laboratory Staff, 1991. Soil survey laboratory methods manual. SCS-USDA. October 1991; 611p.

[35] Soil Survey Staff. 2010. Keys to soil taxonomy. Eleventh Edition. United States Department of Agriculture, Natural Resources Conservation Service. Washington DC.

[36] Tabideh S. N. Hosseinzadeh, and K. R. Nezhad, 2014. Land Preparation: A Strategy for Sustainable Development: A Case Study In Dishmook Area Kuhkiloye City. Indian Journal of Fundamental and Applied Life Sciences ISSN: 2231– 6345 (Online) An Open Access, Online International Journal Available at www.cibtech. org /sp.ed/jls/2014/03/jls.htm. 2014 Vol. 4 (S3), pp. 1717-1728.

Permissions

All chapters in this book were first published in AFF, by Science Publishing Group; hereby published with permission under the Creative Commons Attribution License or equivalent. Every chapter published in this book has been scrutinized by our experts. Their significance has been extensively debated. The topics covered herein carry significant findings which will fuel the growth of the discipline. They may even be implemented as practical applications or may be referred to as a beginning point for another development.

The contributors of this book come from diverse backgrounds, making this book a truly international effort. This book will bring forth new frontiers with its revolutionizing research information and detailed analysis of the nascent developments around the world.

We would like to thank all the contributing authors for lending their expertise to make the book truly unique. They have played a crucial role in the development of this book. Without their invaluable contributions this book wouldn't have been possible. They have made vital efforts to compile up to date information on the varied aspects of this subject to make this book a valuable addition to the collection of many professionals and students.

This book was conceptualized with the vision of imparting up-to-date information and advanced data in this field. To ensure the same, a matchless editorial board was set up. Every individual on the board went through rigorous rounds of assessment to prove their worth. After which they invested a large part of their time researching and compiling the most relevant data for our readers.

The editorial board has been involved in producing this book since its inception. They have spent rigorous hours researching and exploring the diverse topics which have resulted in the successful publishing of this book. They have passed on their knowledge of decades through this book. To expedite this challenging task, the publisher supported the team at every step. A small team of assistant editors was also appointed to further simplify the editing procedure and attain best results for the readers.

Apart from the editorial board, the designing team has also invested a significant amount of their time in understanding the subject and creating the most relevant covers. They scrutinized every image to scout for the most suitable representation of the subject and create an appropriate cover for the book.

The publishing team has been an ardent support to the editorial, designing and production team. Their endless efforts to recruit the best for this project, has resulted in the accomplishment of this book. They are a veteran in the field of academics and their pool of knowledge is as vast as their experience in printing. Their expertise and guidance has proved useful at every step. Their uncompromising quality standards have made this book an exceptional effort. Their encouragement from time to time has been an inspiration for everyone.

The publisher and the editorial board hope that this book will prove to be a valuable piece of knowledge for researchers, students, practitioners and scholars across the globe.

List of Contributors

Elsie Ihuakwu Hamadina
Crop and Soil Science Department, Faculty of Agriculture, University of Port Harcourt, Port Harcourt, Nigeria

Robert Asiedu
International Institute of Tropical Agriculture, IITA, Ibadan, Nigeria

Wuyep Solomon Zitta
Department of Geography, Plateau State University Bokkos, Nigeria and Department of Geography, Environmental Management and Energy Studies, University of Johannesburg, South Africa

Samuel Akintayo Akinseye
Department of Geography, Environmental Management and Energy Studies, University of Johannesburg, South Africa

Yakubu Pwajok Mwanja
Department of Microbiology, Plateau State University Bokkos, Nigeria and Department of Botany and Plant Biotechnology, University of Johannesburg, South Africa

Wabusya Moses and Mugatsia Tsingalia
Department of Biological Sciences, Moi University, Eldoret, Kenya

Humphrey Nyongesa
Department of Sugar Technology, Masinde Muliro University of Science and Technology, Kakamega, Kenya

Martha Konje and Humphrey Agevi
Department of Biological Sciences, Masinde Muliro University of Science and Technology, Kakamega, Kenya

Rayim Wendé Alice Naré, Paul Windinpsidi Savadogo, Zacharia Gnankambary and Michel Papaoba Sedogo
Laboratoire Sol-Eau-Plante Institut de l'Environnement et de Recherches Agricoles (INERA), ouagadougou, Burkina Faso

Hassan Bismarck Nacro
Laboratoire d'Etude et de Recherche sur la Fertilité des Sols (LERF), Université Polytechnique de Bobo Dioulasso (UPB), Bobo Dioulasso, Burkina Faso

Abugre S.
Department of Forest Science, School of Natural Resources, University of Energy and Natural Resources, Sunyani, Ghana

Twum-Ampofo K.
Department of Environmental Management, School of Natural Resources, University of Energy and Natural Resources, Sunyani, Ghana

Oti-Boateng C.
Department of Agroforestry, Faculty of Renewable Natural Resources, KNUST, Kumasi, Ghana

Abiyou Tilahun
Department of Biology, College of Natural Science, Debre Berhan University, Debre Berhan, Ethiopia

Teshome Soromessa and Ensermu Kelbessa
Department of Plant Biology and Biodiversity Management, Science Faculty, Addis Ababa University, Addis Ababa, Ethiopia

Wachira P. M. and Okoth S. A.
School of biological Sciences, University of Nairobi, Nairobi, Kenya

Muindi J. N.
Faculty of Science, Department of Biology, Catholic University of East Africa, Nairobi, Kenya

Kidake K. Bosco, Manyeki K. John, Kirwa C. Everlyne, Ngetich Robert and Mnene N. William
Arid and Range Lands Research Institute-Kiboko, Kenya Agricultural and Livestock Research Organization, Makindu, Kenya

Nenkari Halima
Agricultural Sector Development Support Program, Ministry of Agriculture, Livestock and Fisheries, Kajiado County, Kajiado, Kenya

Faith Ileleji, Elsie I. Hamadina and Joseph A. Orluchukwu
Crop and Soil Science Department, Faculty of Agriculture, University of Port Harcourt, Port Harcourt, Nigeria

Ruzanna Robert Sadoyan
Scientific Center of Agriculture, Ministry of Agriculture, Echmiadzin, Republic of Armenia

Jayanta Kumar Basak
Department of Environmental Science and Hazard Studies, Noakhali Science and Technology University, Noakhali, Bangladesh
Rashed Al Mahmud Titumir
Department of Development Studies, University of Dhaka, Dhaka, Bangladesh

Khosrul Alam
Department of Economics, Noakhali Science and Technology University, Noakhali, Bangladesh

Goué Danhoué
National Polytechnic Institute Felix Houphouët-Boigny of Yamoussoukro, Ivory Coast
Department of training and research in Water, Forests and Environment, Yamoussoukro, Ivory Coast

Yapi Yapo Magloire
National Polytechnic Institute Felix Houphouët-Boigny of Yamoussoukro, Ivory Coast
Department of training and research in Agriculture and Animal Resources, Yamoussoukro, Ivory Coast

Otunaiya Abiodun Olanrewaju and Ologbon Olugbenga A. Chris
Department of Agricultural Economics and Farm Management, College of Agricultural Sciences, Olabisi Onabanjo University, Yewa Campus, Ayetoro, Ogun State, Nigeria

Adigun Grace Toyin
Department of Agricultural Economics and Extension, College of Agricultural Sciences, Landmark University, Omu-Aran, Kwara State, Nigeria

Sammaiah D.
Department of Botany, Govt. Degree College, Huzarabad, (T.S)

Odelu G.
Department of Botany, Govt. Degree College, Jammikunta, (T.S)

Venkateshwarlu M.
Department of Botany, Kakatiya University, Warangal, (T.S)

Srilatha T.
Department of Botany, Govt. Degree & PG College for Women, Warangal, (T.S)

Anitha Devi U.
Department of Botany, Govt. Degree & PG College for Women, Karimnagar, (T.S)

Ugandhar T.
Department of Botany, SRR Govt. Degree & P.G College, Karimnagar, (T.S)

Abiyou Tilahun and Hailu Terefe
Department of Biology, College of Natural Science, Debre Berhan University, Debre Berhan, Ethiopia

Teshome Soromessa
Department of Environmental Science, Science faculty, Addis Ababa University, Addis Ababa, Ethiopia

Halim and Makmur Jaya Arma
Specifications Weed Science, Department of Agrotechnology, Faculty of Agriculture, Halu Oleo University, Southeast Sulawesi, Indonesia

Fransiscus S. Rembon
Specifications Soil Nutrition, Department of Agrotechnology, Faculty of Agriculture, Halu Oleo University, Southeast Sulawesi, Indonesia

Resman
Specifications Soil Science, Department of Agrotechnology, Faculty of Agriculture, Halu Oleo University, Southeast Sulawesi, Indonesia

Safril Kasim
Spesifications Agroforestry, Department of Environmental Science, Faculty of Forestry and Environmental Science, Halu Oleo University, Southeast Sulawesi, Indonesia

Aminuddin Mane Kandari
Spesifications Agroclimatology, Department of Environmental Science, Faculty of Forestry and Environmental Science, Halu Oleo University, Southeast Sulawesi, Indonesia

La Ode Midi
Spesifications Agrohidrology, Department of Environmental Science, Faculty of Forestry and Environmental Science, Halu Oleo University, Southeast Sulawesi, Indonesia

Anita Indriasari
Spesifications Agriculture of Economic and Social, Department of Environmental Science, Faculty of Forestry and Environmental Science, Halu Oleo University, Southeast Sulawesi, Indonesia

Weldemariam Seifu Gessesew
Department of Horticulture, College of Veterinary Medicine and Agriculture, Addis Ababa Univesity, Fiche, Ethiopia

Kebede Woldetsadik and Wassu Mohammed
Department of Plant Sciences, Haramaya University,
Dire-Dawa, Ethiopia

Ibeawuchi I. I., Obiefuna J. C., Tom C. T. and Ihejirika G. O.
Department of Crop Science and Technology, Federal
University of Technology, Owerri, Nigeria

Omobvude S. O.
Department of Crop and Soil Science, University of
Port Harcourt, Rivers State, Nigeria

Mahendra Kumar Trivedi, Alice Branton, Dahryn Trivedi and Gopal Nayak
Trivedi Global Inc., Henderson, USA

Sambhu Charan Mondal and Snehasis Jana
Trivedi Science Research Laboratory Pvt. Ltd., Bhopal,
Madhya Pradesh, India

Francis Sarwuan Agbidye and Thompson Orya Hyamber
Department of Forest Production and Products,
University of Agriculture, Makurdi, Nigeria

Edache Ernest Ekoja
Department of Crop and Environmental Protection,
University of Agriculture, Makurdi, Nigeria

Olufemi Richard Pitan and Folashade Temitope Olaosebikan
Department of Crop Protection, Federal University of
Agriculture, Abeokuta, Nigeria

Bhaskarrao Chinthapalli, Dagne Tafa Dibar, D. S. Vijaya Chitra and Melaku Bedaso Leta
Department of Biology, College of Natural Sciences,
Arba Minch University, Arba Minch, P.O. Box No. 21,
Ethiopia

Mahendra Kumar Trivedi, Alice Branton, Dahryn Trivedi and Gopal Nayak
Trivedi Global Inc., Henderson, USA

Sambhu Charan Mondal and Snehasis Jana
Trivedi Science Research Laboratory Pvt. Ltd., Bhopal,
Madhya Pradesh, India

Faisal Abbas, Nawazish Ali, Naveed Hussain and Tanveer Abbas
Department of Agriculture and Food Technology,
Karakoram International University, Gilgit, Pakistan

Yawar Abbas
Department of Earth & Environmental Sciences, Bahria
University, Islamabad, Pakistan

Attarad Ali
Department of Biotechnology, Quid-i-Azam University,
Islamabad, Pakistan

Abdul-Rehman Phull
Department of Biology, Kongju National University,
Gongju, Republic of Korea

Islamuddin
Rescue 1122 Gilgit, Pakistan

Madete S. K. Pauline, Matofari W. Joseph and Muliro S. Patrick
Department of Dairy and Food Science and Technology,
Egerton University, Nakuru, Kenya

Bebe O. Bockline
Department of Animal Sciences, Egerton University,
Nakuru, Kenya

Mangeni B. Edwin
Livestock Sector, FAO Somalia, Nairobi, Kenya

Mohammed Ariful Islam, Md. Zahidul Islam, Sanzib Kumar Barman and Farjana Morshed
Department of Aquatic Resource Management, Faculty
of Fisheries, Sylhet Agricultural University, Sylhet,
Bangladesh

Sabiha Sultana Marine
Department of Fisheries Technology and Quality
Control, Faculty of Fisheries, Sylhet Agricultural
University, Sylhet, Bangladesh

Agbidye Francis Sarwuan and Emmanuel Zando Angyu
Department of Forest Production and Products,
University of Agriculture, Makurdi, Nigeria

Egbuche Christian. Toochi
Department of Forest and Wildlife Technology, Federal
University of Technology Owerri, Imo State, Nigeria

Aminuddin Mane Kandari
Spesifications Agroclimatology, Department of
Environmental Science, Faculty of Forestry and
Environmental Science, Halu Oleo University,
Southeast Sulawesi, Indonesia

Safril Kasim
Spesifications Agroforestry, Department of
Environmental Science, Faculty of Forestry and
Environmental Science, Halu Oleo University,
Southeast Sulawesi, Indonesia

Muh. Aswar Limi
Spesifications Management, Department of Agribusiness, Faculty of Agriculture, Halu Oleo University, Southeast Sulawesi, Indonesia

Jufri Karim
Spesifications GIS, Department of Geography, Faculty of Science and Mining Technology, Halu Oleo University, Southeast Sulawesi Indonesia

Index